PRINCIPLES OF ENVIRONMENTAL SCIENCE AND TECHNOLOGY

Studies in Environmental Science

Other volumes in this series

Studies in Environmental Science 33

PRINCIPLES OF ENVIRONMENTAL SCIENCE AND TECHNOLOGY

by

S.E. JØRGENSEN

Langkaer Vaenge 9, 3500 Vaerløse, Copenhagen, Denmark

and

I. JOHNSEN

Institut for Økologisk Botanik, Københavns Universitet, Øster Farimagsgade 2D, 1353 Copenhagen, Denmark

ELSEVIER
AMSTERDAM — OXFORD — NEW YORK — TOKYO   1989

ELSEVIER SCIENCE PUBLISHERS B.V.
Sara Burgerhartstraat 25
P.O. Box 211, 1000 AE Amsterdam, The Netherlands

Distributors for the United States and Canada:

ELSEVIER SCIENCE PUBLISHING COMPANY INC.
655 Avenue of the Americas
New York, NY 10010, U.S.A.

First edition 1981

Second revised edition 1989
Second impression 1990

ISBN 0-444-43024-5

Printed in The Netherlands

PREFACE.

This is the second edition of the book "Principles of Environmental Science and Technology". The first edition has been widely used as a textbook at university level for graduate courses in environmental management, environmental science and environmental technology (for non-engineers). As this second edition is significantly improved, it is our hope that it may find an even wider application than the first edition.

The second edition has been improved on the following points:

1) The section on ecotoxicology and effects on pollutants has been expanded considerably.
2) Chapter 4 on ecological principles and concepts has been expanded.
3) A section on ecological engineering - the application of ecologically sound technology *in the ecosystems* - has been added.
4) The problems of agricultural waste have been included in part B and in chapter 6 on waste water treatment several pages have been added about non-point sources and the application of *"soft"* technology.
5) More examples, questions and problems have been added.
6) An appendix on environmental examination of chemicals has been added.
7) All the principles have been shown in the text by use of bold letters.
8) Several figures and tables have been added to illustrate the text better.

The users of the first edition have given many useful advices, which have improved the text of the book. The author would also like to thank Judit Flesborg for a skilled transfer of the text to a Macintosh computer.

S.E. Jørgensen
Copenhagen, May 88.

CONTENTS

CHAPTER 9 (continued)

INTRODUCTION

1.1. ENVIRONMENTAL SCIENCE - AN INTERDISCIPLINARY FIELD.

The past two decades have created a new interdisciplinary field: environmental science, which is concerned with our environments and the interaction between the environment and man. Understanding environmental processes and the influence man has on these processes requires knowledge of a wide spectrum of natural sciences. Obviously biology, chemistry and physics are basic disciplines for understanding the biological/chemical/physical processes in the environment. But environmental science draws also upon geology for an understanding of soil processes and the transport of material between the hydrosphere and lithosphere, on hydrodynamics for an understanding of the transport processes in the hydrosphere, and upon meteorology for an explanation of the transport processes in the atmosphere, just to mention a few af the many disciplines applied in environmental science.

Some also believe that general political decisions are of importance in environmental management and that sociological conditions influence man's impact on the environment.

It is true that a relationship exists between all these factors and that a complete treatment of environmental problems requires the inclusion of political science and sociology in the family of environmental sciences. At present, however, it seems unrealistic to teach all these disciplines simultaneously or to presume that one person has all the background knowledge required for a complete environmental solution taking all aspects into consideration at the same time.

The right solution to environmental problems can be found only by cooperation between several scientists, and it is therefore advantageous if all the members of such a multidisciplinary team know each other's language.

Traditionally, scientists have worked to discover more and more about less and less. In environmental science, however, it is necessary to know more and more about more and more to be able to solve the problems. Environmental science has therefore caused a shift in scientific thinking, by demonstrating that although so much detailed knowledge and so many independent data have been collected, such details cannot be used by man to improve the conditions for life on this earth, unless they can all be considered together.

It is the aim of environmental science to interconnect knowledge from all sciences - knowledge that is required to solve environmental problems.

It is the scope of this book to demonstrate how our present knowledge of natural sciences can be used and interconnected to understand how man influences life on earth. The role of socioeconomic disciplines will only be mentioned briefly. The author feels that it might over-complicate the issue to include these disciplines in the present treatment of environmental science.

The past two decades have seen an unprecedented accumulation of knowledge about the environment and man's impact upon it. Unfortunately, the very mass of this information explosion has created problems for those concerned with the application of this material in research and teaching. University courses structured to meet the growing demand for multidisciplinary treatment of environmental problems became a struggle for teacher and students. The teacher, inevitably a specialist in only one of the number of disciplines, had to gather information from areas remote from his own field. Even after relevant information was obtained the rational organization of the material for meaningful presentation in an environmental course became a massive problem.

Many excellent volumes about environmental science have been published during the last decade, but although some contain a pure interdisciplinary treatment of the subject, most concentrate on one particular aspect. Of course this raises a series of questions. How could one person write on so many diverse topics and still retain a sound factual base? How could so much material be organized to maintain continuity for the reader? How could such a book avoid containing only series of independent facts and problems?, and so on.

This book attempts to solve these problems by focusing on the principles used to solve environmental problems within the framework of natural sciences. The multidisciplinary field of environmental science is growing rapidly: other sciences are bringing new knowledge into the field with an ever increasing velocity, and new problems or new connections between existing problems are continually appearing. Consequently a book dealing with facts and problems will quickly become out of date and the knowledge learned by the students useless. By focusing on the principles instead of events, these obstacles can be overcome, as the same principles are equally valid in the solution of different and new environmental problems.

The purpose of this book is to discover which methods and principles we are using when we wish to understand environmental processes and to use this knowledge to solve concrete environmental problems. Throughout the book the application of methods and principles has been illustrated by examples of real environmental problems, but to give the reader an overview of the problems of today, the last chapter of part A of the book is devoted to a general survey of environmental problems.

1.2. RELATION BETWEEN ENVIRONMENTAL SCIENCE AND TECHNOLOGY, MANAGEMENT, ECOLOGY AND MODELLING.

Concern about the environment has developed from man's ever-increasing impact on the earth. The increasing industrialization, urbanization and population, which we have faced during this century have forced us to consider whether we are changing the very conditions essential to life on the earth? Environmental science is the multidisciplinary field concerned with man's influence on environmental processes, and as such it takes human activity as well as environmental processes into consideration. This relationship is demonstrated in Fig. 1.1, where it can be seen that environmental science is concerned with the interaction between man and the environment. Man's impact on the environment is often termed pollution in its broadest sense.

Fig. 1.1 shows the relationship between environmental science and ecology. Usually ecology is defined as the study of the relationship of organisms to their environment, or the science of the interrelations between living organisms and their environment. Because ecology is concerned especially with the biology of groups of organisms and with functional processes on land, in the oceans, and in fresh water, it is more in keeping with the modern emphasis to define ecology as the study of the structure and function of nature, remembering that man is a part of nature (E.P. Odum, 1971).

One of the definitions of ecology in Webster's Dictionary seems especially appropriate for the closing decades of this century: "The totality or pattern of relations between organisms and their environments".

From these definitions of ecology it is clear that ecology is closely related to environmental science. When man intrudes on an ecosystem (e.g. a lake, a forest, a desert) - he inevitably disturbs the delicate balance of organisms and substances and their activities that have evolved in nature.

The ecosystem can adapt to man's disruptive activities, but only to a certain point. It is therefore crucial to understand the nature of ecosystems to be able to understand the consequences of man's impact on his environment. Today, everyone is acutely aware of the environmental science as indispensable tools for creating and maintaining the quality of human civilization. Consequently ecology is rapidly becoming the branch of sciences that is most relevant to environmental problems or possibly even to the everyday life of every man, woman and child (E.P. Odum, 1971).

But how can we diminish the effect of human activity on the environment? The first step is to understand the relation between the activity and the effect - by means of environmental science - and the second step is to control human impact on ecosystems - by means of environmental management.

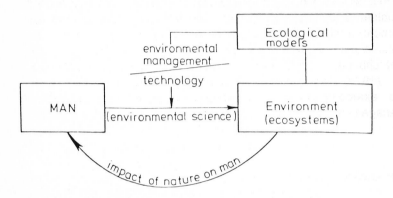

Fig. 1.1. Relations between environmental science, ecology, ecological modelling and environmental management and technology.

Technological development has increased the human impact on the ecosystem, but a new technology aimed at solving the pollution problem has been developed simultaneously. This has occured in the field of environmental technology, in which new methods of purification and recirculation of pollutants are being developed and attempts are being made to change existing technology to reduce pollution. This field must be distinguished from the classical discipline of sanitary engineering, which is limited to water treatment methods, with especial emphasis on design and planning of sewage- and water-supply systems.

Environmental management is often expensive, but many case studies have shown that it can also be economically advantageous to solve environmental problems, e.g. by the recirculation of valuable raw materials. This creates a new and equally important question of how we select the best ecological and economical method for solving a specific environmental problem. Selecting the best ecological solution is a very complex problem. Many organisms and processes interact in the ecosystem, and therefore to map the effects of human activity on ecosystems is a very complicated task.

Ecological modelling, or systems ecology, offers a unique opportunity to screen and select the best methods for pollution control, and as we can see in Fig. 1.1 the circle is closed. Ecosystems are very complex systems and it is not possible to consider all their processes in a management situation. However, for a given problem, it is possible, with a good grounding in chemistry and biology and a good knowledge of the ecosystem considered, to make a simplified model of the ecosystem and its processes, to determine and include the relevant variables and to omit processes of minor importance. Such a model will consist of a mathematical description of processes crucial to the problem, and it is then possible to use the model to simulate different management alternatives. Often the model is based on the principle of mass conservation, in which case a set of differential equations describe the rate of change in concentrations. For an introduction to the subject, see Jeffers (1978), Jørgensen (1981) and Jørgensen (1987) where several concrete examples of the application of ecological models in environmental management, are discussed.

1.3. LEVELS OF ORGANIZATION.

Ecology is considered to be a discipline or division of biology, the science of life. Biology can be divided in two distinct ways. In Fig. 1.2 the divisions form a matrix. The horizontal division is concerned with the fundamentals of biology: morphology, physiology, genetics ecology, molecular biology, evolution theory, etc., while the vertical division is a taxonomic one: bacteriology, ornithology, botany, entomology, etc.

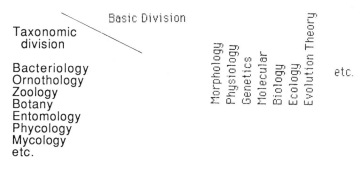

Fig. 1.2.   Division of Biology.

Different levels of biological organization can also be considered to illustrate the content of modern ecology. Fig. 1.3 shows how the biosystem is formed by a combination of a biotic component and an abiotic component.

The biotic components are organized in a hierarchial arrangement from the subcellular level (genes) to communities. The biotic components interact with the physical environment - the abiotic components to form the system, which is meant the regular interaction of interdependent components to form a unified whole. Ecology is concerned with the right hand portion of the spectrum shown in Fig. 1.3. In biology the term population is used to denote groups of individuals of any kind of organism, and community in the ecological sense includes all populations occupying a given area. The community and the non-living environment together form the ecosystem.

Biotic
Component   Genes Cells Organs Organisms Populations Communities

   +

Abiotic
Component   Matter and Energy

   ‖

Systems: Genetic- Cell- Organ- Organismic- Population- Ecosystems

Fig. 1.3. Levels of Biological Organizations.

Environmental science is concerned with the study of the environment of all life forms within the life-bearing layer of the earth. This shallow life layer can be further divided into the gaseous realm (the atmosphere), the liquid water realm (the hydrosphere), the solid realm (the lithosphere) and the living part of the earth (the biosphere). Often the term ecosphere is used to denote the biosphere plus its non-living environment; it refers to the parts of the atmosphere, hydrosphere and lithosphere that bear life.

1.4. THE ENVIRONMENTAL CRISIS.

As mentioned in this chapter three pronounced developments have caused the environmental crisis, we are now facing: the growth in population, industrialization and urbanization.

Fig. 1.4 illustrates world population growth, past and projected. From the graph it can be seen that population growth is experiencing decreasing doubling time, which implies that growth is more than exponential (exponential growth corresponds to a constant doubling time). Fig. 1.4 shows that growth from one billion to two billions took about 100 years, while the next doubling in population took only 45 years.

The net birth rate at present is about 350,000 people per day, while the death rate is 135,000 per day. The population growth is determined by the differences between the two:

population increase = birth rate - death rate.

This implies that the world's population is increasing by more than 200,000 per day, or about 1.5 million per week corresponding to 80 million per year.

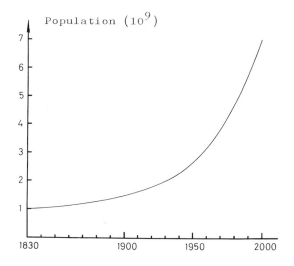

Fig. 1.4. World population growth plotted against time. Notice that the doubling time is decreasing, which implies that the growth is faster than exponential growth.

The need to limit population growth is now more clearly perceived than before. Every living organism requires energy and material resources from its environment. Resources can be classified as renewable and non-renewable (Skinner, 1969). Renewable resources are those that can maintain themselves or be continuously replenished if managed wisely. Food, crops, animals, wildlife, air, water, forest, etc., belong to this class. Land and open space can also be considered as renewable, but they shrink as the population increases. Although we cannot run out of these resources, we can use them faster than they can be regenerated or by using them unwisely affect the environment (Meadows et al., 1972)

Other resources, such as fossil fuels and minerals are non-renewable resources, whose finite supplies can be depleted. Theoretically some of these resources are renewable, but only over hundreds of millions of years, while the timescale of concern to man, is hundreds of years only.

When we talk about finite supplies of resources we should qualify this by discussing finite supplies of substances presently considered to be resources. Often our most important consideration is whether the pollution costs from extraction and use of a resource outweigh its benefit as population or per capita consumption increase.

During the last few decades we have observed a distinct increase in pollution. Many examples illustrate these observations. The concentrations of carbon dioxide, sulphur dioxide and other gaseous pollutants have increased drastically. The concentrations of many toxic substances have increased in soil and water, and the ecological balance has been changed in our ecosystems. In many major river systems oxygen depletion has been recorded, and many recreational lakes are suffering from eutrophication (high concentrations of nutrients - mainly nitrogen and phosphorus).

What has caused this sudden increase in pollution? The answer is not simple, but the growth in population is obviously one of the factors that influences our environment. Other factors include man's rate of consumption and the type and amount of waste, he produces.

Many now recognize two basic causes of the environmental crisis. The first, which occurs in the developing countries, is overpopulation relative to the food supply and the ability to purchase food even if it is available. The second occurs in the technologically advanced countries, in North America, Australia, Japan and Europe. These countries use 80-90% of the world's natural resources, although they only account for about 25% of the world's population. As a result the average consumer in these contries causes 25-50 times as great an impact on our life-supporting system as a peasant in a developing country (Davis, 1970).

The debate centers on the relation between pollution - or environmental impact - and the population, consumption and technology. We can use the crude but useful model proposed by Ehrlich and Holdren (1971). They obtain the environmental impact, I, by multiplying three factors - the number of persons, P, the units of consumptions per capita, C, and the environmental impact per unit of consumption, E,:

$$I = P \cdot C \cdot E \tag{1.1}$$

All three factors are equally important. We can illustrate the importance of all three factors by considering the development in U.S.A. between 1950 and 1970. The population increased 35% during these two decades, while the per capita consumption increased about 51%. In the same period the production of environmentally harmful material has increased by between 40 and 1900%. It is, of course, easy to get the false impression that the percent increases in production as illustrated in Table 1.1 represent percent increases in pollution, whereas in fact only a fraction of these increases actually contributes to environmental pollution.

TABLE 1.1
Changes in per capita consumption (or production) of selected items in the United States between 1950 and 1970.

Item	Percent increase
Synthetic fibers (non-cellulose)	1,890
Air freight (ton miles)	890
Plastics	556
Synthetic organic chemicals	254
All synthetic fibers	220
Total horsepower	212
Phosphates in detergents	210
Electric power	207
Aluminium	182
Natural gas	171
Synthetic rubber	165
Nitrogen fertilizers	143
Synthetic organic pesticides	115
Phosphate rock	110
Motor fuel	74
Motor vehicles registered	65
GNP (per capita 1958 constant dollars)	51
Truck freight (tons)	50
Paper	48
Energy use	46
Petroleum (production)	40
Cement (production)	30
Meat	28
Mercury	20
Protein	8
Steel (production)	1
Food energy	1
Vegetables	0
Fish	-3
Railroad freight (tons)	-4
Coal	-15
Poultry	-17
Lead	-18
Fruits	-20
Synthetic fibers (based on natural cellulose)	-23
Cropland acres	-27
Cigarettes	-28
Natural fibers (cotton, wool, silk)	-43
Natural rubber	-43
DDT	-48
Soap	-52

Indeed, during the same two decades we have learnt to diminish the discharge into the environment. However, the total environmental impact has increased considerable in U.S.A. during these two decades. The two first factors in Equation (1.1) are easy to find from the facts mentioned above, but in such a simple model the third factor will always be a subjective judgement. If we take the rapid growth in production of harmful substances (40-1900%) and also consider the measured increased concentration of such substances in major ecosystems, a reasonable value for the third factor would be around 2. It would give us the following increase in the total impact on the environment:

$$I = 1.35 \cdot 1.51 \cdot 2 = 4 \text{ times} \tag{1.2}$$

This is a very crude simplification, but the increase in environmental impact of 4 times during such a short period as 20 years must give us reason to worry about man's future on this earth, unless we can use all our efforts to manage the problem.

Two major forces can lead to apathy: naive technological optimism - the idea that some technological wonder will always save us regardless of what we do - and the gloom and doom pessimism - the idea that nothing will work and our destruction is assured. The idea behind such a book as this is, of course, that something can be done, but the problem is very complex and difficult to solve. The best starting point must be an understanding of the nature of the problems and the principles and methods that can be used to solve them. It is hoped that by working through this book the reader will be able to grasp these basic concepts.

1.5. FOCUS ON PRINCIPLES OF ENVIRONMENTAL SCIENCE.

An understanding of environmental problems requires only the application of a few principles which must be coupled with environmental data. The principles are discussed in the next three chapters, while the last in Part A, chapter 5, is devoted to a survey of the environmental problems of today.

All life on this planet is dependent on the presence of a number of elements in the right form and concentration. Other elements not used in the life-building processes should not be present in the biosphere or present only in very low concentrations.

If these conditions are not fulfilled, life is either not possible or will be damaged. It is therefore of major importance in understanding environmental deterioration to keep a record of elements and compounds to ascertain whether an abnormal concentration of one or more elements or compounds can explain the environmental problem considered. Therefore the law of mass conservation should be widely used in understanding ecological reactions to pollution. Chapter 2 is devoted to the application of mass balances in the environmental context and the translation of a mass balance results concentration - to environmental effects.

Energy is also required to maintain life. In a thermodynamic sense the earth can be considered a closed, but not isolate, system, which implies that the earth exchanges energy but not matter, with the universe. The same consideratione is somethimes also valid for ecosystems, although the characteristic pollution situation is an input of pollutants, which changes the concentrations in the ecosystem.

The earth's input and output of energy are approximately balanced. Solar radiation is the basic energy requirement for all life on earth, but after this energy has been used to maintain the biological, chemical and physical processes, it is converted to longwave radiation from the earth out to the universe. The balance between input and output assures that a constant average temperature is maintained.

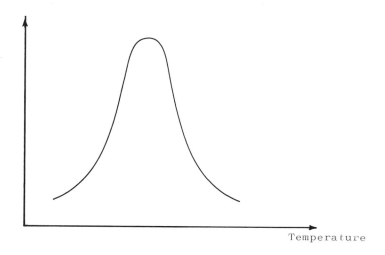

Temperature

Fig. 1.5. Biological processes are strongly dependent on the temperature. A characteristic biological process rate is plotted versus the temperature.

The rate at which biological processes take place is strongly dependent on temperature. In Fig. 1.5 a characteristic response of biological rates to temperature is shown. As seen an optimum temperature exists, at which the biological rate is at its maximum. The present life on earth is dependent on a certain temperature pattern. Changes in the present temperature pattern can therefore have enormous consequences for all life on earth.

Not only is the global energy balance of importance, but also the energy balance within ecosystems can give important information about the life conditions. Chapter 3 is concerned with energy problems related to these environmental issues.

An ecosystem is a complex system that reacts to changed chemical concentrations and temperature in a very complicated way. Sometimes an ecosystem is able to diminish the effect of changed concentrations, in which case the ecosystem is said to have a buffer capacity, but in other ecosystems the effect of pollution is enhanced by "upconcentrating" a harmful component. Chapter 4 on Ecological Principles and Concepts is devoted to an understanding of such ecological reactions to change.

Emphasis will be laid on quantification of environmental problems throughout part A, because it is only through the application of environmental principles for quantification of a problem that we have the right basis for selection of a feasible solution.

A solution might be found in what is named ecological engineering, which implies that the ecosystem is assisted or modified to resist an environmental impact. Ecological engineering requires sound ecological knowledge of the ecosystem reactions. It is only through a profound knowledge of ecosystems and their reactions that good ecological engineering solutions can be found. Whenever such solutions can be used, this will be mentioned in part A, and a survey of these methods will be presented in the last section of chapter 4.

The importance of environmental principles is illustrated throughout Part A by many examples. These examples are not necessarily the most important environmental problems, but they have been selected as those which best illustrate the principles. Hopefully, the use of many examples gives the principles reality and demonstrates that the theory can be used to solve practical problems.

1.6   PRINCIPLES OF ENVIRONMENTAL TECHNOLOGY.

The selection of the rigth technology for the solution of an environmental problem requires a profound understanding of the problem itself. Therefore environmental science is an essential basic field for environmental technology.

Also Part B which deals with environmental technology, attempts to focus on the principles and their application to find technological solutions to environmental problems. Such a solution is not always readily available, but can be obtained by use of environmental legislation. For example considerable lead pollution orginated from the use of lead as an additive in gasoline, but as a result of legislation and the setting of a maximum permitted lead concentration in gasoline, such pollution has now diminished.

Each section in chapters 6-9 discusses a specific environmental problem and all unit processes that are involved in solving that problem will be mentioned. Many processes used in environmental technology are the same processes as are used in nature. References will be given to similar processes already mentioned in part A.

Part B is not written only for the engineer, but rather for all concerned with the solution of environmental problems.
Design criteria are not included, but the following properties of the technological solutions mentioned are considered:

1. Function, expected capacity and efficiency.
2. Advantages and disadvantages.
3. Relation between process variables and capacity/efficiency.
4. Process troubles and how to solve them.
5. Area of application.
6. Environmental evaluation of the method in the broadest possible environmental sense.

The reader is thus given a basic knowledge of environmental technology to enable him to select the best solutions available today and to discuss them with specialists who must design and build the project.
He should obtain a critical view of technological solutions, by considering the environments rather than the economy of prime importance, but also understand when technological solutions have clear advantages over other possibilities. In this context (see above) point 6 is of great importance, as many technological methods solve one problem only to create another. Only a clear environmental analysis of both the problems can reveal whether it is advantageous from an ecological point of view to apply a technological solution.

Environmental technology has developed very rapidly during the last two decades. Many new methods are available today, and for most environmental problems a wide spectrum of methods (processes) are applicable. As a result it is considerably more difficult today to find the very best solution, but at the same time there are better possibilities for an acceptable technological solution.

The problems have been classified in terms of: water pollution, air pollution and solid waste pollution. This is a reasonable classification of environmental technology, as the methods and processes are dependent on the state of the pollutants.

Chapter 6 is devoted to water pollution problems and to water resources, including the technology applied to production of potable and process water from surface and ground water. Chapter 7 discusses solid waste problems and Chapter 8 deals with air pollution problems. Chapter 9 deals with problems related to examination of pollution.

Each chapter is divided into sections dealing with problems, that are considered of importance today.

1.7 HOW TO SOLVE ENVIRONMENTAL PROBLEMS.

Principles and quantifications are used as keywords in our search for solutions to environmental problems.

Fig. 1.6 gives a flow chart of a procedure showing how to go from emission of mass and energy to a solution of the related environmental problems. Emission is translated into imission and concentration. The effect or impact of a concentration of a compound or energy is found by considering all the chemical, physical and biological processes that take place in the ecosystem. This evaluation leads us to an acceptable ecological (and economic) solution by use of ecological engineering or environmental technology. The former attacks the problem in the ecosystem, while the latter attempts to reduce or dilute the emission.

The procedure requires the application of principles and knowledge of environmental processes. Furthermore, the problem must be well understood and quantified so as to be able to find the right solution.

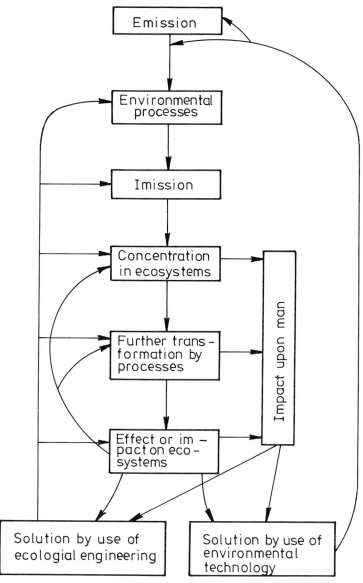

Fig. 1.6. Flowchart illustrating a procedure which can be used to go from emission to solution.
Part B of the book is devoted to how impact on ecosystems or man is translated into an environmental technology solution which attacks the problem by reduction or dilution of the emission. Part A is devoted to all other steps of the procedure.

PRINCIPLES OF ENVIRONMENTAL SCIENCE

As indicated in the introduction, this part consists of four chapters:

Mass Conservation

Principles of energy behaviour applied to environmental issues

Ecological principles and concepts

An overview of the major environmental problems of today

All four chapters are built-of from three elements: Principles, processes of importance in understanding the impact on the environment, and examples to illustrate the application of the principles in an environmental context.

All Principles and processes are designated in the text as with the symbol **P.x.y.**, where x is the chapter number and y the number of a principle or an important process in that chapter.

Specific quantitative examples are named by example + chapter number + a number; but many general examples are also given in the common text.

Each chapter ends with a list of problems, which can be used to discuss the material presented in the chapter.

However, the right way to use the text in learning to apply environmental science to real life problems would be to discuss comprehensive environmental problems, as it is attempted in the text. This requires the combination of data from real case studies with general data, which can be found in tables presented throughout the text or in the appendix.

CHAPTER 2

MASS CONSERVATION

2.1 EVERYTHING MUST GO SOMEWHERE.

According to the law of mass conservation, mass can neither be created nor destroyed, but only transformed from one form to another. Thus everything must go somewhere. The notion of cleaning up the environment or pollution-free products is a scientific absurdity. We can never avoid pollution effects. Nobody - neither man nor nature - consumes anything; we only borrow some of the earth's resources for a while, extract them from the earth, transport them to another part of the planet, process them, use them, discard, reuse or reformulate them (Cloud, 1971).

The law of mass conservation assumes that no transformation of mass into energy takes place, which is formulated in accordance with Einstein as follows:

$$E = mc^2 \tag{2.1}$$

$$c = 3 \cdot 10^8 m \ sec^{-1}$$

If we consider a system which exchanges mass with the environment, then the following equation is valid for an element, e:

$$\frac{dm_e}{dt} = \text{import - export} \tag{2.2}$$

where m_e is the mass of e in the system and t is time.

For a chemical compund, c, the law of mass conservation can be formulated as follows:

$$\frac{dm_c}{dt} = \text{import - export} \pm \text{result of chemical reactions} \tag{2.3}$$

where m_c is the mass of c in the system.

It is possible to compute concentration in ecosystems based upon emission data by use of equations (2.2) or (2.3).

P.2.1. Mass is conserved; it can be neither created nor destroyed, but only transformed from one form to another.

The form and location is of great importance for the effect of pollutants. We should always attempt to discharge our waste in such a way that the change in concentration of the most harmful forms becomes as low as possible. It is therefore noticeable that the four spheres have a completely different composition, as demonstrated in Appendix 1.

TABLE 2.1
Cu-concentrations (characteristic)

Item	Sphere represented	Concentration
Atmospheric particulates (unpolluted area)	Atmosphere	2 mg m^{-3}
Sea-water (unpolluted)	Hydrosphere	$2 \text{ } \mu g \text{ } 1^{-1}$
River water	Hydrosphere	$10 \text{ } \mu g \text{ } 1^{-1}$
Soil	Lithosphere	20 mg kg^{-1}
Fresh-water sediment	Lithosphere	40 mg kg^{-1}
Algae	Biosphere	$20\text{-}200 \text{ } \mu g \text{ } 1^{-1}$

Let us take a concrete example to illustrate these considerations: where should waste containing copper be deposited? To answer this question we need more information than that available in Table 2.1, although this table indicates that the highest concentration of copper is in the lithosphere. Therefore we can assume that the discharge of copper to the lithosphere will produce the smallest change in copper concentration of the four spheres, but need to know something about the effect of copper in its different forms. This information is given in Table 2.2. It is seen that free copper ions are extremely toxic to some aquatic animals. Furthermore, it is demonstrated in Table 2.3, that free copper ions are bound to soil and sediments, which means that the most toxic form will often be present in the environment in low concentrations.

TABLE 2.2

A. Lethal concentrations of copper ions (LC_{50}-values and lethal doses of copper (LD_{50}-values).

Species	Values
Asellus meridianus	$LC_{50*}^{*)} = 1.7 - 1.9$ mg l^{-1}
Daphnia magna	$LC_{50*} = 9.8$ μg l^{-1}
Salmo gairdneri	$LC_{50}^{)} = 0.1 - 0.3$ mg l^{-1}
Rats	$LC_{50} = 300$ mg (kg body weight)$^{-1}$ (as sulphate)

*) dependent on pH, temperature, water hardness and other experimental conditions.

B. Sublethal effects of copper.

Effect	Concentration
Weed, decrease in diversity	$63 - 218$ mg kg^{-1}
Asellus meridianus, decrease in growth rate	0.1 mg l^{-1}
Pimephales promales, decreased number og eggs	0.065 mg l^{-1} causes 40% reduction in number of eggs
Aufwuchs community: 1) Reduced photosynthesis 2) Respiration	0.01 mg l^{-1} 15% reduction 1 mg l^{-1} 50% reduction

The water deposition site is, in the first instance, selected in the sphere where the relative change in concentration is smallest. Furthermore, it is necessary to compare the form and the processes of the waste in the 4 spheres and consider this information for the deposition site specifically.

The values in Tables 2.1 - 2.3 are typical but must not be considered general, as they are strongly dependent on several factors not included in the tables. This does not imply that copper can be deposited in the lithosphere at any given concentration. It is only stated that nature has mainly deposited copper in the lithosphere and that the environmental effect here is smallest. It is necessary to control the concentration of free copper ions in the soil water and guarantee that this is far from reaching a toxic level.

TABLE 2.3
Copper-binding capacities of soil and sediment samples

Sediment Lake Glumsoe	28 mg g^{-1} dry matter at 5 mg l^{-1} in water. pH: 7.6, 20°C (free copper ions)
Humus-rich soil with high ion- exchange capacity	76 mg g^{-1} dry matter at 3.7 mg l^{-1} in soil water.

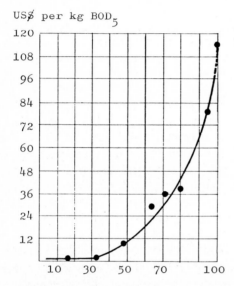

Biological oxygen demand reduction (percent)

Fig. 2.1. Relation between removal of BOD$_5$ (percent) and the cost per kg BOD$_5$ removed from typical industrial water with a high concentration of biodegradable material.

We can collect garbage and remove solid waste from sewage, but they must either be burned, which causes air pollution, dumped into rivers, lakes and oceans, which causes water pollution, or deposited on land, which will cause soil pollution. The management problem is not solved before the final

deposit site for the waste is selected. Furthermore, environmental management requires that *all* consequences of environmental technology be considered. Eliminating one form of pollution can create a new form, as described
above.

Finally, as the production of machinery and chemicals for environmental
technology may also cause pollution, the entire mass balance must be considered in environmental management, including deposit of waste products
and pollution from service industries.

This problem can be illustrated by considering, as an example, the Lake
Tahoe waste water plant, where municipal waste water is treated through
several steps to produce a very high water quality. The conclusion is that it
hardly pays, from an environmental point of view, to make such a comprehensive waste water treatment.

P.2.2. **A total solution to an environmental problem implies that
all environmental consequences are considered by use of a
total mass balance, including all wastes produced and the
service industries.**

Fig. 2.1 shows the relation between the degree of purification (in %) and
the cost of treatment. Increasing the efficiency of the treatment produces an
exponential growth. As the number of possible by-product pollutions often
follow the trend of the cost, the same relation might exist between the environmental side-effects and the percent reduction of pollution.

Consequently our problem is not the elimination of pollution, but its
control. Technology is essential in helping us to reduce pollution levels below a dangerous level, but in the long term pollution control must also include population control and control of the technology including its pattern
of production and consumption. Wise use of existing technology can buy us
some time to develop new methods, but the time we can buy is limited. The
so-called energy crisis is the best demonstration of the need for new and far
more advanced technology. We do not know how much time we have -
probably 20-50 years - so we had better get started now to make sure, we
have the solution to man's many serious problems in time.

The increasing cost of treatment with efficiency (Fig. 2.1) must be taken
into account in urban planning. If it is decided to maintain environmental
quality, increased urbanization (which means increased amounts of wastes)
will require higher treatment efficiency, which leads to a higher cost for
waste treatment per inhabitant or per kg of waste. This fact renders the
solution of environmental problems of metropolitan areas in many developing
countries, economically almost prohibitive.

P.2.3. **The more waste it is required to treat in a given area, the higher the treatment efficiency that is needed and the higher the costs per kg of waste will be to maintain an acceptable environmental quality.**

2.2 THRESHOLD LEVELS.

A pollutant can be defined as any material or set of conditions that created a stress or unfavourable alternation of an individual organism, population, community or ecosystem beyond the point that is found in normal environmental conditions (Cloud, 1971). The range of tolerance to stress varies considerably with the type of organism and the type of pollutant (Berry et al., 1974).

To determine whether an effect is unfavourable may be a very difficult and often highly subjective process. A list of major pollutants can be found in Table 2.4, where the estimated environmental stress now and in the future is shown. Each environmental stress index is obtained by multiplying the weighted factors: **1.** persistance (1 to 5), **2.** geographical range (1 to 5. 1 means only of local interest, 5 that the pollutant is a global problem), and **3.** complexity of interactions and effects (1 to 9). The highest possible index is therefore 225 (the data are taken from Howard Reiguam, 1971).

It is important to recognize that there are both natural and man-generated pollutants. Of course, the fact that nature is polluting does not justify the extra addition of such pollutants by man, as this might result in the threshold level being reached.

In general we can classify pollutants into two groups:

1. non-treshold or gradual agents, which are potentially harmful in almost any amount, and 2. threshold agents, which have a harmful effect only above or below some concentration or threshold level. This classification is illustrated in Fig. 2.2.

For the latter class we come closer to the limit of tolerance for each increase or decrease in concentration, until finally, like the last straw that broke the camel's back, the threshold is crossed. For non-threshold agents, which include several types of radiation, many man-made organic chemicals, which do not exist in nature, and some heavy metals such as mercury, lead and cadmium, there is theoretically no safe level. In practice, however, the degree of damage at very low trace levels is considered negligible or worth the risk relative to the benefits occurred from using the products or processes.

TABLE 2.4
Estimated environmental stress indices for pollutants now and in the future.

	Now	Year 2000-2030
Heavy metals	90	130
Radioactive wastes	35	120
Carbon dioxide	75	75
Solid wastes	35	120
Waterborne industrial wastes	35	80
Oil spills	40	70
Sulphur dioxide and sulphates	20	70
Waste heat	5	70
Nitrogen oxides	20	36
Litter	20	40
Pesticides	30	130
Hydrocarbons in air	10	20
Photochemical oxidants	15	20
Carbon monoxide	10	15
Organic sewage	20	40
Susp. particulates	20	90
Chemical fertilizers	30	50
Noise	5	15

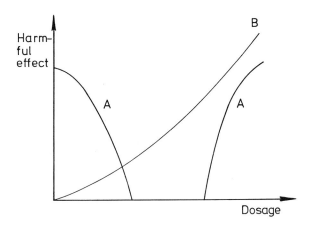

Fig. 2.2. A) Threshold agent. B) Non-threshold or gradual agent. To have a threshold agent it is sufficient that one of the two A-plots is valid.

Threshold agents include various nutrients, such as phosphorus, nitrogen, silica, carbon, vitamins and minerals (calcium, iron, zinc etc.). When they are

added or taken in excess the organism or the ecosystem can be overstimula-
ted, and the ecological balance is damaged. Examples are the eutrophication
of lakes, streams and estuaries from fertilizer run-off or municipal waste
water.

The threshold level and the type and extent of damage vary widely with
different organisms and stresses. The thresholds for some pollutants may be
quite high, while for others they may be as low as 1 part per million (1 ppm)
or even 1 part per billion (1 ppb).

**P. 2.4. The threshold level is closely related to the concentration
found in nature under normal environmental conditions.**

Tables 2.5-2.7 show the concentration of some important elements found
in different parts of the ecosphere. Note that even uncontaminated parts of
the ecosphere contain almost all elements, although some only in very low
concentrations.

In addition, an organism's sensitivity to a particular pollutant varies at
different times of its life cycle, e.g. threshold limits are often lower in the
juvenile (where body defense mechanisms may not be fully developed) than in
the adult stage. This is especially true for chlorinated hydrocarbons such as
DDT and heavy metals - both of which represent some of the most harmful
pollutants.

Pollutants can also be characterized by their longevity in process
organisms and ecosystems. Degradable pollutants will be naturally broken
down into more harmless components if the system is not overloaded. Non-
degradable pollutants or persistent pollutants will, however, not be broken
down or be broken down very slowly and the intermediate components are
often as toxic as the pollutants. Knowledge of the degradability of the different components is naturally of great importance to environmental management, since an undesired concentration of pollutants in the environment is
a function not only of the input and output, but also of the processes which
take place in the environment. These processes must of course be considered
whenever the concentration of pollutants is being computed.

**P. 2.5. Environmental impact assessment requires not only know-
ledge as to the concentration and the form of a pollutant,
but also of the processes which the pollutant might undergo in the environment.**

TABLE 2.5
Background concentration in atmosphere, at stations in unpolluted area.

Element	$\mu g\ m^{-3}$
Ag	0.05 - 0.1
Al	1 - 12
As	0.1 - 1.0
Ca	70 - 250
Cd	0.05 - 1.5
Cu	1 - 4
Fe	3 - 20
H_2	$(250 - 365) \cdot 10^3$
He	$5.2 \cdot 10^3$
K	60 - 180
Kr	$1.1 \cdot 10^3$
Mg	170 - 600
Mn	0.05 - 0.33
Na	1500 - 5500
Ne	$18 \cdot 10^3$
Pb	0.4 - 0.8
S	$(3 - 50) \cdot 10^3$
Sr	0.9 - 4.0
V	0.05 - 0.27
Xe	86

TABLE 2.6
Composition of the sea

Element	mg l^{-1}	Present as	Retention time (y)
H	108,000	H_2O	
He	0.000005	He(g)	
Li	0.17	Li^+	$2.0 * 10^7$
Be	0.0000006		$1.5 * 10^2$
B	4.6	$B(OH)_3, B(OH)_2O^-$	
C	28	$HCO_3^-, H_2CO_3, CO_3^{2-}$	
N	0.5	$NO_3^-, NO_2^-, NH_4^-, N_2(g)$	
O	857,000	$H_2O, O_2(g), SO_4^{2-}$	
F	1.3	F^-	
Ne	0.0001	Ne (g)	
Na	10,500	Na^+	$2.6 * 10^8$
Mg	1350	$Mg^{2+}, MgSO_4$	$4.5 * 10^7$
Al	0.01		$1.0 * 10^2$
Si	3	$Si(OH)_4, Si(OH)_3O^-$	$8.0 * 10^3$
P	0.07	$HPO_4^{2-}, H_2PO_4^-$ HPO_4^{3-}, H_3PO_4	
S	885	SO_4^{2-}	
Cl	19,000	Cl^-	

TABLE 2.6 - continued

Element	mg l^{-1}	Present as	Retention time (y)
A	0.6	A (g)	
K	380	K$^+$	$1.1 * 10^7$
Ca	400	Ca^{2+}, CaSO$_4$	$8.0 * 10^6$
Sc	0.00004		$5.6 * 10^3$
Ti	0.001		$1.6 * 10^2$
V	0.002	VO$_2$(OH)$_2$$^{2-}$	$1.0 * 10^4$
Cr	0.00005		$3.5 * 10^2$
Mn	0.002	Mn^{2+}, MnSO$_4$	$1.4 * 10^3$
Fe	0.01	Fe(OH)$_3$ (s)	$1.4 * 10^2$
Co	0.0005	Co^{2+}, CoSO$_4$	$1.8 * 10^4$
Ni	0.002	Ni^{2+}, NiSO$_4$	$1.8 * 10^4$
Cu	0.003	Cu^{2+}, CuSO$_4$	$5.0 * 10^4$
Zn	0.01	Zn^{2+}, ZnSO$_4$	$1.8 * 10^5$
Ga	0.00003		$1.4 * 10^3$
Ge	0.00007	Ge(OH)$_4$, Ge(OH)$_3$O$^-$	$7.0 * 10^3$
As	0.003	HAsO$_4$$^{2-}$, H$_2AsO_4$$^-$ H$_3$AsO$_4$, H$_3$AsO$_3$	
Se	0.004	SeO$_4$$^{2-}$	
Br	65	Br$^-$	
Kr	0.0003	Kr (g)	
Rb	0.12	Rb$^+$	$2.7 * 10^5$
Sr	8	Sr^{2+}, SrSO$_4$	$1.9 * 10^7$
Y	0.0003		$7.5 * 10^3$
Nb	0.00001		$3.0 * 10^2$
Mo	0.01	MoO$_4$$^{2-}$	$5.0 * 10^5$
Ag	0.0003	AgCl$_2$$^-$, AgCl$_3$$^{2-}$	$2.1 * 10^6$
Cd	0.00011	Cd^{2+}, CdSO$_4$	$5.0 * 10^5$
In	< 0.02		
Sn	0.003		$5.0 * 10^5$
Sb	0.0005		$3.5 * 10^5$
I	0.06	IO$_3$$^-$, I$^-$	
Xe	0.0001	Xe (g)	
Cs	0.0005	Cs$^+$	$4.0 * 10^4$
Ba	0.03	Ba^{2+}, BaSO$_4$	$8.4 * 10^4$
La	0.0003		$1.1 * 10^4$
Ce	0.0004		$6.1 * 10^3$
W	0.0001	WO$_4$$^{2-}$	$1.0 * 10^3$
Au	0.000004	AuCl$_4$$^-$	$5.6 * 10^5$
Hg	0.00003	HgCl$_3$$^-$, HgCl$_4$$^{2-}$	$4.2 * 10^4$
Tl	< 0.00001	Tl$^+$	
Pb	0.00003	Pb^{2+}, PbSO$_4$	$2.0 * 10^3$
Bi	0.00002		$4.5 * 10^5$
Rn	$0.6*10^{-15}$	Rn (g)	
Ra	$1.0*10^{-10}$	Ra^{2+}, RaSO$_4$	
Th	0.00005		$3.5 * 10^2$
Pa	$2.0*10^{-9}$		
U	0.003	UP$_2$(CO$_3$)$_3$$^{4-}$	$5.0 * 10^5$

TABLE 2.7
Characteristic background concentration range in uncontaminated soil.

Element	mg (kg dry matter)$^{-1}$
Ag	0.01 - 0.1
A	10,000 - 300,000
As	0.1 - 40
B	2 - 100
Ba	100 - 3000
Be	0.1 - 40
Br	1 - 10
Ce	1 - 50
Cl	1 - 150
Co	1 - 40
Cr	5 - 3000
Cs	0.3 - 25
Cu	2 - 100
F	2 - 300
Fe	7000 - 550,000
Ga	0.4 - 300
Ge	1 - 50
Hg	0.01 - 0.8
I	1 - 5
K	400 - 30,000
La	1 - 5000
Li	7 - 200
Mg	600 - 6000
Mo	0.2 - 5
N	200 - 2500
Na	750 - 7500
Ni	10 - 1000
P	100 - 4000
Pb	2 - 200
Ra	(3-20) * 10^{-7}
Rb	20 - 600
S	30 - 900
Sc	1 - 10
Se	0.01 - 2
Si	0.2 - 5
Sn	2 - 200
Sr	50 - 1000
Th	0.1 - 12
Ti	1000 - 10,000
Te	0.01 - 0.5
U	0.9 - 9
V	20 - 500
Y	2.5 - 250
Zn	10 - 300
Zr	30 - 2000

2.3 BASIC CONCEPTS OF MASS BALANCE.

The simplest case is an isolated system, where no processes take place, and therefore the concentration of all components is constant. An ecosystem is never an isolated system, but may be either an open system or a closed system. The former exchanges mass as well as energy with the environment. The input of energy to an ecosystem will cause cyclic processes (Morowitz, 1968), in which the important elements will play a part.

Very few pollutants are completely chemically inert, most are converted to other components or degradated. The common degradation process including the degradation of persistent chemicals can be described by a first order reaction scheme:

$$\frac{dC}{dt} = k * C \qquad\qquad (2.4)$$

where C is the concentration of the considered compound, t is the time and k a rate constant. k varies widely from the very easily biodegradable compounds, such as carbohydrates and proteins, to pesticides, such as DDT. The so-called biological half-life time, $t_{1/2}$, is often used to express the degradability. $t_{1/2}$ is the time required to reduce the concentration to half the initial value. The relation between k and $t_{1/2}$ can easily be found:

$$\ln \frac{C_o}{C(t)} = k * t \qquad\qquad (2.5)$$

$$\ln 2 = k * t_{1/2} \qquad\qquad (2.6)$$

Here C_o is the initial concentration.

Appendix 2 shows some characteristic rate constants for biodegradation and biological half-life times. k (and $t_{1/2}$) are dependent on the reaction conditions, pH, temperature, ionic strength, etc. These conditions are indicated in the right-hand column of the tables. As seen, a wide spectrum of values is represented in these tables.

The continuous mixed flow reactor (abbreviated CMF) closely approximates the behaviour of many components in ecosystems. The CMF reactor is illustrated schematically in Fig. 2.3. The input concentration of component i is C_{io} and the flow-rate is Q. The tank (ecosystem) has the constant volume V and in the tank the concentration of i, denoted C_i is uniform. The effluent stream also has a flow-rate Q, and because the tank is considered to be perfectly mixed the concentration in the effluent will be C_i.

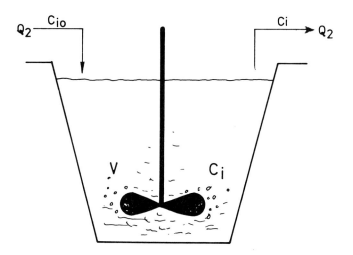

Fig. 2.3. Principle of a mixed flow reactor.

The principles of mass conservation can be used to set up the following simple differential equation: (no reactions take place in the tank)

$$\frac{V dC_i}{dt} = QC_{io} - QC_i \qquad (2.7)$$

In a steady state situation $dC_i = 0$, this means we have:

$$QC_{io} = QC_i \qquad (2.8)$$

If a first order reaction has taken place in the tank (ecosystem) the equation will be changed to:

$$\frac{V dC_i}{dt} = QC_{io} - QC_i - V * k * C_i \qquad (2.9)$$

and in the steady state situation:

$$QC_{io} - QC_i = V * k * C_i \qquad (2.10)$$

By dividing this equation with QC_i the following equation is obtained:

$$\frac{C_{io}}{C_i} - 1 = tr * k \qquad (2.11)$$

where tr is the retention time or the mean residence time in the tank (ecosystem). Rearrangement yields:

$$\frac{C_i}{C_{io}} = \frac{1}{tr * k + 1} \qquad (2.12)$$

For a number of reaction tanks, m, in series, each with volume V, a similar set of equations can be set up:

$$\frac{C_{i\,m}}{C_{io}} = \frac{1}{(1 + tr * k)^m} \qquad (2.13)$$

In an ideal plug flow reactor (PF), bulk flow proceeds through the reactor in an orderly uniform manner. The contents are uniform, and there is no mixing due to longitudinal concentration gradients. Thus, in an ideal PF reactor, composition varies along the axis of flow, and the mass conservation equation must be written for a differential element of volume. The resulting equation is consequently a partial differential equation:

$$\frac{\partial C_i}{\partial t} = - v_x \frac{\partial C_i}{\partial x} \pm k * C_i \text{ (+ generation, - decay)} \qquad (2.14)$$

where x is the axial direction parallel to the flow and v_x is the flow-rate in the direction of the x-axis. A similar equation is also valid for plug flow in an aquatic ecosystem, where diffusion is without influence. The corresponding steady state equation is obtained by setting the left hand side of the equation to zero.

In case where diffusion plays a major role it is necessary to include this term in the equation. The reactants are mixed in the direction of flow as a result of a concentration gradient and diffusion and eddy dispersion. The reactants are perfectly mixed in the radial direction and the rate of the reaction is assumed to follow a first order reaction scheme.

An elemental slice normal to the direction of flow is illustrated in Fig. 2.4. The mass conservation yields the following expression:

$$\frac{\partial C_i}{\partial t} = - v_x \frac{\partial C_i}{\partial x} + D \frac{\partial^2 C_i}{\partial x^2} \pm k * C_i \qquad (2.15)$$

where D is a coefficient of dispersion equal to $D_m + D_e$ - respectively the molecular diffusion coefficient and the eddy dispersion coefficient. For a description of these processes and approximations applied, see textbooks in hydrophysics.

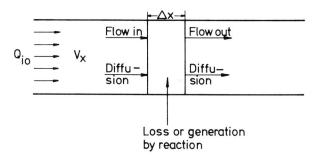

Fig. 2.4. Principle of plug flow with diffusion.

If the component considered is a conservative tracer the effluent response curve can be obtained from the solution of equation (2.15) with $k * C_i$ = 0. This solution has been given by Thomas & McKee (1944) and has the following form: (v_x = a constant)

$$\frac{C_i}{C_{io}} = 2 \sum_{i=1}^{\infty} \mu_i \left[\frac{\beta \sin\mu_i + \mu_i \cos\mu_i}{\beta^2 + 2\beta + \mu_i^2} \right] \exp \left[\beta - \frac{(\beta^2 + \mu_1^2)^c}{2\beta} \right] \quad (2.16)$$

$$\beta = 1/2 \left[\frac{1}{D + V_x * L} \right]$$

$$\mu_i = \cot^{-1} \left[\frac{\mu_i}{\beta} - \frac{\beta/\mu_i}{2} \right] \quad (2.17)$$

$$\Omega = \frac{tv_x}{L}$$

L = length

The equations shown above all assume a first order reaction or no reaction at all. The corresponding steady state solution expressed by means of the mean residence time for 0-mth order reactions are summarized in Table 2.11 for the continuous mixed flow and plug flow situations. The order of reaction is said to be mth order if the following equation is valid for the reaction rate:

$$\frac{dC}{dt} = -k * C^m \tag{2.18}$$

TABLE 2.8
Mean residence time for reactions of different orders.

Reaction order	Mean Residence Times	
	CMF	Plug Flow
0	$\frac{1}{k}(C_{io} - C_i)$	$\frac{1}{k}(C_{io} - C_i)$
1	$\frac{1}{k}(\frac{C_{io}}{C_i} - 1)$	$\frac{1}{k}(\ln \frac{C_{io}}{C_i})$
2	$\frac{1}{kC_i}(\frac{C_{io}}{C_i} - 1)$	$\frac{1}{kC_{io}}(\frac{C_{io}}{C_i} - 1)$
$m (m \neq 1)$	$\frac{1}{kC_i^{m-1}}(\frac{C_{io}}{C_i} - 1)$	$\frac{1}{k(m-1)(C_{io}^{m-1})}((\frac{C_{io}}{C_i})^{m-1} - 1)$

The set of equations presented above consider only the following processes: **continuous or plug discharge, degradation rate and diffusion**. However, in most cases the situation is much more complicated, and only a few cases can be simplified to a description by use of equations (2.4 to 2.17). However, these equations can be used to give a first crude approximation, which in many cases is quite useful in aquatic ecosystems. If the set of equations set up to describe the concentration is far more complicated, more processes must be taken into account, and it is necessary to use an ecological model (see Jørgensen, 1979 and 1980).

Space does not permit a detailed examination of more complicated models, but some processes of interest in an environmental context can be mentioned (in addition to **hydrophysical** and **meteorological** ones):
1. leaching of ions and organic compounds in soil,
2. evaporation of organic chemicals from soil and surface water,
3. atmospheric wash-out of organic chemicals,
4. sedimentation of heavy metals and organic chemicals in aquatic ecosystems,
5. hydrolysis of organic chemicals,
6. dry-deposition from the atmosphere,
7. chemical oxidation,
8. photochemical processes.

Table 2.9 gives some typical examples of the 8 above-mentioned processes and an idea of where these processes are of importance.

The list has not included **biotic** processes, which would further complicate the picture. These processes will be mentioned in 2.8.

All chemical reations proceed until equilibrium in accordance with a nth order reaction scheme, as presented above. At equilibrium no reaction takes place or rather the two opposite reactions have the same rate. If we consider a reaction:

$$aA + bB = cC + dD \qquad (2.18)$$

then the following equation is valid at equilibrium:

$$\frac{[C]^c \, [D]^d}{[A]^a \, [B]^b} = K \qquad (2.19)$$

where [ ] indicates concentration in M (moles per liter).

TABLE 2.9
Some chemicophysical processes of environmental interest

Process	Examples
Leaching of ions and organic compounds in soil	Nutrient run-off from agricultural areas to lake ecosystems
Evaporation of organic chemicals from soil and surface water	Evaporation of pesticides
Atmospheric wash-out of organic chemicals	Wash-out of pesticides
Sedimentation of heavy metals and organic chemicals in aquatic ecosystems	Most heavy metals have a low solubility in sea water and will therefore precipitate and settle
Hydrolysis of organic chemicals	Hydrolytic degradation of pesticides in aquatic ecosystems
Dry deposition from the atmosphere	Dry deposition of heavy metals on land
Chemical oxidation	Sulphides are oxidized to sulphates, sulphur dioxide to sulphur trioxide, which forms sulphuric acid with water
Photochemical processes	Many pesticides are degraded photochemically

In a chemostate or in nature, equilibrium is rarely attained, as one of the reactants is continuously added to the system. However, many processes are rapid and equation (2.19) can be used with good approximation, if the time steps considered are significantly larger than the reaction time. This is often the case for the following environmental processes: adsorption, hydrolysis, chemical (but not biochemical) oxidations and acid-base reactions including neutralization.

P.2.6.   **Concentrations of pollutants as function of time can be found by use of the mass conservation principles. If many processes are involved simultaneously, the use of a computer model is needed. Equilibrium description might be used for rapid processes.**

2.4.   LIFE CONDITIONS.

Section 2.13 describes the effect of toxic substances on animals and plants. This chapter will discuss and illustrate the effect of non-toxic substances on life. Again, the concentration of the component is assumed to be available from computations such as those presented in section 2.3.

P.2.7.   The life building processes require the presence of the element characteristic to the biosphere (see Appendix 1).

The composition of some organisms and their diets are given in Appendix 4 to illustrate these relationships.

P.2.8.   The composition of the biosphere is closely related to the function of the elements.

The high concentration of C, H and O is due to the composition of organic compounds. The   nitrogen concentration results from the presence of proteins, including enzymes and polypeptides, and nucleotides. Phosphorus is used as matrix material in the form of calcium compounds, in phosphate esters and in ATP, which is involved in all energetically coupled reactions. (ATP is an abbreviation for adenosine triphosphate and the molecule is shown in Fig. 2.5).

ATP is an energy-rich compound, because of its relatively large negative free energy of hydrolysis, denoted ΔG:

$$H_2O + ATP = ADP + P + \Delta G_{high}$$

$$\text{(2.20)}$$

$$H_2O + ADP = AMP + P + \Delta G_{high}$$

AMP is not a high energy compound since:

$$AMP = A + P + \Delta G_{low} \qquad \text{(2.21)}$$

These processes do not actually occur in living cells, but ATP participates directly or indirectly in group transfer reactions. A simple example will illustrate this point:

$$ATP + glucose = ADP + glucose\text{-}P \qquad \text{(2.22)}$$

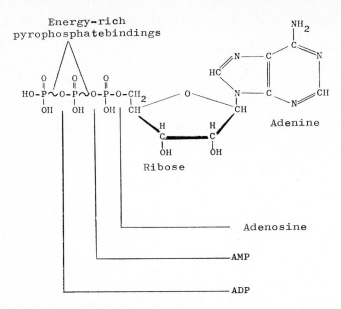

Fig. 2.5. ATP-molecule

Whenever food components are decomposed by a living organism the energy released by the process is either used as heat or stored in the form of a number of ATP molecules.

As with ATP it is possible to indicate a function for other elements needed by living organisms, which is related to the biochemistry of each organism.

P.2.9. **The pattern of the biochemistry determines the relative need for a number of elements (in order of 20-25), while other elements (the remaining 65-70 elements) are more or less toxic. Some elements needed by some species are toxic to others in all concentrations.**
This does not imply that a particular species has a fixed composition of the required elements. The composition might vary within certain ranges.

It is known, for example, that algae species may have from 0.5 to 2.5 g phosphorus per 100 g of dry matter, with an average at about 1 g phosphorus per 100 g dry matter. High phosphorus concentrations are regarded as luxury uptake.

TABLE 2.10
The relation between growth and oxygen concentration for the fish
Salvelinus fontinalis

% of max. growth	Conditions
0	$O_2 = 1.70$ mg l^{-1}, 278 K, weight = 5 g, 8 cm
30	$O_2 = 2.05$ mg l^{-1}, 278 K, weight = 5 g, 8 cm
80	$O_2 = 4.80$ mg l^{-1}, 278 K, weight = 5 g, 8 cm
Max. growth	$O_2 = 6.90$ mg l^{-1}, 278 K, weight = 5 g, 8 cm
0	$O_2 = 1.80$ mg l^{-1}, 283 K, weight = 5 g, 8 cm
30	$O_2 = 2.23$ mg l^{-1}, 283 K, weight = 5 g, 8 cm
80	$O_2 = 5.09$ mg l^{-1}, 283 K, weight = 5 g, 8 cm
Max. growth	$O_2 = 11.40$ mg l^{-1}, 283 K, weight = 5 g, 8 cm
0	$O_2 = 4.24$ mg l^{-1}, 288 K, weight = 5 g, 8 cm
30	$O_2 = 4.64$ mg l^{-1}, 288 K, weight = 5 g, 8 cm
80	$O_2 = 9.71$ mg l^{-1}, 288 K, weight = 5 g, 8 cm
0	$O_2 = 4.33$ mg l^{-1}, 293 K, weight = 5 g, 8 cm
30	$O_2 = 5.36$ mg l^{-1}, 293 K, weight = 5 g, 8 cm
0	$O_2 = 1.00$ mg l^{-1}, 278 K, weight = 400 g
30	$O_2 = 1.12$ mg l^{-1}, 278 K, weight = 400 g
80	$O_2 = 2.41$ mg l^{-1}, 278 K, weight = 400 g
Max. growth	$O_2 = 3.13$ mg l^{-1}, 278 K, weight = 400 g
0	$O_2 = 1.74$ mg l^{-1}, 283 K, weight = 400 g
30	$O_2 = 1.83$ mg l^{-1}, 283 K, weight = 400 g
80	$O_2 = 3.64$ mg l^{-1}, 283 K, weight = 400 g
Max. growth	$O_2 = 6.05$ mg l^{-1}, 283 K, weight = 400 g
0	$O_2 = 2.90$ mg l^{-1}, 288 K, weight = 400 g
30	$O_2 = 3.17$ mg l^{-1}, 288 K, weight = 400 g
80	$O_2 = 6.47$ mg l^{-1}, 288 K, weight = 400 g
0	$O_2 = 3.57$ mg l^{-1}, 293 K, weight = 400 g
30	$O_2 = 4.38$ mg l^{-1}, 293 K, weight = 400 g

Animals require oxygen for respiration, and there is a pronounced relationship between oxygen concentration and growth rate and mortality (see Table 2.10 and Figs. 2.6-2.8). Oxygen has a relatively low solubility in water (see Table 2.11). Low oxygen concentrations in aquatic ecosystems have often caused the destruction of fish due to discharge of otherwise harmless organic material which is biologically decomposed by microorganisms.

Through the breakdown of the organic components oxygen is consumed, causing a critical low oxygen concentration in the water. This is especially dangerous at high temperatures where the biological degradation is fast and the solubility low (see Table 2.11). Section 2.6, which is devoted to oxygen balance in a river system, illustrates the interaction of these processes

Examples of limiting oxygen concentrations for aquatic organisms are shown in Table 2.12.

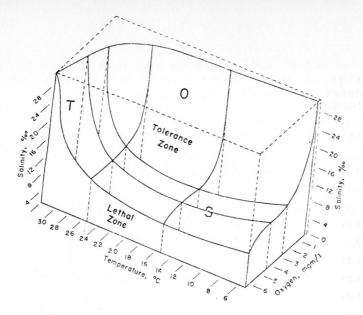

Fig. 2.6. LC$_{50}$ versus temperature, oxygen and salinity for lobster. 48h- LC$_{50}$ test is used.

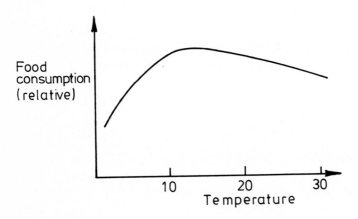

Fig. 2.7. Relative food consumption of large-mouth bass versus temperature

TABLE 2.11
Dissolved oxygen (ppm) in fresh, brackish and sea water at different temperatures and at different chlorinities (%). Values are amount of saturation.

C°	0%	0.2%	0.4%	0.6%	0.8%	1.0%	1.2%	1.4%	1.6%	1.8%	2.0%
1	14.24	13.87	13.54	13.22	12.91	12.58	12.29	11.99	11.70	11.42	11.15
2	13.74	13.50	13.18	12.88	12.56	12.26	11.98	11.69	11.40	11.13	10.86
3	13.45	13.14	12.84	12.55	12.25	11.96	11.68	11.39	11.12	10.85	10.59
4	13.09	12.79	12.51	12.22	11.93	11.65	11.38	11.10	10.83	10.59	10.34
5	12.75	12.45	12.17	11.91	11.63	11.36	11.09	10.83	10.57	10.33	10.10
6	12.44	12.15	11.86	11.60	11.33	11.07	10.82	10.56	10.32	10.09	9.86
7	12.13	11.85	11.58	11.32	11.06	10.82	10.56	10.32	10.07	9.84	9.63
8	11.85	11.56	11.29	11.05	10.80	10.56	10.32	10.07	9.84	9.61	9.40
9	11.56	11.29	11.02	10.77	10.54	10.30	10.08	9.84	9.61	9.40	9.20
10	11.29	11.03	10.77	10.53	10.30	10.07	9.84	9.61	9.40	9.20	9.00
11	11.05	10.77	10.53	10.29	10.06	9.84	9.63	9.41	9.20	9.00	8.80
12	10.80	10.53	10.29	10.06	9.84	9.63	9.41	9.21	9.00	8.80	8.61
13	10.56	10.30	10.07	9.84	9.63	9.41	9.21	9.01	8.81	8.61	8.42
14	10.33	10.07	9.86	9.63	9.41	9.21	9.01	8.81	8.62	8.44	8.25
15	10.10	9.86	9.64	9.43	9.23	9.03	8.83	8.64	8.44	8.27	8.09
16	9.89	9.66	9.44	9.24	9.03	8.84	8.64	8.47	8.28	8.11	7.94
17	9.67	9.46	9.26	9.05	8.85	8.65	8.47	8.30	8.11	7.94	7.78
18	9.47	9.27	9.07	8.87	8.67	8.48	8.31	8.14	7.97	7.79	7.64
19	9.28	9.08	8.88	8.68	8.50	8.31	8.15	7.98	7.80	7.65	7.49
20	9.11	8.90	8.70	8.51	8.32	8.15	7.99	7.84	7.66	7.51	7.36
21	8.93	8.72	8.54	8.35	8.17	7.99	7.84	7.69	7.52	7.38	7.23
22	8.75	8.55	8.38	8.19	8.02	7.85	7.69	7.54	7.39	7.25	7.11
23	8.60	8.40	8.22	8.04	7.87	7.71	7.55	7.41	7.26	7.12	6.99
24	8.44	8.25	8.07	7.89	7.72	7.56	7.42	7.28	7.13	6.99	6.86
25	8.27	8.09	7.92	7.75	7.58	7.44	7.29	7.15	7.01	6.88	6.85
26	8.12	7.94	7.78	7.62	7.45	7.31	7.16	7.03	6.89	6.86	6.63
27	7.98	7.79	7.64	7.49	7.32	7.18	7.03	6.91	6.78	6.65	6.52
28	7.84	7.65	7.51	7.36	7.19	7.06	6.92	6.79	6.66	6.53	6.40
29	7.69	7.52	7.38	7.23	7.08	6.95	6.82	6.68	6.55	6.42	6.29
30	7.56	7.39	7.25	7.12	6.96	6.83	6.70	6.58	6.45	6.32	6.19

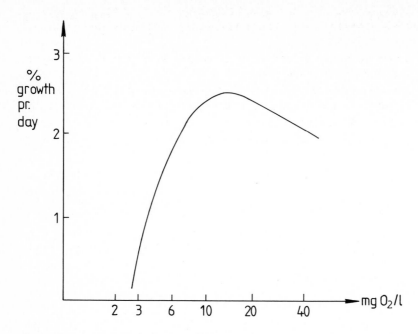

Fig. 2.8. Growth rate of Atlantic Salmon at unrestricted rations versus oxygen concentation. The growth rate is expressed for 50 g fish as % gain in dry weight per day. (Davis et al., 1963)

TABLE 2.12.
Examples of limiting oxygen concentrations for aquatic organisms (Poole et al., 1978)

Organism	Temp. %	mg O_2/l
Brown trout	6-24	1.3-2.9
Coho salmon	16-24	1.3-2.0
Rainbow trout	11-20	1.1-3.7
Worms (*Nereis grubei and capitella capitata*)	22-26	1.5-3.0
Amphipod (*Hyglella azteca*)	-	0.7

The relation between nutrient concentration and growth has long been known for crop yields. Fig. 2.9 illustrates what is called

P.2.10. Liebig's minimum law: if a nutrient is at a minimum relative to its use for growth, there is a linear relation between growth and the concentration of the nutrient.

If the supply of other factors is at a minimum, further addition of the nutrient will, as shown, not influence growth.

In arid climates primary production (the amount of solar energy stored by green plants) is strongly correlated with precipitation, as shown in Fig. 2.10. Here water is obviously "the limiting factor". Among many limiting factors, frequently the most important are various nutrients, water and temperature.

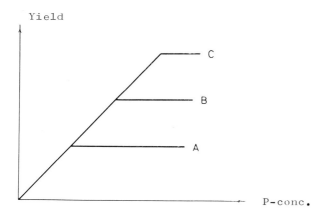

Fig. 2.9. Illustration of Liebig's law. The phosphorus concentration is plotted against the yield and the P-concentration proportional. At a certain concentration another component will be limiting, and a higher P-concentration will not increase the yield. The three levels, A, B and C correspond in this case study to three different potassium concentrations.

P.2.11. The relation between growth and nutrient concentration is described by Michaelis-Menten's equation, which is also used to give the relation between the concentration of a substrate and the rate of a biochemical reaction:

$$\bar{v} = \frac{k * s}{k_m + s} \tag{2.23}$$

where

$\bar{v}$ = the rate (e.g. growth rate)
k = a rate constant

s = the concentration of a substrate
k_m = the so-called half saturation constant

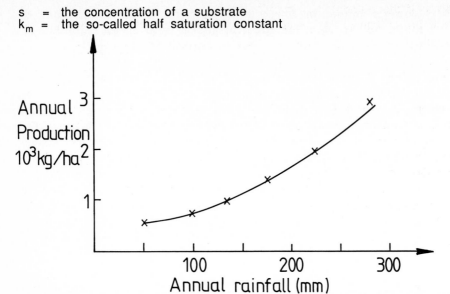

Fig. 2.10. A strong correlation between annual rainfall and primary production along a precipitation gradient in a desert region (South West Africa)
is shown.

The situation in nature is often not so simple, as two or more nutrients
may interact. This can be described by:

$$ v = \frac{k_1 * N_1}{k_{m_1} + N_1} * \frac{k_2 * N_2}{k_{m_2} + N_2} = K_r * \frac{N_1}{(k_{m_1} + N_1)} * \frac{N_2}{(k_{m_2} + N_2)} \qquad (2.24) $$

where
N_1 and N_2 = the nutrient concentrations
K_r = a rate constant
k_{m_1} and k_{m_2} = half saturation constants related to N_1 and N_2

This equation will however, often limit the growth too much and is in
disagreement with many observations (Park et al., 1978).
The following equation seems to overcome these difficulties:

$$ v = K_r * \min \left(\frac{N_1}{k_{m_1} + N_1} , \frac{N_2}{k_{m_2} + N_2} \right) \qquad (2.25) $$

The relationship between nutrient discharge and its effect on an ecosystem is illustrated in section 2.7, where a nutrient balance for a lake is

considered. If the nutrient concentration in a lake is increased the growth of algae will be enhanced. As a result, through photosynthesis, this will produce oxygen:

$$6CO_2 + 6H_2O \rightarrow C_6H_{12}O_6 + 6O_2 \qquad (2.26)$$

However, the increased organic matter produced by the process will sooner or later decompose and thereby cause an oxygen deficit (for further details see section 2.7).

While nutrients are necessary for plant growth, they may produce a deterioration in life conditions for other forms of life. Ammonia is extremely toxic to fish, while ammonium, the ionized form is harmless. As the relation between ammonium and ammonia is dependent on pH: (see also section 6.3.4)

$$NH_4^+ = NH_3 + H^+ \qquad (2.27)$$

$$pH = pK + \log \frac{NH_3}{NH_4^+} \qquad (2.28)$$

where $pK = -\log K$ and $K =$ equilibrium for process (2.27).

The pH value as well as the total concentration of ammonium and ammonia is important. This is demonstrated in Table 2.13.

This implies that the situation is very critical in many hypereutrophic lakes during the summer, when photosynthesis is most pronounced, as the pH increases when the acidic component CO_2 is removed or reduced by this process. The annual variations of pH in a hypereutrophic lake are shown in Fig. 2.11.

pK is about 9.3 in distilled water at 25°C, but increases with increasing salinity. It implies that the concentrations shown in Table 2.13 are higher in sea water.

Temperature is also of great importance to life in ecosystems. The influence of the temperature on some biotic processes is given in Table 2.14 and Fig. 1.5 shows a typical relation between growth and temperature.

The amount of solar energy intercepting a unit of Earth's surface varies markedly with latitude for two reasons. First, at high latitudes a beam of light hits the surface at an angle, and its light energy is spread out over a large surface area. Second, a beam that intercepts the atmosphere at an angle must penetrate a deeper blanket of air, and hence more solar energy is reflected by particles in the atmosphere and radiated back into space. (Local energy that reaches the ground). A similiar result of both these effects is that average annual temperatures tend to decrease with increasing latitude

(Table 2.15). The poles are cold and the tropics are generally warm.

In terrestrial ecosystems, climate is by far the most important determinant of the amount of solar energy plants are able to capture as chemical energy, or the gross primary productivity, see Table 2.16.

TABLE 2.13
Concentration of ammonia (NH_3 + NH_4^+) which contains an un-ionized ammonia concentration of 0.025 mg NH_3 l^{-1}

C°	pH = 7.0	pH = 7.5	pH = 8.0	pH = 8.5	pH = 9.0	pH = 9.5
5	19.6	6.3	2	0.65	0.22	0.088
10	12.4	4.3	1.37	0.45	0.16	0.068
15	9.4	5.9	0.93	0.31	0.12	0.054
20	6.3	2	0.65	0.22	0.088	0.045
25	4.4	1.43	0.47	0.17	0.069	0.039
30	3.1	1	0.33	0.12	0.056	0.035

TABLE 2.14
The influence of temperature on environmental processes

A. $v(T) = v(20°C) * k^{(T-20)}$

Process	k
Nitrification	1.07-1.10
$NH_4^+ \rightarrow NO_2^-$	1.08
$NO_2^- \rightarrow NO_3^-$	1.06
Org-N $\rightarrow NH_4^+$	1.08
Benthic oxygen uptake	1.065
Degradation of organic matter in water	1.02-1.09
Reaeration of streams	1.008-1.026
Respiration og zooplankton	1.05

B. $v(T) = v(T-10°) * k$

Process	k
Respiration of communities	1.7-3.5 (2.5)
Respiration of freshwater fish	2.18-3.28 (2.4)
Sulphate reduction in sediment	3.4-3.9
Excretion of NH_4^+ from zooplankton	2.0-4.3
Excretion of total nitrogen from zooplankton	1.5-2.5
Oxygen uptake of sediment	3.2
Respiration of zooplankton	1.77-3.28

TABLE 2.15
Average Annual Temperature (^{o}C).

Latitude	Year	January	July	Range
90°N	-22.7	-41.1	-1.1	40.0
80°N	-18.3	-32.2	2.0	34.2
70°N	-10.7	-26.3	7.3	33.6
60°N	-1.1	-16.1	14.1	30.2
50°N	5.8	-7.1	18.1	25.2
40°N	14.1	5.0	24.0	19.0
30°N	20.4	14.5	27.3	12.8
20°N	25.3	21.8	28.0	6.2
10°N	26.7	25.8	27.2	1.4
Equator	26.2	26.4	25.6	0.8
10°S	25.3	26.3	23.9	2.4
20°S	22.9	25.4	20.0	5.4
30°S	16.6	21.9	14.7	7.2
40°S	11.9	15.6	9.0	6.6
50°S	5.8	8.1	3.4	4.7
60°S	-3.4	2.1	-9.1	11.2
70°S	-13.6	-3.5	-23.0	19.5
80°S	-27.0	-10.8	-39.5	28.7
90°S	-33.1	-13.5	-47.8	34.3

TABLE 2.16
Net Primary Productivity and World Net Primary Production for the Major Ecosystems

	Area (10^6 km^2)	Net primary productivity per unit area (dry g/m^2/yr)		World net primary production (10^6 dry tons/yr)
		normal range	mean	
Lake and stream	2	100 - 1500	500	1.0
Swamp and marsh	2	800 - 4000	2000	4.0
Tropical forest	20	1000 - 5000	2000	40.0
Temperate forest	18	600 - 2500	1300	23.4
Boreal forest	12	400 - 2000	800	9.6
Woodland and shrubland	7	200 - 1200	600	4.2
Savanna	15	200 - 2000	700	10.5
Temperate grassland	9	150 - 1500	500	4.5
Tundra and alpine	8	10 - 400	140	1.1
Desert scrub	18	10 - 250	70	1.3
Extreme desert, rock and ice	24	0 - 10	3	0.07
Agricultural land	14	100 - 4000	650	9.1
Total land	*149*		*730*	*109.0*
Open ocean	332	2 - 400	125	41.5
Continental shelf	27	200 - 600	350	9.5
Attached algae and estuaries	2	500 - 4000	2000	4.0
Total ocean	*361*		*155*	*55.0*
Total for Earth	*510*		*320*	*164.0*

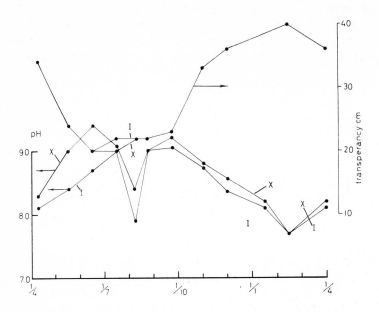

Fig. 2.11. The seasonal variation in the transparency and pH, at station I and X, in a hypereutrophic lake (Lake Glumsoe, Denmark).

2.5. THE GLOBAL ELEMENT CYCLES.

An increase or decrease in the concentration of components or elements in ecosystems are of vital interest, but:

P.2.12. The observation of trends in global changes of concentrations might be even more important as they may cause changes in the life conditions on earth.

Fig. 2.13 illustrates the global carbon cycle. Carbon is used in photosynthesis in form of carbon dioxide and hydrogen carbonate (HCO_3^-), and converted into organic matter, (see equation 2.26).

This is an energy-consuming process and the energy is supplied by solar radiation. The plants supply energy to herbivorous (plant eating) animals, which again are food for carnivorous (meat eating) animals. Every step in this foodchain produces waste (dead animals and plants, faeces, excretions), which is the food of decomposers. Such organisms decompose (mineralize) organic components to inorganic matter, which can then be taken up by plants. By mineralization and respiration carbon dioxide is produced. This recycling of matter is illustrated in Fig. 2.12. The system gets energy from an external source, the sun, which is the basis for the energy consumption of all life - plans, herbivorous and carnivorous animals and decomposers alike.

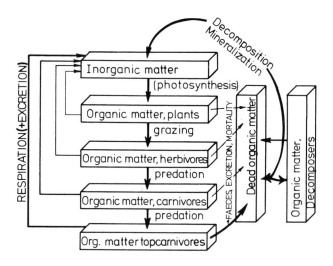

Fig. 2.12.   Biochemical cycling of matter.

This natural recycling of carbon also occurs in the lithosphere and the hydrospere (see Fig. 2.13). As carbon dioxide is a gas the foodchain in the lithosphere is using the atmosphere as a storage for carbon dioxide; hydrogen carbonate is stored in the hydrosphere. Carbon dioxide is exchanged between the atmosphere and the hydrosphere.

P.2.13.   **The solubility of a gas at a given concentration in the atmosphere can be expressed by means of Henry's law:**

$$p = H * x \qquad\qquad (2.29)$$

where
p = the partial pressure
H = Henry's constant
x = molar fraction in solution

H is dependent on temperature (see Table 2.17).

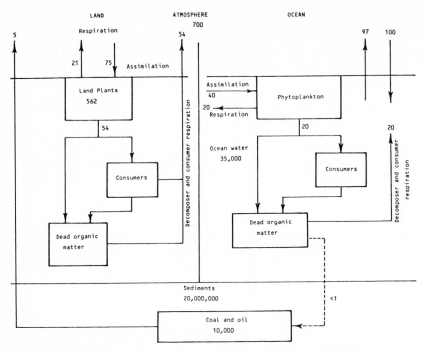

Fig. 2.13. Carbon cycle, global. Values in compartments are in 10^9 tons and in fluxes 10^9 tons/year.

So far we have presented the biochemical part of the global carbon cycle. The rates of action in this cycle are relatively fast, while the process rates in the geochemical cycles are slow. Under the influence of pressure and temperature dead organic matter is slowly transformed to fossil fuel and to carbonate minerals.

These considerations raise the crucial question of whether human activity influences the carbon cycle. It does indeed. Through our use of fossil fuel, which produces carbon dioxide on burning man removes carbon from its store in the lithosphere and transfers it to the atmosphere:

$$C_xH_{2y} + (x + y/2)O_2 \rightarrow xCO_2 + yH_2O \qquad (2.30)$$

It is now for us to consider whether this amount is significant or negligible? As seen in Fig. 2.13 the input to the atmosphere is significant relative to its present concentration. The annual increase would be even higher if the hydrosphere was not taking about 60% of the carbon dioxide produced. However, this input into the hydrosphere is of minor importance (see Fig. 2.13).

TABLE 2.17
Henry's constant (1 atm) for gasses as a function of temperature

Henry's constant (x 10^{-3})

Gas	Temperature (^{o}C)						
	0	5	10	15	20	25	30
Acetylene	0.72	0.84	0.96	1.08	1.21	1.33	1.46
Air (atm)	0.43	0.49	0.55	0.61	0.66	0.72	0.77
Carbon dioxide	0.73	0.88	1.04	1.22	1.42	1.64	1.86
Carbon monoxide	0.35	0.40	0.44	0.49	0.54	0.58	0.62
Hydrogen	0.58	0.61	0.64	0.66	0.68	0.70	0.73
Ethane	0.13	0.16	0.19	0.23	0.26	0.30	0.34
Hydrogen sulphide	26.80	31.50	36.70	42.30	48.30	54.50	60.90
Methane	0.22	0.26	0.30	0.34	0.38	0.41	0.45
Nitrous oxide	0.17	0.19	0.22	0.24	0.26	0.29	0.30
Nitrogen	0.53	0.60	0.67	0.74	0.80	0.87	0.92
Nitric oxide	-	1.17	1.41	1.66	1.98	2.25	2.59
Oxygen	0.25	0.29	0.33	0.36	0.40	0.44	0.48

In Fig. 2.14 the carbon dioxide concentration in the atmosphere is plotted against time for a prediction based on an increase in the consumption of fossil fuel of 2% per annum. Such predictions are only possible by means of ecological models, as many processes interact, as illustrated in Fig 2.13.

The diffusion of carbon dioxide to the deep sea must also be taken into account, as the oceans cannot be considered as mixed tank reactors. Furthermore the solubility of carbon dioxide, varies according to pH which itself will change with the increased uptake of the acidic component of carbon dioxide:

$$CO_2 + OH^- \rightarrow HCO_3^- \quad (2.31)$$

These results create a new question of what effect this charge in the carbon dioxide concentration of the atmosphere will have on the life conditions on earth. This is a difficult question to answer simply, but a further discussion can be found in section 3.3.

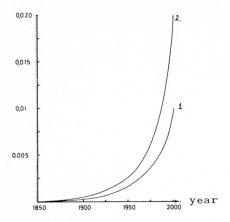

Fig. 2.14 The CO_2-concentration is plotted versus the year. Curve 1 is with and curve 2 without taking into consideration that approximately 60% of the input to the atmosphere is dissolved in the sea.

Concentration of air pollutants are mostly given as %, °/oo, or ppm on a volume/volume basis. The present concentration of CO_2 is about 0.034% (1985) or 340 ppm. It is, however, easy to convert from % or ppm. to mass per volume.

Example 2.1.
Convert 0.034% CO_2 to mg/m^3 at 0°C and 1 atm.

Solution.
Molecular mass is 12+2*16 = 44 g/mol
1 m^3 contains 0.034 * 1/100 m^3 = 0.00034 m^3 = 0.34 l

p * v = nRT   1 * 0.34 = n * 0.082 * 273
    n = 1.52 * 10^{-2} mol ≈ 0.668 g = 668 mg

668 mg/m^3 at 0°C and 1 atm. ≈ 0.034% CO_2

Global cycles for all elements can be set up, although our knowledge of some elements is rather limited. However, in principle good examples exist of the cycles of the more important elements.

Fig. 2.15 shows the sulphur cycle, while Fig. 2.16 illustrates the global nitrogen cycle. Both cycles are out of equilibrium as a result of human activity. The production of nitrogen fertilizer has meant that gaseous nitrogen from the atmosphere is converted to ammonia or nitrate, which are deposited in the lithosphere, while the major part is washed out to the hydrosphere, where it may cause eutrophication problems on a local scale.

There is a similar inbalance in the phosphorus cycle. Phosphorus minerals are also used in the production of fertilizers, which are applied in the lithosphere from which the major part is later transported to the hydrosphere.

For a more comprehensive introduction to the subject of global element cycles, see Svensson and Söderlund (1976) and Bolin (1979).

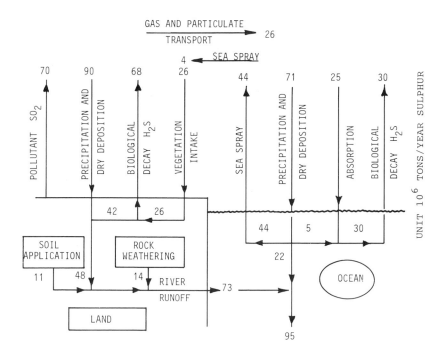

Fig. 2.15. The global sulphur cycle.

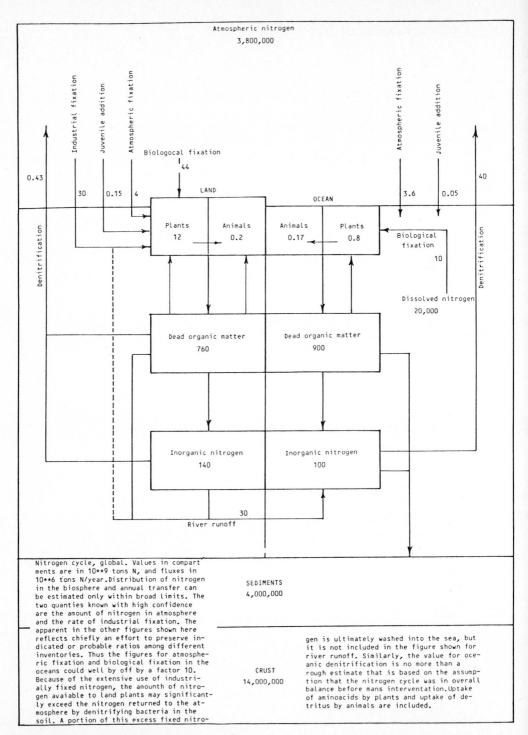

Atmospheric nitrogen
3,800,000

Industrial fixation

Juvenile addition

Atmospheric fixation

Biologocal fixation
44

0.43

30   0.15   4

Denitrification

LAND

OCEAN

Plants
12

Animals
0.2

Animals
0.17

Plants
0.8

Atmospheric fixation

Juvenile addition

3.6   0.05

40

Denitrification

Biological
fixation

10

Dissolved nitrogen
20,000

Dead organic matter
760

Dead organic matter
900

Inorganic nitrogen
140

Inorganic nitrogen
100

30

River runoff

Nitrogen cycle, global. Values in compart
ments are in 10**9 tons N, and fluxes in
10**6 tons N/year.Distribution of nitrogen
in the biosphere and annual transfer can
be estimated only within broad limits. The
two quanties known with high confidence
are the amount of nitrogen in atmosphere
and the rate of industrial fixation. The
apparent in the other figures shown here
reflects chiefly an effort to preserve indicated or probable ratios among different
inventories. Thus the figures for atmospheric fixation and biological fixation in the
oceans could well by off by a factor 10.
Because of the extensive use of industrially fixed nitrogen, the amounth of nitrogen avaiable to land plants may significantly exceed the nitrogen returned to the atmosphere by denitrifying bacteria in the
soil. A portion of this excess fixed nitro-

SEDIMENTS
4,000,000

CRUST
14,000,000

gen is ultimately washed into the sea, but
it is not included in the figure shown for
river runoff. Similarly, the value for oce-
anic denitrification is no more than a
rough estimate that is based on the assumption that the nitrogen cycle was in overall
balance before mans intervention.Uptake
of aminoacids by plants and uptake of detritus by animals are included.

Fig. 2.16. The global nitrogen cycle.

2.6. OXYGEN BALANCE OF A RIVER.

Maintenance of a high oxygen concentration in aquatic ecosystems is crucial for survival of the higher life forms in aquatic ecosystems. At least 5 mg l^{-1} is needed for many fish species (see also Table 2.12). At 20-21 °C it corresponds to (see Table 2.11) $5/9 = 56\%$ saturation. The oxygen concentration is influenced by several factors, of which the most important are:

1. **P.2.14. The decomposition of organic matter, can often be described by use of first order kinetic equation:** (see 2.3)

$$\frac{dL}{dt} = -K_1 * L \tag{2.32}$$

$$\int_0^t \frac{dL}{L} = \int_0^t -K_1 * dt \tag{2.33}$$

$$\ln \frac{L_t}{L_o} = -K_1 * t \tag{2.34} \text{ or}$$

$$L_t = L_o * e^{-K_1 t} \tag{2.35}$$

where
L_t = concentration of organic matter at time $= t$
L_o = initial concentration of organic matter
K_1 = rate constant

The concentration of organic matter can be expressed by means of the oxygen consumption in the decomposition process. This is usually BOD_5, which is the oxygen consumption measured over 5 days. The oxygen consumption after 5 days is therefore not included, but it will rarely be of importance, as all easily decomposed matter is broken down within 5 days.

BOD_5 of municipal waste water is usually 80-250 mg O_2 l^{-1} and for waste water after mechanical-biological treatment (concerning these processes, see section 6.2) 10-20 mg O_2 l^{-1}.

2. **The nitrification of ammonia (ammonium)** in accordance with the following process:

$$NH_4^+ + 2O_2 \rightarrow NO_3^- + H_2O + 2H^+ \qquad (2.36)$$

Ammonia is formed by decomposition of organic matter. Proteins and other nitrogenous organic matter are decomposed to simpler organic molecules such as amino acids, which again are decomposed to ammonia. Urea and uric acid, the waste products from animals, are also broken down to ammonia. Nitrifying microorganisms can use ammonia as an energy source, as the oxidation of ammonia is an energy-producing process. This decomposition chain is illustrated in Fig. 2.17, where it can be seen that the energy (chemicals) is decreased throughout the chain.

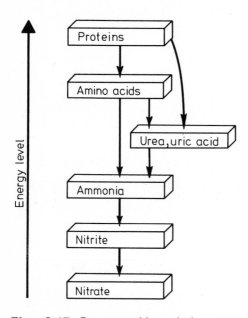

Fig. 2.17 Decomposition chain: protein to nitrate.

P.2.16. The nitrification process can be described by the following first order kinetic expression:

$$\frac{dN}{dt} = K_N * t \qquad (2.37)$$

$$\ln \frac{N_t}{N_o} = -K_N * t \qquad (2.38) \text{ or}$$

$$N_t = N_o * e^{-K_N * t} \qquad (2.39)$$

where
N_t = concentration of ammonium at time = t
N_o = concentration of ammonium at time = 0
K_N = rate constant, nitrification

N_t and N_o are here expressed by the oxygen consumption corresponding to the ammonium concentration in accordance with (2.41).

Values for K_1, K_N, L_o and N_o are given for some characteristic cases in Table 2.18.

P.2.16. K_1 and K_N are dependent on the temperature as illustrated in the following expression:

$$K_1 \text{ or } K_N \text{ at } T = (K_1 \text{ or } K_N \text{ at } 20°C) * K_T^{(T-20)} \qquad (2.40)$$

where T = the temperature (°C), K_T = a constant, see Table 2.19.

TABLE 2.18
Characteristic values, K_1, K_N, L_o and N_o (20 °C)

	K_1(day^{-1})	K_N(day^{-1})	N_o	L_o
Municipal waste water	0.35-0.40	0.15-0.25	80-130	150-250
Mechanical treated municipal waste water	0.35	0.10-0.25	70-120	75-150
Biological treated municipal waste water	0.10-0.25	0.05-0.20	60-120	10-80
Potable water	0.05-0.10	0.05	0-1	0-1
River water	0.05-0.15	0.05-0.10	0-2	0-5

The relation between ammonium concentration and oxygen consumption in accordance with (2.36) is calculated to be (2 * 32)/14 = 4.6 mg O_2 per mg NH_4^+ - N, but due to bacterial assimilation of ammonia this ratio is reduced to 4.3 mg O_2 per mg NH_4^+ - N in practice.

TABLE 2.19
Temperature dependence K_1 and K_N

	K_1	K_N
K_T	1.05	1.06-1.08

3. **Dilution**. The dilution capacity of a stream can be calculated using the principles of mass balance (principles 2.1 and 2.7):

$$L_S * Q_S + L_W * Q_W = L_m * Q_m \qquad (2.41)$$

where L represents the concentration (mass/volume) and Q the flow rate (volume/time). The subscripts S, W and m designate the stream, waste and mixture conditions, respectively.

4. **Settling or sedimentation** is nature's method of removing particles from a water body. Large solids will settle out readily, while colloidal particles can stay in suspension for a long period of time.

5. **Resuspension** of solids is common in aquatic ecosystems due to strong wind stress, flooding or heavy run-off.

6. **Photosynthesis**, which produces oxygen, see equation (2.26).

7. The use of oxygen for **respiration** by plants and animals.

8. **Oxidation of organic matter in the sediment**, which also causes consumption of oxygen and release af ammonia in accordance with the following equation:

Organic matter $+ O_2 - CO_2 + H_2O + NH_4^+$ \qquad (2.42)

9. If the oxygen concentration is below the saturation (see Table 2.11) **reaeration** from the atmosphere will take place. The equilibrium between the atmosphere and the water can be expressed by use of Henry's law (see equation 2.29, section 2.5).

P.2.17. The aeration is dependent on 1. the difference between the oxygen concentration at equilibrium and the present oxygen concentration, 2. the temperature, 3. the flow rate of water, and 4. the water depth.

These relationships are formulated quantitatively in the following equations:

$$K_a(20°C) = \frac{2.26 * v}{R^{2/3}} \quad (day^{-1}) \tag{2.43}$$

$$K_a(T) = K_a(20) * e^{\Omega(T-20)} \tag{2.44}$$

$$R_a = K_a(T)(C_s - C_t) \tag{2.45}$$

where

$K_a(20°C)$ = reaeration coefficient at 20°C (day^{-1})
$K_a(T)$ = reaeration coefficient at T°C (day^{-1})
v = average flow ($m * s^{-1}$)
R = depth (m)
R_a = rate of reaeration ($mg\ l^{-1}\ day^{-1}$)
Ω = constant = 0.024 $°C^{-1}$, 15 °C < T < 25 °C
C_s = oxygen concentration at saturation ($mg\ l^{-1}$)
C_t = actual oxygen concentration at time = t ($mg\ l^{-1}$)

O'Connor and Dobbins have found following alternative expression satisfactory:

$$K_a(20°C) = 649 * \frac{\sqrt{D} * S^{1/4}}{R^{5/4}} \tag{2.46}$$

where

D = coefficient of molecular diffusion (liquid film $m^2\ day^{-1}$)
S = slope (m/m)
Many other alternative expressions can be found in the literature; see for instance the review by Gromiec (1983).

If all these processes are considered together the following equation for the oxygen concentration, C_t, is obtained:

$$\frac{dC_t}{dt} = P - R - K_1 * L_t - K_N * N_t + K_a (C_s - C_t) - S \qquad (2.47)$$

where
P = oxygen production by photosynthesis
R = oxygen consumption by respiration
S = oxygen consumption of sediment

Dilution can be considered by calculation of L_t and N_t. Sedimentation and resuspension may be included in the rate constants K_1 and K_N or by dividing L_t and N_t into dissolved and suspended material.

If we omit the biochemical oxygen demand and S, the equation is reduced to:

$$\frac{dC_t}{dt} = P - R + K_a (C_s - C_t) \qquad (2.48)$$

Based upon measurements of the oxygen concentration as a function of the time, it is possible to calculate P, R and K_a. If measurements at two stations available are C_1 and C_2, we get:

$$\frac{\Delta C}{\Delta t} = \frac{C_2 - C_1}{t_2 - t_1} = K_a (\bar{C}_s - \bar{C}_t) + \bar{P} - \bar{R} \qquad (2.49)$$

where $\bar{C}_s$, $\bar{C}_t$, $\bar{P}$ and $\bar{R}$ represent average values and $t_2 - t_1$ = flow time between stations 1 and 2.

For the time between sunset and sunrise, P can be omitted and we get:

$$\frac{\Delta C}{\Delta t} = \frac{C_2 - C_1}{t_2 - t_1} = K_a (\bar{C}_s - \bar{C}_t) - \bar{R} \qquad (2.50)$$

From a set of measurements taken over 24 hours it is now possible to find estimates of K_a, P and R.

However, if P, L and R are considered as state variables and not as constant contributions, and temperature variations are included, it is necessary to use a more complicated model, which requires computer analysis.

50 years ago Streeter and Phelps were using a simpler approach, as they only considered two processes to be of importance in oxygen balance - biological decomposition and reaeration. They used the following differential

equation:

$$\frac{dD}{dt} = K_1 * L_t - K_a * D \tag{2.51}$$

where $D = C_s - C_t$ (2.52)

By use of equation 2.35, equation 2.51 can be reformulated:

$$\frac{dD}{dt} + K_a * D = K_1 * L_o * e^{-K_1 * t} \tag{2.53}$$

This equation can be solved analytically. When $D = D_o$ at $= 0$, we get:

$$D = \frac{K_1 * L_o}{K_a - K_1} (e^{-K_1 * t} - e^{-K_a * t}) + D_o * e^{-K_a * t} \tag{2.54}$$

For $K_1 - K_a$ the solution is: ((2.53) is not valid in this case)

$$D = (K_1 * t * L_o + D_o) * e^{-K_1 * t} \tag{2.55}$$

From equation (2.54) it is possible to plot D, C_t and the reaeration against time.
The point where the oxygen concentration is at a minimum is termed the critical point (see Fig. 2.18). This point can be found from:

$$\frac{dD}{dt} = 0 \qquad \frac{d^2D}{dt^2} < 0 \text{ (minimum)} \tag{2.56}$$

$$t_c = \frac{1}{K_a - K_1} \ln \left(\frac{K_a}{K_1} - \left(\frac{K_a}{K_1} - 1 \right) \frac{D_o}{L_o} \right) \tag{2.57}$$

-71-

$$D_c = \frac{K_1 * L_o}{K_a} * e^{-K_1 * t_c} \qquad (2.58)$$

If a nitrification term is added to the Streeter-Phelps equation, we get the following solution:

$$D = \frac{K_1 * L_o}{K_a - K_1} * (e^{-K_1 * t} - e^{-K_a * t}) + \frac{K_N * N_o}{K_a - K_N} (e^{-K_N * t} - e^{-K_a * t})$$

$$+ D_o e^{-K_a * t} \qquad (2.59)$$

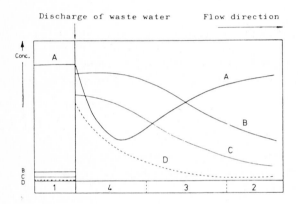

Fig. 2.18. A. Oxygen concentration, B. BOD_5, C, Nutrients, D. Suspended matter, versus distance in running water. The number corre- spond to the classification in the saprobic system (see text).

In equation (2.52) to (2.59) time is used as independent variable, but if a river is considered it might be useful to translate time to distance in flow direction, x, by use of the flow velocity, v:

$$x = v * t \qquad (2.60)$$

The equations can be used to describe the oxygen profile in the flow direction, but if so, some hydraulic assumptions must be introduced: 1. that by discharge of polluted waste water into the river complete mixing with the flow will take place, 2. that the flow rate is the same throughout the entire cross-section of the river.

This means that the concentration, C, just after discharge of waste water can be calculated from the following simple expression:

$$C = \frac{Q_w * C_w + Q_r * C_r}{Q_w + Q_r}$$

(2.61)

where

Q_w = waterflow, waste water (1 s^{-1})

C_w = concentration in waste water $(\text{mg } 1^{-1})$

Q_r = waterflow, river (1 s^{-1})

C_r = concentration in river water $(\text{mg } 1^{-1})$

Example 2.2.

A municipal waste-water plant discharges secondary effluent to a surface stream. The worst conditions occur in the summer months, when stream flow is low and water temperature high. The waste water has a max. flow of 12000 m^3/day, a BOD_5 of 40 mg l^{-1} (at 20°C), a dissolved oxygen concentration of 2 mg l^{-1} and a temperature of 25°C. The stream has a minimum flow of 1900 m^3 h^{-1}, a BOD_5 of 3 mg l^{-1} (at 20°C) and dissolved oxygen concentration of 8 mg l^{-1}. Use K_1 = 0.15 (see Table 2.18 for the mixed flow). The temperature in the stream reaches a maximum of 22°C. Complete mixing is almost instantaneous. The average flow (after mixing) is 0.2 m sec^{-1} and the depth of the stream 2.5 m.

Solution: Find the critical oxygen concentration.

$$K_a (20°C) = \frac{2.26 * 0.7}{2.5^{2/3}} = 0.25 \text{ d}^{-1} \quad \text{(equation 2.45)}$$

Mixture: 500 m^3/h (ww) + 1900 m^3/h (stream) = in total 2400 m^3/h

BOD_5 of mixture : $\dfrac{500 * 40 + 1900 * 3}{2400}$ = 10.7 mg l^{-1}

$L_5 = L_o * e^{-0.15*5}$ $*$ $BOD_5 = L_o - L_5 = 10.7$

$10.7 = L_o (1 - e^{-0.15*5})$

-73-

$L_o = 20.3$ mg l^{-1}

Dissolved oxygen $= DO_{mixture} = \dfrac{500*2+1900*8}{2400} = 6.7$ mg l^{-1}

Temperature$_{mixture} = \dfrac{500*25+1900*22}{2400} = 22.6$

K_1 at $22.6 = 0.15 * 1.05^{22.6-20} = 0.17$ (equation 2.43)

K_a at $22.6 = 0.25 * e^{0.024(22.6-20)} = 0.27$ (equation 2.46)

Initial oxygen deficit $= 8.7 - 6.7 = 2$ mg l^{-1} (use Table 2.14)

Critical location (time) $= \dfrac{1}{0.27-0.17}$ ln $(\dfrac{0.27}{0.17} - (\dfrac{0.27}{0.17} - 1) \dfrac{2}{20.3})$

$\qquad\qquad\qquad = 4.25$ d.

(equation 2.57)

$D_c = \dfrac{0.17 * 20.3}{0.27} e^{-0.17*4.25} = 6.2$ mg l^{-1}

This conditions occurs at $0.2 * 3600 * 24 * 4.25$ m $= 73440$ m

Oxygen concentration at critical point: $8.7 - 6.2 = $ **2.5 mg l^{-1}**

Oxygen concentrations in riverwater can be determined by means of more complex ecological models, which take into account hydraulic components, growth of phytoplankton and zooplankton, oxygen consumption of the sediment, the spectrum of biodegradability of the components present in discharged waste water, and the presence of toxic matter affecting the biodegradability. For further information on these more complicated models, see Jørgensen (1980), Rinaldi et al. (1979), Orlob (1981) and Jørgensen (1981).

In section 2.4 we mentioned the importance of oxygen concentration for the ecological equilibrium and survival of aquatic ecosystems. Using a biologico-chemical classification of running waters; based on the oxygen profile, it is easy to demonstrate how the oxygen concentration can be used to assess the ecological quality of the ecosystem.

As the variation in oxygen concentration need only be small to significantly alter the ecology of running water, it could be concluded that an ecological examination of the aquatic ecosystem is a more sensitive instrument for assessing water quality than a chemical examination, although naturally a close relationship exists between the two results. Another advantage of the ecological examination is its ability to provide a picture of the long-term conditions, while the chemical examination always gives a momentary picture.

One of the most commonly used ecological examinations is the application of the saprobien-system, which classifies running waters in 4 classes (Hynes, 1971):

1. *Oligosaprobic water* , in which the water is unpolluted or almost unpolluted.
2. *Beta-mesoaprobic water* , in which the water is slightly polluted.
3. *Alfa-mesosaprobic water* , in which the water is polluted.
4. *Poly-saprobic water* , in which the water is very polluted.

In american literature the following names are used for the same 4 classes:
1. Clear Water Zone
2. Recovery Zone
3. Active Decomposition Zone
4. Degradation Zone

The classification is based on an examination of plant and animal species present in the water at a number of locations. The species can be divided into four groups:

1. Organisms characteristic of unpolluted water
2. Species dominating in polluted water
3. Pollution indicators
4. Indifferent species

Figs. 2.19 to 2.23 illustrate some of the species characteristic of each group. Table 2.20 outlines the relationships between the groups of species and the classification of running water. Figs. 2.18 and 2.24 to 2.28 are illustrative examples of the relationship between the oxygen profile, the BOD_5, some chemical parameters and the species found in a running water. It can be seen that the oxygen concentration decreases after the discharge point and reaches a minimum, after which it again increases due to re-aeration. The BOD_5 decreases steadily due to the biodegradation of organic matter. Ammonium reaches a maximum as the nitrogenous material is decomposed, and then decreased due to nitrification. Phosphates are also released by the decomposition of organic matter and are thereafter taken up along with nitrate by algae. The species dominant in polluted waters are seen at their maximum where the oxygen concentration is low. As the oxygen concentration increases the species characteristic of unpolluted water become more dominant.

The classification presented above includes only four classes. For more complicated systems, see Hynes (1971).

Normally the assessment of the ecological conditions of running waters is a complicated procedure, and the application of the saprobic system requires some experience and is time consuming. Although a direct translation of chemical parameters to ecological conditions is not possible, it is often advantageous to measure chemical parameters, because it is a much faster procedure. A tentative translation of the chemical analysis to the saprobic system is shown in Table 2.21, but the table should be used very cautiously. As seen in this table it is not sufficient to measure the BOD_5 and the oxygen concentration alone: the ammonium and nitrate concentration must also be included in the examination to obtain a full picture of the conditions. Furthermore

P.2.18. assessment of the diurnal variations is important, as it is the minimum oxygen concentration, which is crucial.

this can usually be recorded shortly before sunrise.

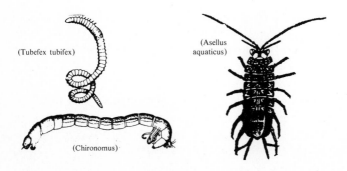

Fig. 2.19. Insects dominant in polluted water (5 x).

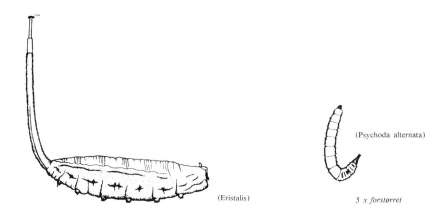

(Psychoda alternata)

(Eristalis)

5 x forstørret

Fig. 2.20 Pollution indicators (5 x).

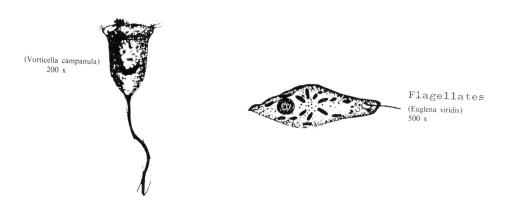

(Vorticella campanula)
200 x

Flagellates
(Euglena viridis)
500 x

Fig. 2.21 Protozoans.

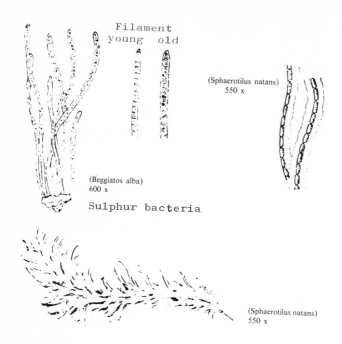

Filament
young old

(Sphaerotilus natans)
550 x

(Beggiatos alba)
600 x

Sulphur bacteria

(Sphaerotilus natans)
550 x

Fig. 2.22.   Bacteria and fungi.

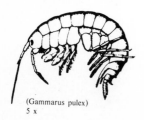

(Gammarus pulex)
5 x

Fig. 2.23   Characteristic of unpolluted water.

TABLE 2.20
Classification of running water

Class	Species	
Oligo-saprobic	Bacteria:	<100 germs pr. ml
	Algae:	Cladophora typical
	Similium:	Few Gammarus pulex, Hydropsyche
	Fish:	Salomoid species, trout
Beta-meso-saprobic	Bacteria:	<100,000 germs pr. ml
	Algae:	e.g. Cladophora, Spirogyra
	Plants:	Potamogeton, Elodea, Batrachium and others
	Insects:	Tubifex, Chironimus and Asellus Gammarus pulex. Baëtis
	Fish:	Eel
		Helobdella, Glossiphonia, Sphaerium Pisidium, Planorbis, Ancyclus
Alfa-meso-saprobic	Aerobic bacteria:	>100,000 germs pr. ml (Sphaerotilus), Protozoans
	Algae:	Green, diatoms, bluegreen
	Plants:	Potamogeton crispus
	Insects:	Tubifex, Chironimus, Asellus aquaticus, Sialis
	Fish:	Stickleback, Sphaerium, Herbobdella
Poly-saprobic	Bacteria: Anaerobic species	>1,000,000 germs pr. ml
		Beggiatoa, Sphaerotilus, Apodyalactea, Fusarium aqueductum, Psotozaus: Carchesium, Vorticella, Bodo Euglena, Colpidium, Glaucoma
	Algae:	Bluegreen e.g. Oscillatoria
	Insects:	Tubifex, Chironimus dominating, Eristatis, Ptychoptera, Psycoda
	Fish:	Nil

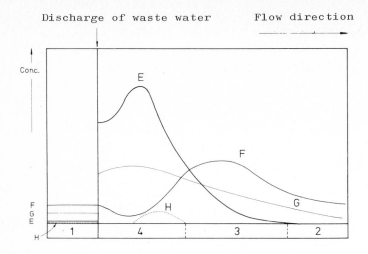

Discharge of waste water   Flow direction

Conc.

Fig. 2.24 E. $NH_3 + NH_4^+$, F. NO_3^-, G. PO_4^{3-}, H. NO_2^- - concentrations versus distance in running water. The numbers correspond to the classification in the saprobic system.

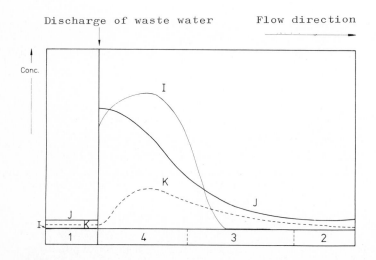

Discharge of waste water   Flow direction

Conc.

Fig. 2.25. I. Pollution indicator, J. Bacteria, K. Protozan - concentrations versus distance in running water. The classification is indicated with numbers (see text).

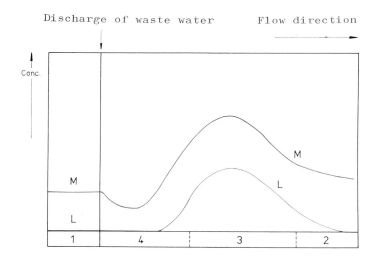

Fig. 2.26. L. Cladophora, M. Other algae -concentrations versus distance in running water. Classification in accordance with saprobic system (see text).

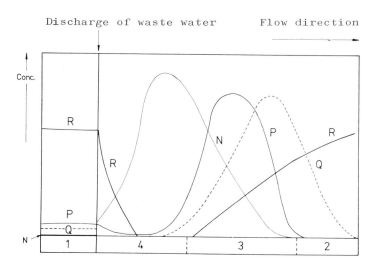

Fig. 2.27. N. Tubifex, P. Chironomus, Q. Gammarus pulex - concentrations and R. Species characteristic of unpolluted water versus distance in running water. Classification with numbers in accordance with the saprobic system.

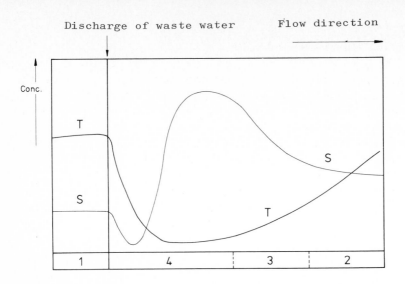

Fig. 2.28. S. Number of higher organisms, T. Number of species versus distance in running water. Classification with numbers in accordance with saprobic system.

TABLE 2.21
The saprobic-system and physiochemical parameters

Physico-chemical parameter	Poly-saprobic	Alfa-meso-saprobic	Beta-meso-saprobic	Oligo-saprobic
Dissolved oxygen	0-3 mg l^{-1} <50% saturation	BOD_5 increasing >70% saturation BOD_5 decreasing <30% saturation	>60% saturation	>90% saturation
BOD_5	high	increasing <5 mg l^{-1} decreasing >20 mg l^{-1}	<5 mg l^{-1}	<3 mg l^{-1}
NH_4^+	0.5-2 mg N l^{-1}	0.3-1.2 mg N l^{-1}	<0.2 mg N l^{-1}	<0.1 mg N l^{-1}
NO_2^-	0-0.2 mg N l^{-1}	0-0.2 mg N l^{-1}	~0.2 mg N l^{-1}	<0.05 mg N l^{-1}
NO_3^-	very low	1-2 mg N l^{-1}	2-6 mg N l^{-1}	high
Turbidity	high	low	very low	very clean

2.7. THE EUTROPHICATION PROBLEM.

The word eutrophic generally means "nutrient rich". Naumann introduced in 1919 the concepts of oligotrophy and eutrophy. He distinguished between oligotrophic lakes containing little planktonic algae and eutrophic lakes containing much phytoplankton.

The eutrophication of lakes in Europe and North America has grown rapidly during the last decade due to the increased urbanization and the increased discharge of nutrient per capita.

The production of fertilizers has grown exponentially in this century as demonstrated in Fig. 2.29 and the concentration of phosphorus in many lakes reflects this growth as shown in Fig. 2.30 (from Ambühl, 1969).

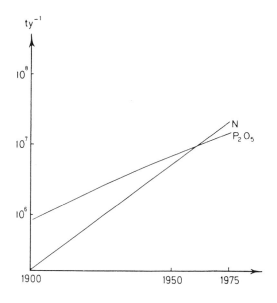

Fig. 2.29. The production of fertilizers (t y^{-1}), as demonstrated for H and P$_2$O$_5$, has grown exponentially (the y-axis is logarithmic).

The word eutrophication is used increasingly in the sense of the artificial addition of nutrients, mainly nitrogen and phosphorus, to water. Eutrophication is generally considered to be undesirable, although it is not always so.

The green colour of eutrophic lakes makes swimming and boating more unsafe due to increased turbidity. Furthermore, from an aesthestic point of view the chlorophyll concentration should not exceed 100 mg m^{-3}. However, the most critical effect from an ecological viewpoint is the reduced oxygen content of the hypolimnion, caused by the decomposition of dead algae. Eutrophic lakes might show high oxygen concentrations at the surface during the summer, but low oxygen concentrations in the hypolimnion, which may cause fishkill, see Fig. 2.31 - the oxygen profile. The zones of deep lakes are shown in Fig. 2.32.

On the other hand an increased nutrient concentration may be profitable for shallow ponds used for commercial fishing, as the algae directly or indirectly form food for the fish population.

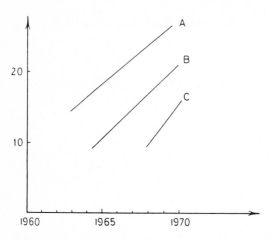

Fig. 2.30.   The total P-concentration (μg/l) in 3 segments of Vierwalds-tättersee as a function of time. A Kreutzrichter, B Lake Gersauer, C Lake Urner.

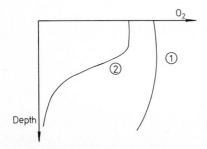

Fig. 2.31. Oxygen profile in a stratified lake: winter condition (1), summer condition (2).

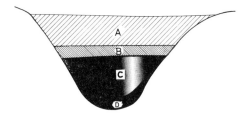

Fig. 2.32. Thermal stratification. A) Epilimnion, B) Thermocline, C) Hypo-limnion, D) Mud.

About 16-20 elements are necessary for the growth of freshwater plants, as demonstrated in Table 2.22, where the relative quantities of essential elements in plant tissue are shown. Compare with the table in appendix 4, where elements in *dry* plant tissues are given.

TABLE 2.22
Average fresh-water plant composition on *wet* basis

Element	Plant content (%)
Oxygen	80.5
Hydrogen	9.7
Carbon	6.5
Silicon	1.3
Nitrogen	0.7
Calcium	0.4
Potassium	0.3
Phosphorus	0.08
Magnesium	0.07
Sulphur	0.06
Chlorine	0.06
Sodium	0.04
Iron	0.02
Boron	0.001
Manganese	0.0007
Zinc	0.0003
Copper	0.0001
Molybdenum	0.00005
Cobalt	0.000002

The present concern about eutrophication relates to the rapidly increasing amount of phosphorus and nitrogen, which are normally present at relatively low concentrations. Of these two elements phosphorus is considered the major cause of eutrophication, as it was formerly the growth-limiting factor for algae in the majority of lakes but, as demonstrated in Fig. 2.29, its usage has greatly increased during the last decades. Nitrogen is a limiting factor in number of East African lakes as a result of the nitrogen depletion of soils by intensive erosion in the past. However, today nitrogen may become limiting to growth in lakes as a result of the tremendous increase in the phosphorus concentration caused by discharge of waste water, which contains relatively more phosphorus than nitrogen. While algae use 4-10 times more nitrogen than phosphorus, waste water generally contains only 3 times as much nitrogen as phosphorus. Furthermore, nitrogen accumulates in lakes to a lesser extent than phosphorus and a considerable amount of nitrogen is lost by denitrification (nitrate to N_2).

The importance of phosphorus in the eutrophication of lakes is shown in Fig. 2.33, where the algal concentration, expressed as µg chlorophyll-a per l, is plotted against the phosphorus concentration for several lakes. The correlation is obvious.

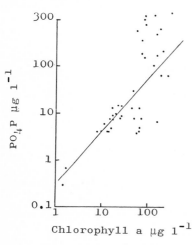

Fig. 2.33. Algae biomass (max) versus ortho-P for several lakes in England.

P.2.19. From a thermodynamic point of view a lake can be considered an open system which exchanges material (waste water, evaporation, precipitation) and energy (evaporation, radiation) with the environment.

However, in many great lakes the input of material per year is unable to change the concentration significantly. In such cases the system can be considered as (almost) closed, which means that it is exchanging energy but not material with the environment.

P.2.20. **The flow of energy through an ecosystem lead to at least one cycle of material in the system.**
(provided that the system is in steady state, see Morowitz, 1968).

As demonstrated in Figs. 2.34 to 2.38 the important elements all participate in process cycles. These cycles contain the processes that determine eutrophication.

The growth of phytoplankton is the key process in eutrophication and it is therefore of great importance to understand the interacting processes regulating its growth.

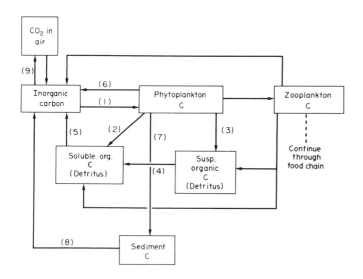

Fig. 2.34. The carbon cycle. Processes: (1) Photosynthesis, (2) Excretion, (3) Mortality, (4) and (5) Decomposition (mineralization), (6) Respiration, (7) Settling, (8) Mineralization of sediment, (9) Exchange of CO_2 with atmosphere, (10) Grazing.

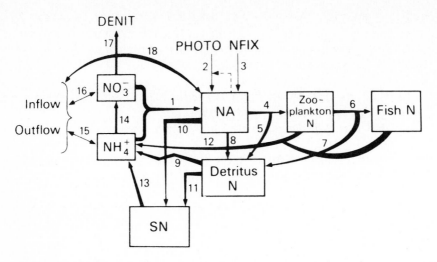

Fig. 2.35. The nitrogen cycle. The processes are: (1) Uptake of NO_3^- and NH_4^+ by algae, (2) Photosynthesis, (3) Nitrogen fixation, (4) Grazing with loss of undigested matter, (5), (6) and (7) correspond (4) and (5), (8) Mortality, (9) Mineralization, (10) Settling, (11) Settling of ditritus, (12) Settling, (13) Release from sediment, (14) Nitrification, (15), (16) and (18) Input/output, (17) Denitrification.

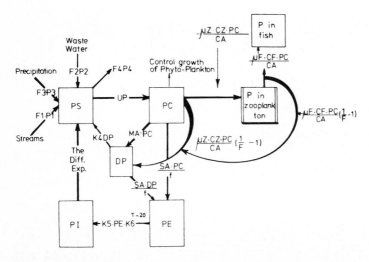

Fig. 2.36. The phosphorus cycle. The equations for the process are included in the diagram.

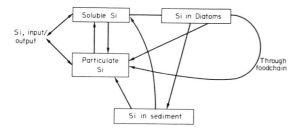

Fig. 2.37. The silica cycle.

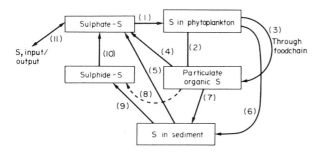

Fig. 2.38. The sulphur cycle. The oxidation from S in sediment and particulate organic sulphur to sulphate-S has sulphide as an intermediate, but as the oxidation from sulphide-S to sulphate-S is a very rapid process, the kinetic equation of the processes (4) and (5) can be described as a one-step process, as is also indicated on the diagram. The processes shown with dotted lines are taking place under anaerobic conditions, which means that sulphide-S is accumulating.

Primary production has been measured in great detail in many great lakes. This process represents the synthesis of organic matter, and can be summarized as follows:

$$\text{light} + 6CO_2 + 6H_2O \longrightarrow C_6H_{12}O_6 + 6O_2 \tag{2.62}$$

This equation is necessarily an oversimplification of the complex metabolic pathway of photosynthesis, which is dependent on sunlight, temperature and the concentration of nutrient. The composition of phytoplankton is not constant (note that Table 2.30 only gives an average concentration), but to a certain extent reflects the chemical composition of the water. If, e.g. the phosphorus concentration is high, the phytoplankton will take up relatively more phosphorus - the luxury uptake.

P.2.21. **As seen Table 2.22 phytoplankton consists mainly of carbon, oxygen, hydrogen, nitrogen and phosphorus, and without these elements no algae growth will take place. So each of these elements represents a limiting factor on algae growth.**

For further details, see section 2.4.

P.2.22. **Another side of the problem is the consideration of nutrient sources. It is important to set up mass balances for the most essential nutrients.**

The sequence of events leading to eutrophication often occurs as follows. Oligotrophic waters often have a ratio of N:P greater than or equal to 10, which means that phosphorus is less abundant relative to the needs of phytoplankton than nitrogen. If sewage is discharged into the lake the ratio will decrease since, the N:P ratio for municipal waste water is about 3:1, and consequently nitrogen will be less abundant than phosphorus relative to the needs of phytoplankton. Municipal waste water contains typically 30 mg l^{-1} N and 10 mg l^{-1} P. In this situation, however, the best remedy for the excessive algal growth is not necessarily to remove nitrogen from the sewage, because the mass balance might show that nitrogen-fixing algae would produce an uncontrollable input of nitrogen into the lake.

It is necessary to set up a mass balance for the nutrients. This will often reveal, that the input of nitrogen from nitrogen-fixing blue green algae, precipitation and tributaries is already contributing too much to the mass balance for any effect to be produced by nitrogen removal from the sewage. On the other hand the mass balance may reveal that most of the phosphorus input (often more than 95%) comes from the sewage, and so demonstrates that it is better management to remove phosphorus from the sewage rather than nitrogen. It is, therefore

P.2.23. **not important which nutrient is limiting, but which nutrient can most easily be made to limit the algal growth.**

These considerations have implied that the eutrophication process can be controlled by a reduction in the nutrient budget. For this purpose a number of eutrophication models have been developed, which take a number of processes into account. For details, see Jørgensen (1979), Jørgensen (1978), Jørgensen (1980), Orlob (1981) and Jørgensen (1981).

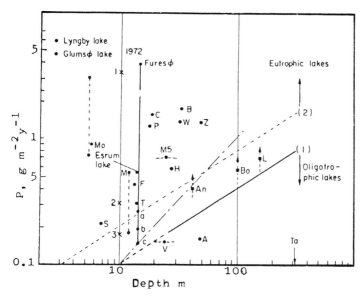

Fig. 2.39. a, b and c correspond to removal of 90%, 95% and 99% (Furesoe) of P input respectively. Glumsoe, 1972 (Jørgensen, Jacobsen and Høi, 1973):, Lyngby lake, 1972 (F.L. Smidth/MT, 1973); Esrom lake, 1972.

G - Greifensee	L - Lac Leman	An - Lac Annecy
P - Pfäffikersee	V - Vänern	Ta - Lake Tahoe
B - Baldeggersee	M - Lake Mendota	Bo - Bodensee
F - Furesoe (1954)	T - Türlersee	Mo - Lake Moses
H - Halwillersee	A - Aegerisee	
W- Lake Washington	Z - Zürichsee	

It will suffice here to consider the so-called Vollenweider plot (Vollenweider, 1969), which is much simpler to use than ecological models, but as it does not consider the dynamics of the phytoplankton population, the annual variation, the sediment and its interaction with the water body, it can only give a crude picture of the possible control mechanisms in existence. The plot is shown in Fig. 2.39, where the phosphorus loading in g/m^2 year is related to the depth of water. The diagram consists of three areas corresponding to oligotrophic, mesotrophic and eutrophic lakes.

A similar plot can be constructed for nitrogen and a comparison of the two diagrams can approximately show whether a possible reduction in the nitrogen loading would be a better management solution than a reduction in the phosphorus loading. Vollenweider has later (1975) improved these considerations by taking input, output and the net loss to the sediment into consideration and by using a correction factor for stratified lakes, but in

cases where these improvements are required it is better to use an ecological model. Under all circumstances Vollenweider's plot should be used as a
first approximation.

Lakes can be classified in accordance with their primary production - the
so-called oligotrophic-eutrophic series - which is shown in Table 2.23.

Typical oligotrophic lakes are deep, with the hypolimnion larger than the
epilimnion (for zonation, see Fig. 2.32). Littoral plants are scarce and the
plankton density is low, although the number of species can be large. Due to
the low productivity the hypolimnion does not suffer from oxygen depletion.
The nutrient concentration is low and plankton blooms are rare, so the water
is highly transparent. This relationship between productivity and transparency is shown in Fig. 2.40. Here the transparency in m is plotted against
the maximum G_{24}-value (the maximum production in mg per 24h and per m^3).

This relationship does not always general apply, as the transparency is
dependent not only on the phytoplankton concentration, but also on the
concentration of inorganic suspended matter (e.g. clay) and the colour of the
water (very humic rich lakes are brownish), but in most lakes the
transparency is mainly determined by the phytoplankton.

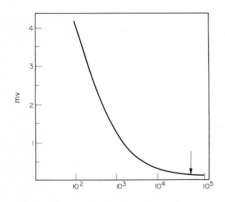

Fig. 2.40. The transparency, v, plotted to G_{24max}. The arrow indicates the
values for Lake Glumsoe.

Eutrophic lakes are generally shallower and have a higher phytoplankton
concentration, and thus a generally lower transparency. Littoral vegetation
is abundant and summer and spring algal blooms characteristic (see Figs.
2.41 and 2.42).

Another characteristic difference between oligotrophic and eutrophic lakes is the profile of the change in photosynthesis with depth (see Fig. 2.43).

Table 2.23
Classification of lakes

Trophic type	Mean primary productivity (mg C/m^2/day)	Phytoplankton density (cm^3/m^3)	Phytoplankton biomass (mg C/m^3)	Chlorophyll (mg/mg^3)	Dominant phytoplankton
Ultra-oligotrophic	< 50	< 1	< 50	0.01-0.5	
Oligotrophic	50-300		20-100	0.3-3	Chrysophyceae,
					Cryptophyceae,
Oligo-mesotrophic		1-3			Dinophyceae,
					Bacillariophyceae
Mesotrophic	250-1000		100-300	2-15	
Meso-eutrophic		3-5			
Eutrophic	> 1000		> 300	10-50	Bacillariophyceae
					Cyanophyceae,
Hypereutrophic		> 10			Chlorophyceae,
					Euglenophyceae
Dystrophic	< 50-500		< 50-200	0.1-10	

TABLE 2.23, continued

Trophic type	Light extinction coefficients (n/m)	Total organic carbon (mg/l)	Total P (μg/l)	Total N (μg/l)	Total inorganic solids (mg/l)
Ultra-oligotrophic	0.03-0.8		< 1-5	< 1-250	2-15
Oligotrophic	0.05-1.0	< 1-3			
Oligo-mesotrophic			5-10	250-600	10-200
Mesotrophic	0.1-2.0	< 1-5			
Meso-eutrophic			10-30	500-1100	100-500
Eutrophic	0.5-4.0	5-30			
Hypereutrophic			30- 5000	500- 15000	400-60000
Dystrophic	1.0-4.0	3-30	< 1-10	< 1-500	5-200

Fig. 2.41. Eutrophic lake.

Fig. 2.42. Littoral vegetation is abundant in eutrophic lakes.

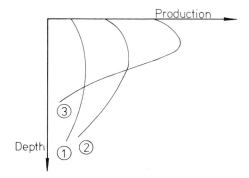

Production →

Depth ① ②

③

Fig. 2.43. Productivity of phytoplankton per unit of volume versus the depth in a series of lakes. (1) oligotrophic, (2) mesotrophic, (3) eutrophic.

2.8.   MASS CONSERVATION IN A FOOD CHAIN.

The mass flow through a food chain has been discussed in 2.2. The food taken in by one level in the food chain is used in respiration, waste food, undigested food, excretion, growth and reproduction. If the growth and reproduction are considered as the net production, we can state that:

net production = intake of food - respiration - excretion - waste food

$$(2.63)$$

and we can call the ratio of the net production to the intake of food the net efficiency (for further details about these considerations, see 3.1 and 3.2).

The net efficiency is dependent on several factors, but is often as low as 10%.

P.2.24. **Any toxic matter in the food is unlikely to be lost through respiration and excretions, because it is much less bio-degradable than the normal components in the food.**

This being so, the net efficiency of toxic matter is often higher than for normal food components, and as a result some chemicals, such as chlorinated hydrocarbons including DDT and some heavy metals, can be magnified in the food chain.

This phenomenon is denoted **biological   magnification** and is illustrated for DDT in Table 2.24.

TABLE 2.24
Biological magnification **(Data after Woodwell et al., 1967)**

Trophic level	Concentration of DDT (mg/kg dry matter)	Magnification
Water	0.000003	1
Phytoplankton	0.0005	160
Zooplankton	0.04	~ 13,000
Small fish	0.5	~ 167,000
Large fish	2	~ 667,000
Fish-eating birds	25	~ 8,500,000

P.2.25. **DDT and other clorinated hydrocarbons have a specially high biological magnification, because they have a very low biodegradability and are only excreted from the body very slowly, because they are dissolved in fatty tissue.**

These considerations can also explain why

P.2.26. **pesticide residues observed in fish increase with the increasing weight of the fish** (see Fig. 2.44).

TABLE 2.25
Concentration of DDT (mg per kg)

Atmosphere	0.000004
Rain water	0.0002
Atmospheric dust	0.04
Cultivated soil	2.0
Freshwater	0.00001
Seawater	0.000001
Grass	0.05
Aquatic macrophytes	0.01
Phytoplankton	0.0003
Invertebrates on land	4.1
Invertebrates in sea	0.001
Freshwater fish	2.0
Sea fish	0.5
Eagles, falcons	10.0
Swallows	2.0
Herbivorous mammals	0.5
Carnivorous mammals	1.0
Human food, plants	0.02
Human food, meat	0.2
Man	6.0

As man is on the last level of the food chain, relatively high DDT concentrations have been observed in the human body fat (see Table. 2.25). The sources within the dietary intake can be found in Table 2.26.

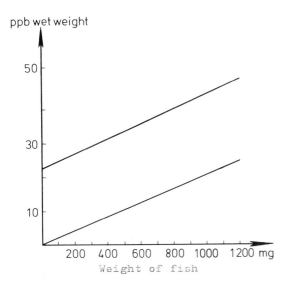

Fig. 2.44. Increase in pesticide residues in fish as weight of the fish increases. Top line = total residues; bottom line = DDE only. (After Cox, 1970).

TABLE 2.26
Uptake efficiencies by oral intake

Compound	Species	Efficiency (%)
Biphenyl	Rat	98.4-98.9
Cadmium	Goat	2
Cobalt	Rat	0.6
Copper	Fish	2.9
DDT, fruits	Man	9.9
DDT, meat, fish	Man	40.8
Mercury, inorganic	Man	6-8
Mercury, organic	Man	15-100
Mercury, organic	Rat	90
Lead	Man	5-17
Lead	Rabbit	0.8-1
Lead	Sheep	1.3
2-chlorobiphenyl	Rat	98.4-98.9
2,2', 4,4', 5,5' hexachlorobiphenyl	Rat	95.1-95.3

In addition to pesticides man is exposed to a wide variety of heavy metals. Generally heavy metals are excreted faster than pesticides, and fortunately their uptake efficiency is generally lower than that of organic matter (see Table 2.26).

2.9.   MASS CONSERVATION IN A SOCIETY.

It is, as stressed in 2.6, of great importance to quantify any inbalance in the global cycles of elements. We have pointed out in 2.4. and 2.5 that determination of the concentrations of life-essential components, as well as toxic components, is crucial for the assessment of the life conditions in the ecosystem. This paragraph is devoted to regional pollution problems.

P.2.27.   Many toxic substances are widely dispersed and a global increase in the concentration of heavy metals and pest-icides has been recorded (see, e.g. Fig. 2.45).

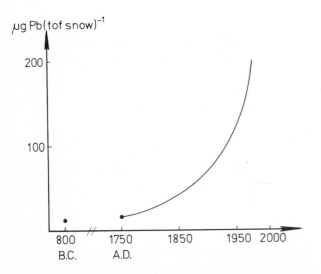

Fig. 2.45. Accumulation of lead in the Greenland ice pack from 800 B.C. to the present. (After Chemistry, 1968).

However, such pollution problems are often far more severe on the regional level. This is illustrated in Table 2.27, where the concentrations of several heavy metals found in the river Rhine are compared with their concentration in the North Sea.

P.2.28. **It is often appropriate to start the solution of a regional pollution problem by setting up a mass balance for the considered component or element to clarify the sources of pollution and to state the most effective means of solving the problem.**

TABLE 2.27
Heavy metal pollution in the river Rhine

	The river Rhine t/year	Ratio	$\dfrac{\text{conc. in the Rhine}}{\text{conc. in the North Sea}}$
Cr	1000		20
Ni	2000		10
Zn	20,000		40
Cu	200		40
Hg	100		20
Pb	2000		700

To illustrate these principles it is appropriate to use an example. What is causing the lead contamination in Denmark? What is the effect of this contamination, and if there is a problem, how can it be solved? The methods used to find the answer to these questions are not limited to lead or to any specific region (Denmark), but could, as principles, be applied to other similar regional pollution problems.

The lines we will follow can be summarized as follows:

1. *From the regional statistical information a mass balance can be set up, including a statement of import, export, mining, production, recirculation and loss.*
2. *The loss is the source of pollution, and the next issue to consider is whether the pollution is widely dispersed in the region or localized,* and how it is dispersed. This can often be answered when the nature of loss is known.

   As the loss will often take place in very diluted form, it might be advantageous to determine the concentration of the pollutant, in order to find the total loss to the environment, by multiplying the concentration with the total amount of "carrier", which is often known or is measurable.

3. The further fate of the pollutant in the environment is the most crucial detail from an ecological point of view, but it is also a very difficult problem to solve. Although intensive research has been carried out, our knowledge in this field is limited. Some crude guidelines with relation to heavy metals can however be given:
   a) *A particulate pollutant in the atmosphere will settle, the rate of settling being dependent on the particle size.*
   b) *Several chemical processes can take place in the hydrosphere. Some of these we are able to predict and can give rate constants for and other relevant information.* (See also 2.8, 2.13, 3.1 and 3.2).
   c) *Similar processes can take place in soil and soil water. Some knowledge is available about the uptake of pollutants by plants in the lithosphere, but this process is very much dependent upon the soil properties,* see Jørgensen (1975) and (1976). These processes in the environment have also been discussed.
4. It is, of course, of special interest *to obtain knowledge about the overall impact on man.* The dispersion of pollutants in the atmosphere, hydrosphere and lithosphere will cause contamination of the air, drinking water, portable water and food. The crucial question is, therefore, what the overall effect of these contaminations has on man?

As mentioned above, these considerations can be illustrated by the regional lead pollution in Denmark.

Fig. 2.46 summarizes the information about the lead mass balance in Denmark. Export, import and recirculation can be found from current statistics. The loss to environment is determined for each application by measurements or calculation of the loss percentage. Table 2.28 gives some of the details not included in Fig. 2.46.

TABLE 2.28
Discharge of lead to the environment

Source	t Pb/year	
Wasted batteries	600	
Wasted by industrisl use	500	*)
Wasted by regeneration	100	*)
Wasted to atmosphere (gasoline-75%)	900	
Paints and pigments	600	
By hunting	600	
By use of fossil fuel	100	
Solid waste	1000	
Total	4400	

*) approximately 360 tons can be found in waste water

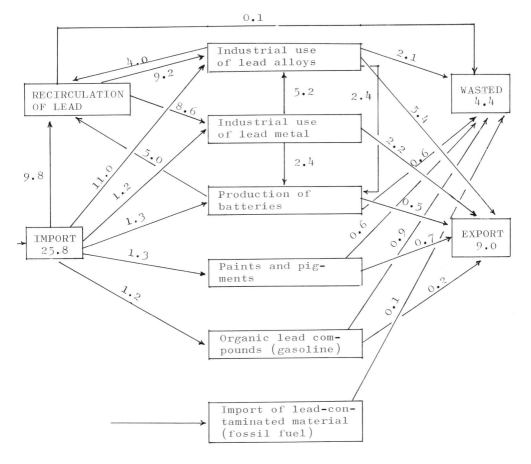

Fig. 2.46. The Danish lead balance. The loss to the environment is found as referred in the text to 4.4 * 10³ t/year (see table 2.28). The difference 12.4 * 10³ t/year corresponds an increase of the lead capital. As unit is used 1000 t Pb per year. Denmark 1969.

Fig. 2.47 shows the dispersion of the lead losses to the environment. The amount lost to the atmosphere, hydrosphere and lithosphere is indicated.

Furthermore, it shows to a certain extent, where the final deposition will take place. We can distinguish between what is deposited locally and what is evenly dispersed. Approximately 90% of the particulate lead in the atmosphere will not settle close to the source, but will be more or less evenly dispersed over the entire region, while the remaining 10% will settle close to the source. As much as approximately 1000 t of atmospheric lead is evenly dispersed. If we consider that all 1000 t will settle on land with none in the marine environment, the lead deposition will average about 220 g ha[-1] (Denmark covers an area of 43,000 km^2). This assumption is, of course, not quite correct, but an actual measurement is close to this figure: 200 g ha[-1] (Stads - og Havneingeniøren 1973).

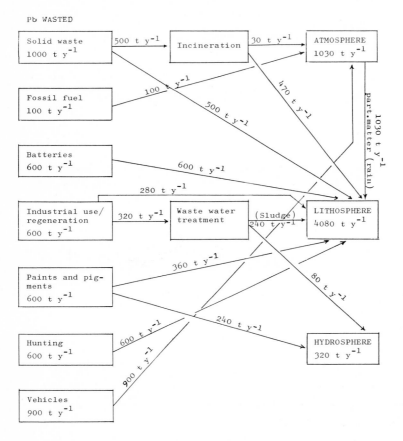

Fig. 2.47. The dispersion of lead losses to the environment. (Denmark 1969).

There is a close relationship between dispersion of lead in the environment and the contamination of potable water, food and air. The World Health Organization (WHO) recommends 0.1 mg l[-1] as the maximum permissible concentration in potable water, but this is probably too high. WHO uses such a high concentration because it would be impossible to meet a lower standard in some countries where lead water-pipes are used.

The average Dane consumes 0.3 mg of lead with food and water but at least 90% of this is not assimilated and lost in the excrement.

0.025 mg is contained in the daily inhaled air, of which 50% or 0.0125 mg is taken up by the blood. A model of a Dane's intake of lead is shown in Fig. 2.48. As can be seen the total uptake per day is 42.5 µg, of which 30 µg is excreted, while 12.5 µg is accumulated in bone tissues.

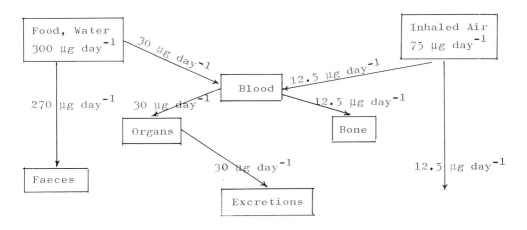

Fig. 2.48. Model of an average Dane's uptake of lead.

The excretion can appropriately be described by a first order reaction (see also 2.4):

$$\frac{d[Pb]}{dt} = -k * [Pb] \qquad (2.64)$$

where
[Pb] = concentration of lead (mg (kg dry matter)$^{-1}$)
k   = excretion rate constant (day^{-1})

$\dfrac{d[Pb]}{dt}$ = excretion rate (mg (kg dry matter * day)$^{-1}$)

This implies that the daily excretion, 30 µg, is a result of a certain accumulation, and Fig. 2.48 describes a steady state situation where the intake through the food and excretion is balanced.

The intake of lead with food (300 µg per day) is an average value, which is strongly dependent upon the food items, as illustrated in Table 2.29, where the lead concentration in different foods is listed. Another source of lead, which must not be forgotten in this context, is the application of white and red lead as dyestuffs in ceramic household articles. Recent legislation in many countries has minimized or banned the application of lead dyestuffs in household articles, as they were discovered to be an essential source of lead contamination in food.

TABLE 2.29
Lead in food

| Food items | Typical lead concentration *) in mg/kg fresh weight | | |
	England	Holland	Denmark
Milk	0.03	0.02	0.005
Cheese	0.10	0.12	0.05
Meat	0.05	<0.10	<0.10
Fish	0.27	0.18	0.10
Eggs	0.11	0.12	0.06
Butter	0.06	0.02	0.02
Oil	0.10	-	-
Corn	0.16	0.045	0.05
Potatoes	0.03	0.1	0.05
Vegetables	0.24	0.065	0.15
Fruits	0.12	0.085	0.05
Sugar	-	0.01	0.01
Soft drinks	0.12	0.13	-

*) The number of analyses are limited. The values cannot be considered as averages.

2.10. THE HYDROLOGICAL CYCLE.

Water is the most abundant chemical compound on earth (see Table 2.30), and also has some unique properties. Its importance for all life on earth can be demonstrated as follows:

1. *Our body consists of 70% water* and we need at least 1.5 litres per day to survive. We can survive without food for perhaps 80 days, but only a few days without water.
2. *Water serves as a basic transport medium for lifegiving nutrients.*
3. *Water removes and dilutes many natural and man-made wastes.*
4. *Water has a great ability to store heat energy and to conduct heat, and has an extremely high vaporization temperature compared with the*

molecular weight. These thermal properties are major factors of influence in the climatic pattern of the world, and in minimizing sharp changes in temperature on the earth.

5. *Water has its maximum density at 4 °C above its freezing point,* so solid water, ice, is less dense than liquid water. This is the reason why a water body freezes only on the top. If ice was denser than liquid water, lakes, rivers and oceans would freeze from the bottom up, killing most higher forms of aquatic life.

P.2.29. **Water shows a physical cycling** (compare with the chemical cycling of the elements, see section 2.5), **as demonstrated in Fig. 2.49. In this vast cycle, driven by solar energy, our supply of water is renewed again and again.**

TABLE 2.30
Water resources and annual water balance of the Continents of the World

Component	Europe	Asia	Africa	N. America	S. America	Australia	Total
Area (1E6 km^2)	9.8	45	30.3	20.7	17.8	8.7	132.3
Precipitation (km^3)	7165	32,690	20,780	13,910	29,355	6405	110E3
Total river runoff (km^3)	3110	13,190	4225	5960	10,380	1965	38,830
Underground runoff (km^3)	1065	3410	1465	1740	3740	465	11,885
Infiltration (km^3)	5120	22,910	18,020	9690	22,715	4905	83,360
Evaporation (km^3)	4055	19,500	16,555	7950	18,975	4440	71,475
% underground runoff of total	34	26	35	32	36	24	31

Water evaporates from the oceans, rivers, lakes and continents, and gravity pulls it back down as rain. Some of the water falls on the land sinks or percolates into the soil and ground to form ground water. The soil, can like a sponge, hold a certain amount of water, but if it rains faster than the rate at which the water percolates, water begins to collect in puddles and ditches and runs off into nearby streams, rivers and lakes. This run-off causes erosion. The water runs eventually into the ocean, which is the largest water storage tank. Because of this cycle, water is continually replaced, as indicated in Table 2.31.

Long-term average water-budget equations for extensive hydrological systems can be expressed as

P = E   (2.65)

where P is the precipitation inflow and E the evaporation outflow. The storage change, S, is zero.

Water-budget equations for entire land and water masses must also contain the total water discharge from land to ocean, Q:

$$P + Q + E \qquad \text{for oceans} \qquad\qquad (2.66)$$
$$P + Q + E \qquad \text{for land} \qquad\qquad (2.67)$$

The numerical equality is illustrated in Table 2.32.

The short-term water-budget equation for a terrestrial ecosystem must include a storage term, S:

$$P = Q + E + S \qquad\qquad (2.68)$$

If subsurface flows are included, we have

$$P + Q_i + L_i = E + Q_o + L_o + S \qquad\qquad (2.69)$$

where Q_i and Q_o are surface inflow and outflow, respectively, and L_i and L_o are the corresponding subsurface flows.

TABLE 2.31
The water cycle

Water in	Is replaced every
Human body	month
The air	12 days
A tree	one week
Rivers	a few days
Lakes	0.1 - 100 years
Oceans	3600 years
Polar ice	15,000 years

TABLE 2.32
Mean annual water balance components for the earth

Item	Land	Ocean	Earth
Area (10^6 km^2)	148.9	361.1	510.0
Volume (10^3 km^3)			
Precipitation	111	385	496
Evaporation	- 71	- 425	- 496
Discharge	- 40	40	0
Mean depth (mm)			
Precipitation	745	1066	973
Evaporation	- 477	- 1177	- 973
Discharge	- 269	111	0

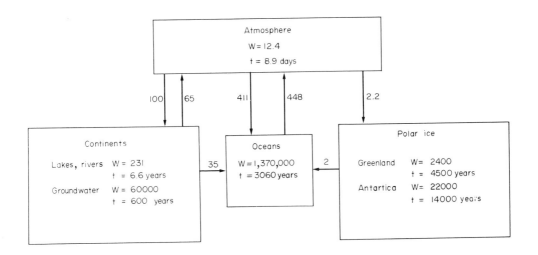

Fig. 2.49. Water cycle. W = Water content in compartment in $10^3 km^3$ * t = retention time. Numbers on fluxes represent $10^3 km^3 y^{-1}$.
Estimate of ground water is to a depth of 5 km in the earth's crust; much of this water is not actively exchanged.

2.11.   PLUME DISPERSION.

P.2.30.   Determination of the atmospheric concentration of pollutants emitted from point sources is an important example of the use of the mass conservation principle.

Fig. 2.50 illustrates a plume from a source at x = 0,   y = 0. As the plume moves downwind it grows through the action of turbulent eddies. The instantaneous plume has a high concentration over a narrow width. Over for instance, ten minutes the plume will touch a much broader area but the concentration will of course be correspondingly lower. Likewise in two hours the plume will stretch out in the cross wind direction and its concentration will be further reduced. The concentration distribution perpendicular to the axis of the wind appears to be gaussian or, rather, the gaussian distribution seems to be a useful model for the calculation of plume concentration. The shortcomings of this model will be discussed later.

The scope of the plume dispersion model is to determine, C, the concentration of the pollutant, as a function of its position downwind from the source. C is inverse proportional to the wind speed, U (m s^{-1}):

$$C \propto \frac{1}{U} \tag{2.70}$$

where U is the wind speed.

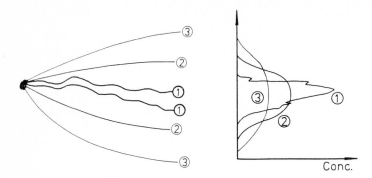

Fig. 2.50. Plume from a source (1) corresponds to the instantaneous plume. (2) is 10 minutes average plume and (3) is the 2 hour average plume. The diagram represents the cross-plume distribution patterns.

Thus we might model the plume as a gaussian function in vertical, z, and horizontal, y, co-ordinates. The behaviour of the wind direction, the x co-ordinate, should be directly proportional to the source strength and to I/U, see equation (2.65).

The gaussian function in the y direction is expressed as follows:

$$C \propto A * \exp \left[- \frac{1}{2} \left(\frac{y}{\partial_y} \right)^2 \right] \tag{2.71}$$

where A is a constant and ∂_y is the standard deviation.

If the gaussian function is normalized, so that the area under the curve has a unit value, by taking A to be $1/\sqrt{2\pi} * \partial_y$, the integral of concentration must be equal to the total amount of pollutant emitted.

Thus:

$$C \propto \frac{1}{\sqrt{2\pi} * \partial_y} \left[- \frac{1}{2} \left(\frac{y}{\partial_y} \right)^2 \right] \tag{2.72}$$

(2.72) gives, however, the distribution in one direction, and we are concerned with the distribution in three dimensions, as shown in Fig. 2.51.

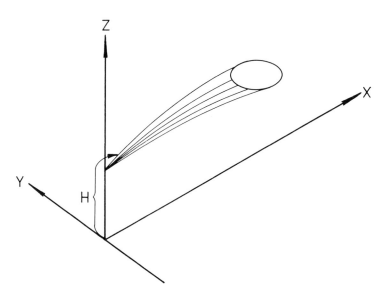

Fig. 2.51. Co-ordinate system applied in the plume model.

The gaussian function in 3 dimensions, considering that the integral of concentration over x, y and z coordinates must be equal to the total emission, takes the form:

$$C(x,y,z) = \frac{Q}{2\pi\partial_y\partial_z \cdot U} \; \exp\left[-1/2(\frac{y}{\partial_y})^2\right] \; \left[\exp(-1/2(\frac{z-H}{\partial_z})^2)+\exp(-1/2(\frac{z+H}{\partial_z})^2)\right]$$

$$(2.73)$$

where

C = the concentration (g m^{-3})
Q = source strength (g sec^{-1})
U = average wind speed (m sec^{-1})
H = the effective stack height (m)
    (see Fig. 8.4 for explanation)

Notice that z - H accounts for the real above-ground source and the term z + H accounts for the reflection of the plume, corresponding to an imaginary source below the ground.

The effective stack height is the physical stack height (h) + the plume (Δh) rise. In section 8.2.4 it will be demonstrated how Δh can be found.

For z = 0, at ground level, we find:

$$C(x,y,0) = \frac{Q}{\pi \partial_y \partial_z * U} \exp\left[-\left(\frac{y^2}{2\partial_y^2} + \frac{H^2}{2\partial_z^2}\right)\right] \tag{2.74}$$

For the concentration along the centreline (y=0) and at ground level, we have:

$$C(x,0,0) = \frac{Q}{\pi \partial_y \partial_z * U} \exp\left[-\frac{H^2}{2\partial_z^2}\right] \tag{2.75}$$

Compare (2.75) with (2.72).

U is a function of z but usually the average wind speed at the effective stack height is used. If the wind speed U_1 at level z_1 is known, U can be estimated by the following equation:

$$U = U_1 \left(\frac{H}{z_1}\right)^n \tag{2.76}$$

Smith (1968) recommends n = 0.25 for unstable and n = 0.50 for stable conditions. For further discussion on this subject see Turner (1970).

∂_y and ∂_z can be considered as diffusion coefficients and can be found as functions of atmospheric conditions (Smith, 1968).

These relationships are demonstrated in Table 2.33 and Figs. 2.52 and 2.53.

TABLE 2.33
Key to stability categories

Surface wind speed (at 10 m) m sec^{-1}	Day			Night	
	Incoming solar radiation			Thiny overcast or	
	Strong	Moderate	Slight	>4/8 low cloud	<3/8 cloud
< 2	A	A-B	B		
2-3	A-B	B	C	E	F
3-5	B	B-C	C	D	E
5-6	C	C-D	D	D	D
> 6	C	D	D	D	D

The neutral class, D, should be assumed for overcast conditions during day or night. Source: Turner, 1970.

Uncertainties in the estimation of ∂_y are generally fewer than those of ∂_z. However, wide errors in the estimate of ∂_z occur over longer distances. In some cases ∂_z may be expected to be correct within a factor of 2. These cases are 1) stability for distance of travel to a few hundred metres, 2) neutral to moderately unstable conditions for distances to a few kilometres, 3) unstable conditions in the lower 1000 metres of the atmosphere with a marked inversion above for a distance of 10 km or more.

It can be shown that the maximum ground-level concentration occurs where $\partial_2 = 0.707$ H, provided that the ratio ∂_4/∂_z is constant with downwind distance X.

Turner (1970) discusses a procedure for handling diffusion, when the plume expansion is limited by an upper level inversion. The principle of the metod is to calculate the concentrations as if they were distributed uniformly throughout the layer of height H_i - the distance from ground level to the inversion. In this case the expression for C at ground level becomes:

$$C = \frac{Q}{\sqrt{2\pi}\ UH_i\partial_y}\ \exp\left[-\ 1/2\ \left(\frac{y}{\partial_y}\right)^2\right] \qquad (2.77)$$

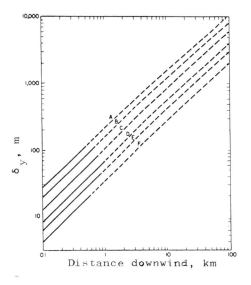

Fig. 2.52. Horizontal dispersion coefficient as a function of downwind distance from the source. (Turner, 1970).

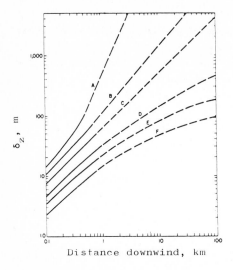

Fig. 2.53. Vertical dispersion coefficient as a function of downwind distance from the source. (Turner, 1970).

This equation can be applied downwind, of when the plume first reaches the H_i elevation, which can be estimated when 2.15 $\partial_z = H_i - H$. Turner (1970) recommends using equation (2.77) for $x > 2x_i$, where x_i is the point at which the plume reaches the inversion. Before x_i the regular diffusion result applies and between x_i and $2x_i$ an interpolation between the results at x_i and $2x_i$ is recommended.

The same equation can be applied under fumigation conditions, in which case H_i is the height to which the unstable air risen. (For an explanation of inversion and fumigation, see Fig. 8.2). However, Q must then be corrected to the fraction of the plume which can be carried down to the ground. Note that the stable classifications used to calculate ∂_y and ∂_z are also used to predict fumigation conditions. Turner, however, uses a correction factor to account for the plume spreading in the y direction:

$$\partial_{y,fum} = \partial_{y,stable} + H/8 \qquad (2.78)$$

As the example below demonstrates it is easy to use the plume model in practice, but, of course, such a relatively simple approach has some limits:

1. the diffusion coefficients applied are not very accurate,
2. the turning of the wind with height owing to friction effects is neglected,
3. adsorption or deposition of pollutants is neglected, but could easily be included, if required,
4. chemical reactions along the plume path have been omitted, but might be included if necessary knowledge were available,
5. shifts in wind direction are not taken into consideration.

In view of these limitations, the plume model should only be used as a first estimate of pollutant concentration (see Scorer, 1968).

Example 2.3.

Estimate the concentration of SO_2 downwind of a power plant burning 12,000 tons of 1.5% sulphur coal per day. The effective stack height is 200 m. The wind speed has been measured on a clear sunny day as 4 m/sec at the top of a 10 m tower. Find the concentration at ground level at x = 1 km and at 5 km, if it is estimated that 25% of the sulphur remains in the ash and is collected. At what height would an inversion affect the ground level concentration at x = 10 km?

Determine the maximum ground-level concentration of SO_2 and the distance from the stack at which the maximum occurs.

Solution:

From Table 2.32: stability category B
From Fig. 2.52 and 2.53

x	∂_y	∂_z
1 km	150 m	100 m
5 km	700 m	700 m
10 km	1000 m	1200 m

From equation (2.76):

$$U = U_1 \left(\frac{H}{z_1}\right)^n = 4\left(\frac{200}{100}\right)^{0.5} = 18 \text{ m sec}^{-1}$$

S-emission: $\dfrac{0.75 \ * \ 12 \ * \ 10^9 \ * \ 1.5}{100 \ * \ 24 \ * \ 3600} = 1562.5 \text{ g sec}^{-1}$

SO_2-emission: $1562.5 \ * \ \dfrac{64}{32} = 3125 \text{ g sec}^{-1}$

$x = 1$ km: $\quad C_{SO_2} = \dfrac{3125}{\pi \ * \ 150 \ * \ 100 \ * \ 18} \ \exp\left(-\dfrac{200^2}{2 \ * \ 100^2}\right) = 500 \ \mu g \ m^{-3}$

$x = 5$ km: $\quad C_{SO_2} = \dfrac{3125}{\pi \ * \ 700 \ * \ 700 \ * \ 18} \ \exp\left(-\dfrac{200^2}{2 \ * \ 700^2}\right) = 108 \ \mu g \ m^{-3}$

$2.15 \ * \ 1200 = H_i - H = H_i - 200$

$H_i = 2780$ m

$S_2 = 0.707 \ * \ 200 = 141$ m

which is obtained at $x = 1.1$ km

$C_{SO_2} = \dfrac{3125}{\pi \ * \ 200 \ * \ 100 \ * \ 18} \ \exp\left(-\dfrac{200^2}{2 \ * \ 141^2}\right) = 1000 \ \mu g \ m^{-3}$

2.12. EFFECTS OF AIR POLLUTANTS.

The effect of toxic substances on plants, animals and human health is obviously dependent on a number of factors, which will be mentioned in section 2.13.

Here only the closed interrelationship between the effect and the concentration versus time will be touched on.

Exposure time is an important piece of information when results of toxicity examinations have to be interpreted (see also section 2.13), but in the context of air pollutants the relationship between concentration and time is of significant importance, probably due to the following factors:

1. The effect of air pollutants on human health is very well examined for a number of combinations, concentration/exposure time.

2. The uptake through the respiratory system is proportional to the exposure time, if all other conditions are the same.

3. A number of examinations of indoor air pollution have been carried out, and here the exposure time (short exposure versus constant exposure) plays an important role.

4. The air-quality standards reflect the interrelation between concentration and exposure time. See Table 2.34, which shows one example out of many.
   For water-quality standards this is not the case, see Table 9.10 and Appendix 6.
   Figs. 2.54-2.56 demonstrate this expected 2-dimensional interrelationship between concentration and time to give a certain effect.
   To summarize:

P.2.31. **The effect of toxic substances is dependent on an interrelationship between concentration and time which can be clearly demonstrated for air pollutants. Air-quality standards reflect this interrelationship between concentration and exposure time.**

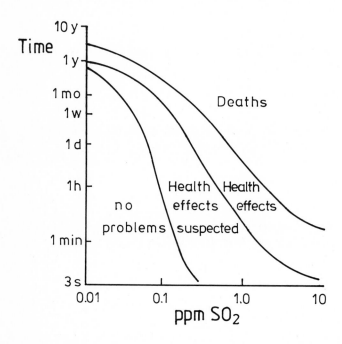

Fig. 2.54. Effects of sulphur pollution on human health (Air Quality, 1967).

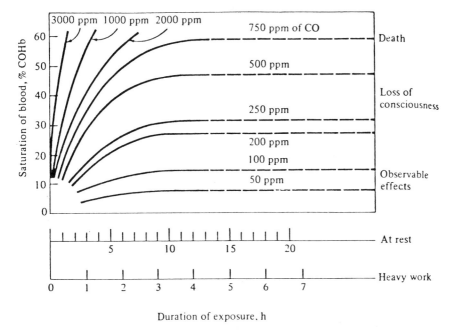

Fig. 2.55. COHb levels in blood: correlation to atmospheric CO, duration of exposure, and type of physical activity (Wolf, 1971).

TABLE 2.34
An example of air-quality standards (clean air acts amendments, U.S. 1970 (see Council of Environmental Quality, 1982)

Pollutant	Time indications	Concentration $\mu g/m^3$	ppm
Particulate matter	Annual geometric mean	75	-
Particulate matter	Maximum 24-h conc.	260	-
Hydrocarbons	Maximum 3-h conc.	160	0.24
Carbon monoxide	Maximum 8-h conc.	10	9
Carbon monoxide	Maximum 1-h conc.	40	35
Sulphur oxides	Annual arithmetic mean	80	0.03
Sulphur oxides	Maximum 24-h conc.	365	0.14
Sulphur oxides	Maximum 3-h conc.	1300	0.5
Nitrogen oxides	Annual arithmetic mean	100	0.05
Photochemical oxidants	Maximum 1-h conc.	240	0.12

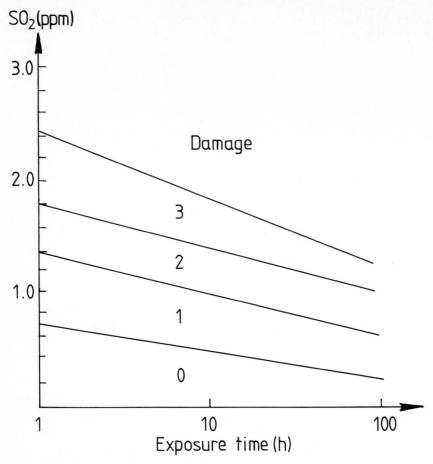

Fig. 2.56. SO_2-concentrations that can produce visible injury to vegetation with continuous exposure (Zahn, 1961). (3) is resistance group 3 (truck crops, cabbage species), (2) is resistance group 2 (cereals, beans, strawberries, roses), (1) is resistance group 1 (clover-like forage species) and (0) corresponds to "damage is unlikely".

2.13 PRINCIPLES OF ECOTOXICOLOGY.

Assessment of the environmental impact of pollutants requires extensive knowledge of dose and effect of a variety of receptors and a variety of effects in those receptors. In other words, we need to be able to translate the concentration in the ecosystem to an effect at the organism level. Most discussions of the results of environmental pollution concentrate on effect and neglect the problem of estimating dose. Without accurate knowledge of the dose there can be no quantitative assessment of the effect.

This problem of dose estimation is one of the characteristic features that distingiushes ecotoxicology from classical toxicology.

The concentration can either be measured or assessed by use of the principles of mass conservation, see section 2.3. In both cases, however, we need information on the dynamics of the pollutant. What is the fate of the pollutant? In which form does it occur? Will it be transformed to other forms and at what rate? It seems necessary to set up a model to get an overview of the final concentration as a result of the many processes in which the pollutant might take part.

An attempt is made below to give a general view of the many possibilities that must be considered to arrive at an answer as to: What will the concentration of a specific pollutant be in the environment as a function of time? Information on a number of processes is required:

1. Evaporation
2. Adsorption
3. Adsorption to and accumulation in soil and sediment
4. Biodegradation
5. Chemical oxidation (by air, nitrate, sulphate etc.)
6. Photodegradation
7. Complex formation (several possible inorganic and organic ligands are present in all ecosystems)
8. Hydrolysis

Fig. 2.57 visualizes how complex it is to assess the fate and effect of toxic substances. Fig. 2.58 shows how the principal forms and conversion pathways might be extended when trace metals are considered.

As is seen from Fig. 2.57, in addition to a knowledge of physical, chemical and biological processes it is required to relate the concentration to the effect on all levels in the biological hierarchy (see Fig. 2.59).

At *the cellular level* it is necessary to know whether the toxic substances might cause formation of mutation, which is possible by either alteration of the genes (the base sequence in DNA) or by changes in the chromosomes. The toxic substances might also cause an increased concentration of enzymes, or a competive or non-competitive inhibition. Other biochemical effects are included in the following more comprehensive list of responses:

enzyme activity (3 possibilities, see above)
change of steroid hormones
activation or suppression of biochemical pathways
alteration of membrane properties
changes of chromosomes
alteration of genes.

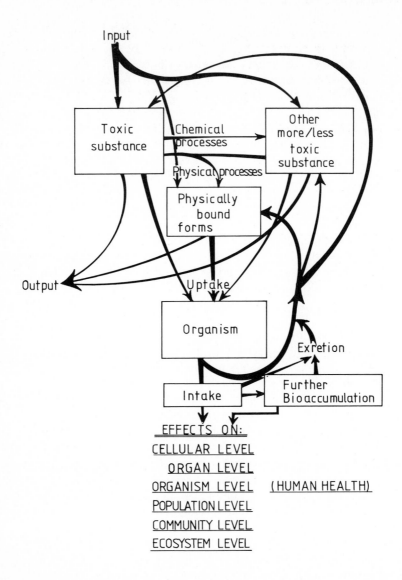

Input

Toxic
substance

Chemical
processes

Other
more/less
toxic
substance

Physical processes

Physically
bound
forms

Output

Uptake

Organism

Exretion

Intake

Further
Bioaccumulation

EFFECTS ON:
CELLULAR LEVEL
ORGAN LEVEL
ORGANISM LEVEL   (HUMAN HEALTH)
POPULATION LEVEL
COMMUNITY LEVEL
ECOSYSTEM LEVEL

Fig. 2.57. The diagram gives a conceptual model of the fate and effect of toxic substances in the environment.

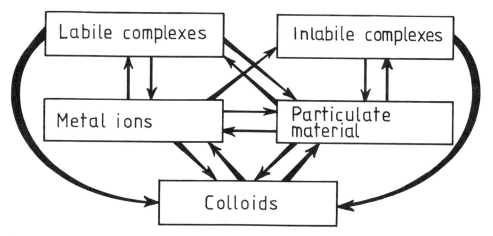

Fig. 2.58. The principal forms and conversion pathways of trace metals in natural waters.

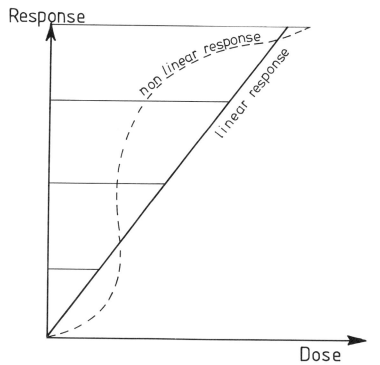

Fig. 2.59. Model of response/dose relationship.

Our knowledge in this biochemical field is at present very limited, although intensive research is being performed all over the world to fill the gaps.

Toxic substances soluble in lipids are transported in blood bound to lipoproteins and are taken up by tissue rich in lipids.

Another significant binding mechanism is the formation of chemical bonds with specific groups. For instance, mercury and organic lead are bound by SH-groups in proteins. Similarly, lead is accumulated in the skeleton by replacing calcium. Cadmium is accumulated in kidneys due to unknown reasons. In the first instance such accumulation effects naturally cause damage to the organ, which might imply a sublethal effect already at a very low intake.

Response to a toxic substance at *the organism level* can be categorized according to the dose rate and to the severity of damage:

1. Acute toxicity causing mortality: LD_{50}- and LC_{50}-values, see also section 2.1.

2. Chronically accumulating damage ultimately causing death.

3. Sublethal effects related to physiology: oxygen consumption, osmotic and ionic balance, feeding and digestion, assimilation, excretion, photosynthesis, N-fixation. The response can often be assessed as an increased or decreased rate.

4. Sublethal effects related to morphology: histological changes in cell and tissue, tumours, gross anatomical deformity.

5. Sublethal behavioural effects: locomotory activity, motivation and learning, equilibrium and orientation, migration, aggregation, reproductive behaviour, prey vulnerability, predator inability, chemoreception, photo-geo-taxes.

Chemical injury to individuals resulting in premature death or reduced reproductive success and recruitment is ultimately refelcted in lower abundances of exposed populations. Adaptation to toxic substances causes, furthermore, a change in the gene pool of the population. Further treatment of these important issues will be found in Chapter 4 on ecological principles, but it should already be stressed here that a minor reduction in reproductivity might have a pronounced long-term effect on the population, as is observed, for instance, in lakes exposed to acid rain.

The influence on *the population level* will, of course, affect the entire community: population extinction, changes in species composition, switches of dominance, changes in diversity and similarity patterns, reduction in biomass, alterations in spatial structure, and reduced stability, including the ecological buffer capacity and successional influences. An understanding of these issues requires a treatment of the related concepts, which can be found in Chapter 4.

An effect on *the ecosystem level* is observed in several cases by characterization of the nutrient, energy and water budgets of the ecosystem and this explains their dynamic behaviour through an understanding of the basic mechanisms governing the internal processes of the system. An eco-system can minimize and counteract the influence of environmental stress through population shifts and interactions. Therefore, the ecosystem is characterized as persisting, in spite of pertubations, through dynamic shift in its nutrient and energy metabolism. Populations are sacrificed to preserve the integrity and function of the ecosystem, although there is a limit to the buffer capacity of the system. Another aspect of the effect on ecosystem level is the shift in the basic composition of the ecosystem and its mass-cycles.

For instance, acid rain might cause another ionic composition of the entire ecosystem by release of previously bound ions to the environment. Here, the interaction with other pollutants plays an important role as the binding capacity of soil and sediments for heavy metals is highly dependent on pH. These problems will also be covered further in Chapter 4.

It seem clear from this overview of ecotoxicological problems, that their management requires the implementation of models.

The predictive values of toxic-substance models are generally less certain than those of other ecological models. First of all, because the process and corresponding parameters have been less well studied than such common processes as photosynthesis, nutrient uptake, grazing, etc.; also because the amount of data available for each case study is considerably more limited for toxic-substance models than for many other management models. On the other hand, the need for high accuracy in predictions is smaller. LD_{50} and LC_{50} values have been published for most toxic compounds which have environmental considerations, but less is known about the sublethal effects of these compounds.

Consequently, the management aim is not to obtain a concentration below a certain value in the lake water or at a given trophic level, but rather to assure a concentration of a given order of magnitude lower than the LD_{50} or LC_{50} value. Such statements as the WHO's maximum permissible concentrations in drinking water (see appendix 6) or recommended concentration in food (including fish) are based on the same principle - the uncertainty means that a large safety margin should be allowed for when human health or ecological belance is threatened.

Thus, in spite of its uncertainty, a toxic-substance model will still be a very valuable management tool.

P. 2.32. Due to the many unknown parameters, it is important to make toxic-substance models as simple as possible.

This requires a comprehensive knowledge of the chemistry and biology of the toxic compounds. Two examples will illustrate these considerations in relation to lakes.

The free copper ions are very toxic to fish and zooplankton. The LC_{50} value for Daphnia magna is as low as 20 µg l^{-1} and for Salmonoid species 100-200 µg l^{-1}. A copper model should consequently focus on the concentration of free copper ions in the water, and include the processes determining this concentration.

In the case of copper, a lethal concentration will be reached before the concentration in fish become toxic to humans, and as the uptake and excretion of copper by plants and animals are insignificant for the free copper ion concentration, these processes can be omitted.

This means that a model as simple as that shown in Fig. 2.60 can be applied to give at least a first estimate of how much copper and in what form can be discharged into an aquatic ecosystem.

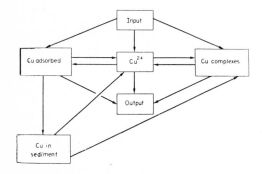

Fig. 2.60. A simple copper model.

The input of copper is the forcing function. The partition between free copper ions and copper adsorbed on suspended matter should also be included as the forcing function.

The equilibrium: copper ions + ligands = copper complexes, can be described to a certain extent by use of known equilibrium constants (Jørgensen et al., 1979 and Jørgensen 1979).

The adsorption process of copper ions on suspended matter requires laboratory investigation to find the adsorption capacity of the suspended matter. The release of copper from the sediment should also be studied, although some information is available (Lu et al., 1977).

As can be seen from this study, a rather simple approach which, however, is complex enough to require application of a model, can be used as a management tool, although the amount of data necessary to calibrate and validate the model is limited.

A useful, but simple model for the distribution and effect of DDT in aquatic ecosystems can also be built. Here, the problem is the DDT concentration in fish at the highest trophic level in the lake, as DDT is accumulated mainly through the food chain. The WHO has recommended the maximum permissible concentration of DDT in human food as 1-7 mg per kg net weight, which corresponds to a daily intake of 0.005 mg/kg body weight.

The management problem could be to keep the DDT concentration in all fish species below this value, divided by a safety factor of, say, 10, which means below a concentration of 0.1 mg per kg net weight. The model shown in Fig. 2.61 is suggested for this purpose.

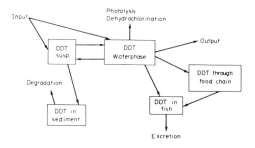

Fig. 2.61. A simple DDT model.

The direct uptake rate from water and the excretion rate coefficient are known for some fish (Jørgensen et al., 1979) and are not significantly different from species to species. The DDT accumulation through the food chain and the rate of photolysis and dehydrochlorination are known with acceptable accuracy. As with the copper problem, the equilibrium between DDT in the water phase and adsorbed on suspended matter must be studied in the laboratory, while the degradation rate in the sediment can be approximated with data from wet soil.

As can be seen from these two cases of medelling the distribution of copper and DDT, it is possible to make simple, workable models which might provide the answer to specific management problems related to discharge of toxic materials into the aquatic ecosystem.

The growth of application of models in environmental management is very pronounced, see for instance Jørgensen (1984).

However, the need for some general lines became clear in the mid seventies, when some modellers started to model the fate and effect of toxic substances in ecosystems (see, e.g., Thomann, 1978; Jørgensen, 1979; Hill et al.,1976). The overwhelming amount of parameters needed for a general application of such models makes it impossible to find these parameters in the litterature or by one's own laboratory experiments. Mankind uses as much as 50,000 different chemicals in industries or in everyday life, and if we were to model the fate and effect of all these chemicals, knowledge would be required of uptake rates, efficiencies from feed and direct contact influence on growth rates, excretion rates, decomposition rates, etc. - a totally of perhaps 30 parameters. This information would be needed for at least, let us say, 1000 different species of animals and plants representing the more than 1 million different species on earth. Consequently, we are talking about 1.5 billion pieces of information, if we are to have a good parameter estimation for all possible toxic substances which might harm the environment.

How can we get out of this dilemma? One attempt would be to find some basic scientific principles to make a holistic approach to ecological modelling.

Physics and chemistry have attempted to solve this problem by setting up some general relationships between the properties of the chemical compounds and their composition. If needed data cannot be found in the literature, such relationships are widely used as the second best approach to the problem.

In many cases, the application of such general relationships in chemistry gives a quite acceptable estimation. In many ecological models used in environmental context the required accuracy is not very high. In many toxic-substance models we need, for instance, only to know whether we are far from or close to the toxic levels. It is, therefore, understandable that a wider application of these relationships must be foreseen in the nearest future, although they will only give a first approximation of the parameters. However, more experience with the application of such general relationships is needed before a very general use can be recommended.

In this context it should be emphasized that in chemistry such general relationships are used only very carefully.

A workable example of this approach will be given in Chapter 3, dealing with energy behaviour, as it is possible to "translate" biological processes of toxic substances into energy terms. An example of parameter estimation based on chemical solubility data will be given here. Lipid solubility seems to be an effective predictor of bioconcentrations in environmental situations.

As a first estimation one can estimate the n-octanol-water partition coefficient of compounds, which can be reliably predicted by several simple measurements (see Fig. 2.62). Furthermore, it has been shown that the n-octanol:water partition coefficient is related to water solubility (see Fig. 2.63). Such approaches clearly give only a first approximation.

In view of the many data needed to assess the fate and effect of toxic substances, the OECD has set up guidelines to measure the properties of pollutants. A summary of the data is given in appendix 10.

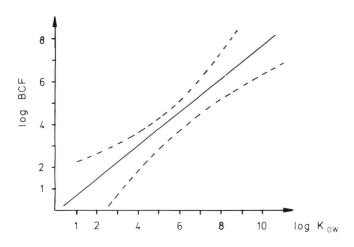

Fig. 2.62. Biological concentration factor versus the n-octanol-water partition coefficient for various fish species.

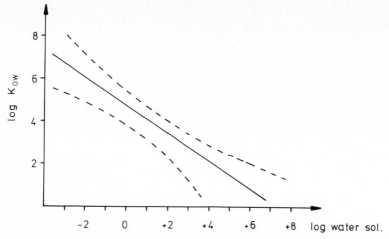

Fig. 2.63. A linear logaritmic regression of the water solubility versus the n-octanol-water partition for a range of organic compounds. The water solubility is expressed as μmol l^{-1} and determined at 10-30 C.

It is important at this stage to distinguish between intake and uptake.

P. 2.33. Intake is the entry of a substance into the lungs, the gastrointestinal tract or subsutaneous tissues of animals. The fate of the intake will be governed by processes of absorption. Uptake is the absorption of a substance into extracellular fluid and the fate of material taken up will be governed by metabolic processes.

The information about intake and uptake is different for various biota and will be presented below for four classes: mammals, fish, terrestial and aquatic plants, but first some typical quantitative, mainly empirical descriptions of the processes involved will be presented.

P. 2.34. The rate of uptake of a pollutant by ingestion of contaminated food is dependent on both the organism's maintenance metabolic rate and the rate of growth (Nordstrøm et al., 1975).

For freshwater fish the following expression for this relationship (at 20°C) is often used:

$$I = C_f \left(0.25 \, m^{0.8} + 2\frac{dm}{dt}\right) \, f(g \; day^{-1}) \tag{2.79}$$

where

I = rate of uptake
C_f = concentration of pollutant in food (gg^{-1})
m = body mass (g)

$$\frac{dm}{dt}$$ = growth rate $(g\ day^{-1})$

f = fractional absorption from gastrointestinal tract

P. 2.35. The respiratory uptake of a pollutant through the gills of freshwater fish was shown experimentally to be dependent on the metabolic rate (de Freitas et al., 1975) (at 20°C)

$$I_g = 1000\ m^{0.8} * C_w * f_g\ (g\ day^{-1}) \tag{2.80}$$

where
I_g = rate of uptake through the gills $(g\ day^{-1})$
m = body mass (g)
C_w = concentration of pollutant in water (g/g)
f_g = fractional absorption through the gills

The uptake of pollutants by terrestrial plants has been studied because it is the first step in the transport to humans through the food chain. Burton et al. (1970) suggest the following equation:

$$C = P_d * F_d + P_r * F_r \tag{2.81}$$

where
C = yearly average uptake/ g of plant material
P_d = soil factor
F_d = concentration in soil $(g\ km^{-2})$
P_r = rate factor
F_r = yearly fall-out rate of pollutant $(g\ km^{-2}\ y^{-1})$

In higher aquatic plants the uptake of waterborne pollutants by stems and leaves is often much more important than the absorption from the sediments by the roots (Erikson and Mortimer, 1975). The rate of uptake from water of both inorganic and methylmercury by growing aquatic plant can be expressed as:

$$\frac{[\text{Hg - plants}]}{[\text{Hg - water}]} = 3000 \cdot t \qquad (2.82)$$

where t is the duration of growth in days.

This relation was found to hold for water concentrations ranging from 1 to 10,000 mg l^{-1} (Mortimer and Kudo, 1976).

P. 2.36. The uptake from water can often be expressed in the same simple manner for both animals and plants.

With a good approximation we have:

$$\frac{C_b}{C_w} = CF \qquad (2.83)$$

where
C_b = the biotic concentration (g kg^{-1})
C_w = the concentration in water (g l^{-1})
CF = a concentration factor

Appendix 3 gives a list of characteristic concentration factors. The retention of toxic substances is also determined by the excretion rate, which can be approximated by means of the following equation:

$$r_e = k_e * C_b \qquad (2.84)$$

where
r_e = excretion rate (g day^{-1} (body weight)$^{-1}$)
k_e = excretion rate coefficient (day^{-1})
C_b = concentration of toxic substances (g (body weight)$^{-1}$)

P. 2.37. The excretion rate coefficient, k_e, can again be approximated as:

$$k_e = a \cdot m^b \qquad (2.85)$$

where a and b are constants (b is close to 0.75), and m is the body weight.

The retention can now be calculated as:

$$\frac{dC_b}{dt} = U - r_e \qquad (2.86)$$

where U = uptake from food + uptake from air + uptake from water + uptake from soil.

This model of the concentration of toxic material in plants and animals is extremely simple and should only be used to give a first rough estimate. For a more comprehensive treatment of this problem, see Butler (1972); ICRP (1977); Marshall et al. (1973); de Freitas et al. (1975); Mortimer and Kudo (1975); Seip (1979).

Table 2.35 and 2.36 give some characteristic excretion rates and uptake efficiencies. Note that the uptake efficiency is dependent on the chemical form of the component and on the composition of the food.

TABLE 2.35
Excretion rates

Species	Component	Excretion rate (% of abs. amount day^{-1})
Rat (urine)	Cd	1.25
Homo sapiens (urine)	Hg	0.01
Rat (urine)	Hg	1.0
Sheep (urine)	Pb	0.5 - 1.0
Homo sapiens	Zn	8.0

TABLE 2.36
Uptake efficiencies

Component	Species	Uptake efficiency	
DDT	Homo sapiens	14.4%	(dairy product)
DDT	Homo sapiens	40.8%	(meat product)
DDT	Homo sapiens	9.9%	(fruit)
Hg	Monkey	90.0%	(methyl-Hg)
Hg	Rat	90.0%	(methyl-Hg)
Hg	Rat	20.0%	(Hg-acetate)
Pb	Rabbit	0.8-1%	(in food)
Pb	Sheep	1.3%	(in food)
Zn	Pinfish	19.0%	(in food)

A wide variety of terms is used in an inconsistent and confusing manner to describe uptake and retention of pollutants by organisms by different paths and mechanisms. However, three terms are now widely applied and accepted to these processes:

1. *Bioaccumulation* is the uptake and retention of pollutants by organisms via *any* mechanism or pathway. It implies that both direct uptakes from air and water and uptake from food are included.

2. *Bioconcentration* is uptake and retention of pollutant by organisms directly from the water through gills or epithelial tissue. This process is often described by means of a concentration factor, see above.

3. *Biomagnification* is the process whereby pollutants are passed from one trophic level to another and it exhibits increasing concentrations in organisms related to their trophic status. This process has been touched on in section 2.8.

An enormous amount af data has been published on chemical analyses of plants and animals, but much is of doubtful scientific value. The precise questions to be answered through a given examination need to be clearly formulated at the initial stage (Holden, 1975). Again the problem is very complex. It is not sufficient to set up computations for the retention of toxic substances, it is necessary to ascertain the distribution in the organism, the lethal concentration, the effect of sublethal exposure and the effects on populations over several generations. Some idea of the complexity of the problem is given in Tables 2.37 and 2.38.

TABLE 2.37
Cd: Blood glucose - Pleuronectes flesus

Value (mg/100 ml)	Conditions (W = weeks, D = days)
27.8 (± 1.2)	Cd = 5 ppb, 4W, 9-10 Fish, 283K
29.2 (± 1.2)	Cd = 5 ppb, 9W, 9-10 Fish, 283K
31.7 (± 1.2)	Cd = 50 ppb, 4W, 9-10 Fish, 283K
41.0 (± 2.7)	Cd = 50 ppb, 9W, 9-10 Fish, 283K
34.1 (±1.4)	Cd = 50 ppb, 4W, 9-10 Fish, 283K
28.4 (± 3.6)	Cd = 50 ppb, 9W, 9-10 Fish, 283K
25.9 (±0.8)	Cd = 0 ppb, 4W, 9-10 Fish, 283K
24.7 (± 0.8)	Cd = 0 ppb, 9W, 9-10 Fish, 283K
20	Cd = 0 mg/l, 15D
21	Cd = 0.1 mg/l, 15D
29	Cd = 1.0 mg/l, 15D
35	Cd = 10 mg/l, 15D

TABLE 2.38

Cd: Decrease in yield

Item	Value (% of control)	Conditions Cd added to soil
Rice	0	0.001 % of CdO
Rice	8	0.003 % of CdO
Rice	8	0.01 % of CdO
Rice	7	0.03 % of CdO
Rice	31	0.1 % of CdO
Rice	68	0.3 % of CdO
Rice	81	0.6 % of CdO
Rice	99	1.0 % of CdO
Wheat	28	0.003 % of CdO
Wheat	84	0.01 % of CdO
Wheat	87	0.03 % of CdO
Wheat	97	0.1 % of CdO
Wheat	97	0.3 % of CdO
Wheat	98	0.6 % of CdO
Wheat	99	1.0 % of CdO

Our knowledge in the field of ecotoxicology is rather limited and further research in the area is urgently needed. The following issues should be studied (Moriarty, 1972).

1. Development of compartment models as a means of understanding the factors that determine the amount and distribution of pollutants within animals and plants.

2. Understanding the pollutant modes of action and the consequent sub-lethal effects on the whole animal's functions.

3. Studies of relatively simple ecosystems, with simulations of significant aspects in laboratory experiments in an attempt to improve our ability to understand and predict ecological effects.

Apart from environmental concentration/effect relationships, time/effect relationships are also important in understanding toxic effect. This has already been mentioned in section 2.12 for air pollutants where it is reflected in the air quality standards.

Fig. 2.64 shows a toxicity curve of log exposure time versus log LC_{50}. Threshold LC_{50} indicates when the curve becomes asymptotic to the time axis. *The threshold* LC_{50} is magnitudes greater than the concentration found in nature or the threshold values reflected in environmental standards, see section 2.2.

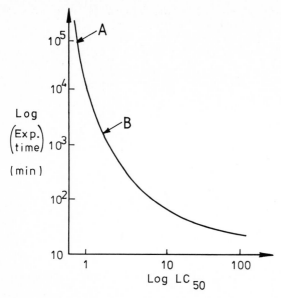

Fig. 2.64. Log exposure time versus log LC_{50}. A indicates threshold LC_{50} and B 24-hr LC_{50}.

P. 2.38. **We distinguish between 5 classes of dose/response curves, Fig. 2.65: (Bridges, 1980 and Brown, 1978)**

1. **a threshold dose** exists. It is simply indicated by dose levels at which there are no abnormal responses over controls. These threshold values are comparable with those mentioned in section 2.2.,

2. **a cumulative response**, where the response increases more than linearly with the dose and any dose gives an certain response,

3. **a quasi threshold relationship**, where no effect is observed below a certain value, but the response above the threshold dose is cumulative,

4. **a linear relationship** between response and dose,

5. **a saturating effect** is observed. In this case, the curve has a steep slope at low doses, but tends to flatten out at higher

levels. The following mechanisms could explain this relationship: an inducible pathway of detoxification, a progressively saturating activation pathway or differential repair processes.

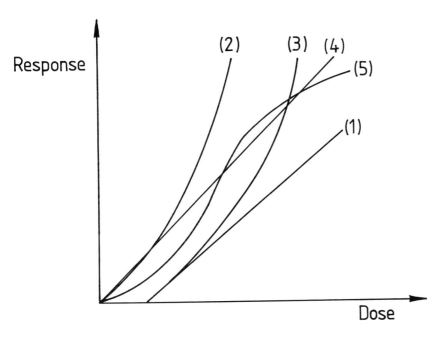

Fig. 2.65. Types of responses are shown: (1) threshold, (2) cumulative, (3) quasi threshold, (4) linear and (5) saturating.

Mechanisms of interactions are as follows:

1. The kinetic phase, by altering mechanisms of toxicant uptake, distribution, deposition, metabolism and excretion.

2. The dynamic phase, by altering toxicant receptor binding affinity and activity.

3. Chemical interactions between pollutants which produce new compounds, complexes or changes of chemical state; interactions between pollutants and substrates which alter physicochemical forms of pollutants and their toxicities (Anderson and D'Appolinia, 1978).

Some important types of pollutant interactions and related responses are

illustrated in Figs. 2.66 and 2.67. These involve synergistic and antagonistic interactions for combinations of two substances.

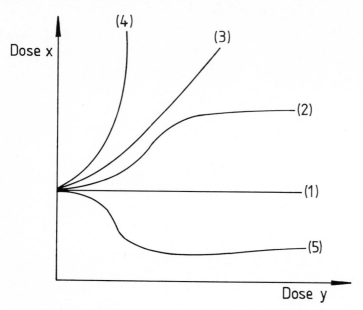

Fig. 2.66. Isoboles (curves of equal biological response) for combinations
of a substanse X, that is active on its own and a substance Y, that is inactive
when given alone, but which influences the action of X. The following
possibilities are shown (1) Y is inert, (2) antagonistic effect by physiological function, (3) antagonistic effect by chemical competition, (4) antagonistic effect by a non-competitive irreversible process, (5) Y sensitizes
for X ≈ synergistic effect.

Our present knowledge about teratogenesis, mutagenesis and carcino-
genesis is especially limited, and very poor when it comes to exposure of
two or more chemicals simultaneously.

Figs. 2.68 and 2.69 illustrate cases of two interacting parameters.

The uncertainty and limited knowledge in this avenue of environmental
science makes it necessary to treat such problems cautiously. This does not
mean, however, that we have to abandon their solution - on the contrary, we
have to find one although it can only be preliminary and will be very
uncertain.

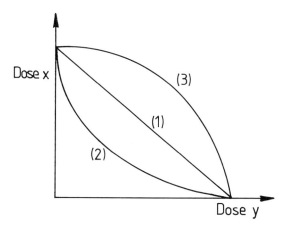

Fig. 2.67. Isoboles for combinations of two substances, X and Y, each of which is active by itself. (1) corresponds to additive action, (2) synergism and (3) antagonism.

If the uncertainty is too great it might be necessary to avoid the use of chemicals at all, at least in certain areas. This is, however, a political decision, and one which traditionally has come down on the side of economic rather than ecological advantage.

Fig. 2.70 provides a tentative procedure for the assessment of the environmental impact of toxic pollutants. It can be followed generally, but it is limited in application by our insufficient knowledge.

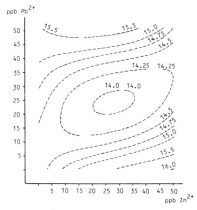

Fig. 2.68. The influence of Pb and Zn on the larval development until the megalopa stage (time in days). Experiment with R. harristi (crab larvae). (Benijts and Benijts, 1975).

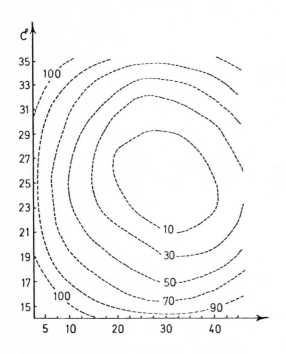

Fig. 2.69. Mortality (%) of crab larvae at various combinations of temperature and salinity (°/oo). (Sprague, 1970)

1. **Basic properties of the toxic pollutant:**
   solubility (water, fat), chemical forms, specific gravity, molecular weight, melting and boiling point.

2. **What concentrations occur in the abiotic environments:**
   water, atmosphere, soil etc.   If possible should mass balances be developed?

3. **Is there any possibility of accumulations in sediment (adsorption) or by biomagnifications? (see section 2.9).**

4. **What is known about uptake, concentration factors, uptake efficiencies, excretion rates etc.?**   Do we have enough knowledge to set up **a model** for the concentration in organisms as function of time and concentration in the environment?

5. **What do we know about lethal concentrations, sublethal effects, synergisms, antagonisms?**

6. **What do we know about the distribution in organisms?   Are any organs very sensitive to toxic pollutants?**

7. **What can we conclude about the possible environmental impact?**

Fig. 2.70. Tentative procedure: assessment of environmental impact of toxic pollutants.

QUESTIONS AND PROBLEMS

1. Where should a) Calcium sulphate, b) Nickel, c) Manganese dioxide be deposited?
   Use the same considerations as those made for copper in 2.1 and set up tables similar to 2.2, 2.3 and 2.4.
   The relevant information can be found in the literature, e.g. Jørgensen et al. (1979).

2. Indicate some elements which could be discharged into a bay in the order of few mg per litre without harmful effect.

3. Indicate, by use of table Appendix 2, some factors, that influence the degradation rate of pesticides.

4. Consider a lake with a volume of 500,000 m^3. 2000 m^3 24 h^{-1} waste water containing 0.5 mg l^{-1} methoxychlor is discharged into the lake. The natural flows to the lake corresponds to 4 months retention time. Methoxychlor follows a first order decomposition rate, and the approximate half-life time in water can be found in Appendix 2 (no H_2O_2 is added). Find the equlibrium concentration in the lake water. Assume that precipitation and evaporation are balanced.

5. Approximately what mercury concentration of zooplankton would be expected in a lake with 10 μM mercury per litre?

6. A person lives in an industrial area, where the Cd-concentration is as high as 5 μg per m^3. He smokes 20 cigarettes per day each containing 1 μg Cd. His food contains 75 μg Cd. The uptake efficiency of Cd is: by lungs 20% and by digestion 7%. What is his daily intake? The excretion rate constant is 0.002 day^{-1}. Set up an equation to determine his Cd-concentration as a function of time. What will his maximum concentration be?

7. If iodine is the limiting factor for brown algae what is the approximate maximum concentration of brown algae in water with a concentration of 0.01 mg per litre iodine?

8. Explain why a biological waste water treatment plant shows less nitrification during the winter than during the summer.

9. Explain the relationship between DDT in freshwater, phytoplankton, freshwater fish and eagles, using the values from Table 2.25.

10. Calculate the emission of carbon dioxide to the atmosphere in the year 2000 assuming that the global C-cycle in Fig. 2.13 is valid for 1975 and the consumption of fossil fuel increases at 3% per annum.
What will the carbon dioxide concentration be in the atmosphere in the year 2000 under these conditions? It is assumed that 2/3 of the emission will be discharged in the sea, in accordance with Fig. 2.13.

11. Give the consequences of a further inbalance in the phosphorus and nitrogen cycles.

12. Find the critical point, by use of the Streeter-Phelps equation including nitrification, for the following cases:
100 l sec^{-1} waste water with BOD_5 = 30 mg l^{-1} and NH_4^+-concentration = 20 mg l^{-1} is discharged into a river with a flow of 1200 l sec^{-1}. The flow rate in the river is 0.5 m sec^{-1}, the depth is 4 m, and the temperature is 20°C.
Is the saturation at the critical point acceptable?

13. How would you classify the river at the critical point?

14. A lake has a phosphorus concentration of 80 µg l^{-1}. What is the expected maximum chlorophyll concentration?

15. A lake is 50 m deep and has an area of 10 km^2. The phosphorus input to the lake comes mainly from waste water. Approximately 50,000 m^3 pr 24h of municipal waste is discharged into the lake. How would you classify the lake? The waste water is only treated by a mechanico-biological plant. What level of efficiency of phosphorus removal is necessary to improve the conditions of the lake to the mesotrophic and oligotrophic state, respectively?

16. The transparency in a lake is 0.8 m, estimate G_{24max}.

17. Write the reaction between acetic acid and nitrate respectively sulphate under anaerobic conditions.

18. Show that the conversion from ppm for a gaseous pollutant to µg m^{-3} at 25°C and 760 mm Hg is given by:
10^3 (ppm) molecular weight of pollutant/24.5.

19. The average concentration of particulate lead is 6 µg/m^3, of which 75% is less than 1 µm in size. If a person respires 15 m^3 air daily and the uptake efficiency is 50% of the particles below 1 µm and negligible for those above this size, calculate how much lead is absorbed in the lungs

each day.

20. The quilibrium: $2NO_2$ (g) = N_2O_4 (gas) is exothermic with the formation of about 60 kJ per mole. While nitrogen dioxide is brown, dinitrogen tetraoxide is colourless. Would you expect the colour of the photo-chemical smog to be deeper on a warm day or a cool day?

21. The distribution coefficient for DDT between olive oil and water is 925. The solubility or DDT in water is 1.2 μg l^{-1}.
Calculate the solubility of DDT in olive oil in mg l^{-1} and molarity. The molecular weight of DDT is 354.5 g mol^{-1}.

22. Iron(II) ion released by weathering is stable at low pH, but is easily oxidized to iron(III) ion at higher pH. Explain why this is so.

23. What happens to the oxidizing ability of water at high pH?

24. The earth's rivers add to the oceans 30 * 10^{15} kg water per annum. Calculate, using tables from the text, the residence time of K and Na, assuming a constant concentration of these elements in the oceans.

25. The average concentration of dissolved iron ion in river water is 1 ppm, but in seawater it is only 8 ppb. What happens to iron(Fe^{3+})-bearing river water when it enters the sea?

26. A waste-water treatment plant disposes its effluent in a surface stream. What would be the lowest oxygen concentration as a result of the waste-water discharge and what would be the dissolved oxygen concentration in the stream after 2 days, when the following characteristics are valid:

	Waste water	Stream
BOD_5	48 mg/l	3 mg/l
Temperature (°C)	16	20
Dissolved oxygen	1.0 mg/l	8.2 mg/l
Flow	0.2 m^3/s	5.8 m^3/s

Nitrification can be omitted.

27. Find the NH_3-concentration at pH = 7.6, when the total concentration of $NH_3 + NH_4^+$ has been found to be 4.6 mg/l.

28. Find solubility of hydrogen sulphide in water at 20°C.

29. What is the solubility of methane in water at 5°C?

CHAPTER 3

PRINCIPLES OF ENERGY BEHAVIOUR
APPLIED TO ENVIRONMENTAL ISSUES

3.1. FUNDAMENTAL CONCEPTS RELATED TO ENERGY.

Energy is defined as the ability to do work, and the behaviour of energy can be described by the first and second laws of thermodynamics.

P.3.1. **The first law of thermodynamics states that energy may be transformed from one type to another but is never created or destroyed.**

It can also be applied in a more ecological way as follows: you can not get something for nothing -   there is no such thing as a free lunch (Commoner, 1971).

Thus when a change of any kind occurs in a closed system (see 2.3 for definition) an increase or decrease in the internal energy occurs, heat is evolved or absorbed and work is done. Therefore:

$$\Delta E = Q + W \tag{3.1}$$

where
ΔE = change in internal energy
Q = heat absorbed
W = work done on the system

As mentioned in 2.1 a relationship exists between mass and energy, which dictates that energy is produced as a result of nuclear processes. Equation (3.1) assumes that such processes have not taken place.

In environmental science we are primarily concerned with the quantity of incident solar energy per unit area in an ecosystem and the efficiency with which this energy is converted by organisms into other forms. This situation is illustrated in Fig. 3.1, where the fate of solar radiation upon grass-herb vegetation of an old field community in Michigan is shown (Golley, 1960).

The transformation of solar energy to chemical energy by plants conforms with the first law of thermodynamics:

| Solar energy assimilated by plants | $-$ | chemical energy of plant tissue | $+$ | heat energy of respiration (3.2) |

For the next level in the foodchain, the herbivorous animals, the energy balance can also be set up:

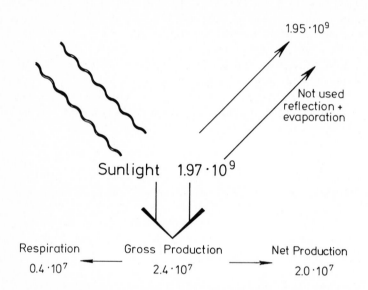

Fig. 3.1. Fate of solar energy incident upon the perennial grass-herb vegetation of an old field community in Michigan. All values in J m^{-2} y^{-1}.

$$F = A + UD = G + H + UD \qquad (3.3)$$

where
F = the food intake converted to energy (Joule)
A = the energy assimilated by the animals
UD = undigested food or the chemical energy of faeces
G = chemical energy of animal growth
H = the heat energy of respiration

These considerations pursue the same lines as those mentioned in 2.8, where the mass conservation principle was applied. The conversion of biomass to chemical energy is illustrated in Table 3.1.

P.3.2. In nature we can distinguish two processes: spontaneous processes which occur naturally without an input of energy from outside and non-spontaneous processes, which require an input of energy from outside. These facts are included in the second law of thermodynamics, which states that processes involving energy transformations will not occur spontaneously unless there is a degradation of energy from a non-random to a random form, or from a concentrated into a dispersed form. In other words all energy transformations will involve energy of high quality being degraded to energy of lower quality (e.g. potential energy to heat energy). The quality of energy is measured by means of the thermodynamic state variable entropy (high quality ~ low entropy).

TABLE 3.1
A.   Heat combustion of animal material

Oragnism	Species	Heat of combustion (kcal/ash-free gm)
Ciliate	Tetrahymena pyriformis	-5.938
Hydra	Hydra littoralis	-6.034
Green hydra	Chlorohydra viridissima	-5.729
Flatworm	Dugesia tigrina	-6.286
Terrestrial flatworm	Bipalium kewense	-5.684
Aquatic snail	Succinea ovalis	-5.415
Brachiipod	Gottidia pyramidata	-4.397
Brinc shrimp	Artemia sp.(nauplii)	-6.737
Cladocera	Leptodora kindtii	-5.605
Copepod	Calanus helgolandicus	-5.400
Copepod	Trigriopus californicus	-5.515
Caddis fly	Pycnopsyche lepido	-5.687
Caddis fly	Pycnopsyche guttifer	-5.706
Spit bug	Philenus leucopthalmus	-6.962
Mite	Tyroglyphus lintneri	-5.808
Bettle	Tenebrio molitor	-6.314
Guppie	Lebistes reticulatus	-5.823

TABLE 3.1 (continued)
B. Energy values in an Andropogon virginicus Old-Field Community in Georgia

Component	Energy value (kcal/ash-free gm)
Green grass	-4.373
Standing dead vegetation	-4.290
Litter	-4.139
Roots	-4.167
Green herbs	-4.288
Average	-4.251

C. Heat combustion of migratory and nonmigratory birds

Sample	Ash-free material (kcal/gm)	Fat ratio (% dry weight as fat)
Fall birds	-8.08	71.7
Spring birds	-7.04	44.1
Nonmigrants	-6.26	21.2
Extracted bird fat	-9.03	100.0
Fat extracted: fall birds	-5.47	0.0
Fat extracted: spring birds	-5.41	0.0
Fat extracted: nonmigrants	-5.44	0.0

D. Heat of combustion of components of biomass

Material	ΔH protein (kcal/gm)	ΔH fat (kcal/gm)	ΔH carbohydrate (kcal/gm)
Eggs	-5.75	-9.50	-3.75
Gelatin	-5.27	-9.50	
Glycogen			-4.19
Meat, fish	-5.65	-9.50	
Milk	-5.65	-9.25	-3.95
Fruits	-5.20	-9.30	-4.00
Grain	-5.80	-9.30	-4.20
Sucrose			-3.95
Glucose			-3.75
Mushroom	-5.00	-9.30	-4.10
Yeast	-5.00	-9.30	-4.20

Source Morowitz, 1969

The second law of thermodynamics can be more precisely expressed as the existence of a state variable S, defined by:

$$dS = d_eS + d_iS \qquad (3.4)$$

where $d_eS = dQ/T$ and $d_iS \geq 0$ (= 0 for reversible processes), which will change during any process such that dS (universe ≥ 0).

Organisms, ecosystems and the entire ecosphere possess the essential thermodynamic characteristic of being able to create and maintain a high state of internal order or a condition of low entropy (entropy can be said to measure disorder, lack of information on molecular details, or the amount of unavailable energy). Low entropy is achieved by a continual dissipation of energy of high utility - light or food - to energy of low utility - heat. Order is maintained in the ecosystem by respiration, which continually produces disorder (heat).

P.3.3. **The second law of thermodynamics explains why ecosystems can maintain organization or order. A system tends spontaneously toward increasing disorder (or randomness), and if we consider the system to consist of an ecosystem and its surroundings, we can understand that order (negative entropy) can be produced in the ecosystem if more disorder (entropy) is produced in its surroundings. In the ecosystem the ratio of total community respiration to the total community biomass (denoted as the R/B ratio) can be regarded as the maintenance to structure ratio or as a thermodynamic order function for homogeneous systems.**

The behaviour of energy in an ecosystem can conveniently be termed the energy flow, because the energy transformation occurs only in one direction towards lower quality energy. However, as the chemical compounds carry chemical energy and the elements cycle (see 2.5), energy will also cycle. Solar radiation is the inflowing energy, which is used by phototrophic organisms to produce biomass. The chemical energy in the biological matter produced is of a non-random character, but at the same time heat is produced by respiration and becomes the outflowing energy. The chemical cycling of the elements is a result of this unidirectional flow of energy. Without solar radiation there would be no order in the ecosystem and no cycling of elements would take place.

P.3.4. **From a physical standpoint the environmental crisis is an entropy crisis, as pollution makes disorder.**

An example of this is given in Fig. 2.45, which illustrates the accumulation of lead in the Greenland ice pack from 800 B.C. to the present. This steadily increased accumulation demonstrates that lead released to the atmosphere is distributed worldwide and that entropy is correspondingly increased. That entropy is increased by distribution of pollutants, can be demonstrated by a simple model consisting of two bulbs of equal volume, connected with a stop cock. If one chamber contains one mole of a pure gas and the second is empty, then opening the valve between the two chambres causes an increase in entropy of:

$$\Delta_e S \equiv \int \frac{dQ}{T} = \frac{Q}{T} = -W = R * \ln \frac{V_2}{V_1} = R * \ln 2 \tag{3.5}$$

where $\Delta_e S$ = the increase in entropy, V_2 = the volume occupied by the mole of gas after the valve was opened, while V_1 is the volume before the valve was opened.

Thus paradoxically, the more we attempt to maintain order, the more energy we require and the greater stress we inevitably put on the environment.

3.2. ENERGY USE AND ENERGY RESOURCES.

A human needs about 9,000 kJ per day to survive, but today the average inhabitant in the most developed countries uses more than 900,000 kJ, 100 times more than the survival level. Most energy is consumed by the industrial nations. They have 30% of the world's population but consume 80% of the world's energy.

Paragraph 2.8 has shown that only 10-20% of the initial biomass can be used by the organisms on the next level in the foodchain. As biomass can be translated into energy (see Table 3.1), this is also true of energy transformation through a foodchain. This implies that the short foodchain of grain — human should be prefered to the longer and more wasteful grain — animal — human.

Of course the problem of food shortage cannot be solved so simply, since animals produce proteins with a more favourable amino acid composition for human food (lysine is missing in plant proteins) and eat plants that cannot all be used as human food at present. But to a certain extent food production can be increased by making the foodchain as short as possible.

This concept also applies to the use of fossil fuel or other energy sources. Fig. 3.2 shows such an energy chain where the more links there are the greater the waste of energy.

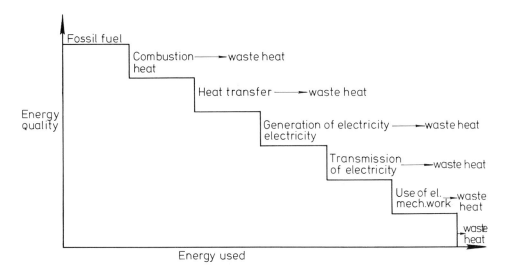

Fig. 3.2. Degradation of energy through an energy chain.

P.3.5. **At each link in the chain energy is transferred from one form to another and heat is wasted as a consequence of the second law of thermodynamics.**

In Table 3.2 is listed the net energy efficiency for some of the most commonly used energy systems. We cannot escape the second law of thermodynamics, but we can minimize energy waste:

1. by keeping the energy chain *as short as possible,*
2. *by increasing the efficiency,* i.e. the ratio of useful energy output to the total energy input,
3. *by wasting as little heat to the surroundings as possible,* e.g. by *insulation,* and *using heat produced by energy transfer* heat produced at power stations can be used for heating purposes).

The first law of thermodynamics can also be applied to the energy situation. The only free energy available is solar radiation, because the earth is an open system that gets energy from external sources; if we have to get energy from other sources it will always cost energy to provide energy. For example, at present only one third of the oil in an average reservoir is recovered, and the energy needed to produce the steam that is rejected into the borehole to increase the recovery may exceeds the energy recovered as oil. As a result, although total energy production is declining total energy consumption is increasing.

TABLE 3.2
The energy efficiencies of common individual systems

Conversion system	Percentage efficiency approximately
Processing of natural gas	97
Mining of nuclear fuel (uranium)	95
Processing of coal	92
Processing of oil	88
Home natural gas furnace	85
Electric car storage battery	80
Surface mining of coal	78
Extraction of natural gas	73
Propeller driven wind turbine	70
Home oil furnace	65
Fuel cell	60
Processing of nuclear fuel	57
Deep mining of coal	55
Fossil fuel power plant with proposed MHD topping cycle	50
Steam turbine engine	45
Diesel engine	42
Offshore oil well	40
Fossil fuel power plant and proposed nuclear breeder plant	38
Today's nuclear power plant	31
Internal combustion engine	30
Fluorescent lamp	28
Wankel engine	22
Advanced solar cells (if developed)	15
Today's solar cells	10
Incandescent lamp	5

Fig. 3.3 demonstrates how steeply energy consumption has been in this century, and that the gap in per capita energy consumption between developed and developing nations is widening.

Based on the best possible information available today, it seems clear that the major supplies of oil and gas will probably run out somewhere between the years 2025 and 2050, while there is still sufficient coal for 200-400 years. Coal, however, cannot replace oil and gas completely because of pollution problems (see 2.6 and 3.5). Since it takes 25 to 50 years to develop a new energy system we have no time to loose in seeking replacements for oil and gas.

We will not go further into energy problems, since this book is concerned with the principles of environmental science, but it is worth mentioning that the world's greatest energy problem today is food: about 40% of the world's population do not receive the basic minimum quantity or quality of energy needed to maintain good health.

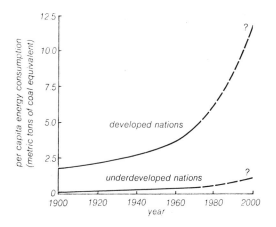

Fig. 3.3. The gap in per capita energy consumption between the developed nations and the underdeveloped nations is widening. (Source: United Nations)

3.3.   PRODUCTIVITY.

The *primary productivity* of an ecological system is defined as the rate at which radiant energy is stored by the photosynthetic and chemosynthetic activity of producer organisms, chiefly green plants, as chemical energy in the form of organic matter. As unit is often used g (kcal, kJ) or g dry matter per m^2 og m^3 per 24h or per year. It is essential to distinguish between *gross primary productivity,* which is the total rate of synthesis including the organic matter used up in *respiration*, and *net primary productivity,* which is the rate of storage of organic matter. The rate of energy storage at consumer levels is referred to as *secondary productivity.*

The standing crop of biomass present at any given time must not be confused with net productivity, which is the change in biomass per unit of time:

$$P = \frac{dB}{dt} = f(B, \text{temp, environmental factors}) \qquad (3.6)$$

where P = productivity, B = biomass.

Table 3.3 gives some characteristic figures for gross and net primary productivity and Table 3.4 gives the estimated annual gross primary production of the biosphere and its distribution among major ecosystems.

TABLE 3.3
Characteristic figures for gross and net primary producion

Ecosystem	Gross prim. production	Net prim. production
Agricultural land	650 g dry weight $m^{-2}y^{-1}$ (range 100-4000)	
Alfalfa field	24000 kcal $m^{-2}y^{-1}$	15200 kcal $m^{-2}y^{-1}$
Desert	200 kcal $m^{-2}y^{-1}$	150 kcal $m^{-2}y^{-1}$
Oak-Pine forest	11500 kcal $m^{-2}y^{-1}$	5000 kcal $m^{-2}y^{-1}$
Pine forest	12200 kcal $m^{-2}y^{-1}$	7500 kcal $m^{-2}y^{-1}$
Rain forest	45000 kcal $m^{-2}y^{-1}$	13000 kcal $m^{-2}y^{-1}$
Coal reef	18.2 g dry weight $m^{-2}day^{-1}$	
Grassland	1200 g dry weight $m^{-2}y^{-1}$ (range 250-2500)	500 g dry weight $m^{-2}y^{-1}$ (range 150-1500)
Eutrophic lakes	2.1 g dry weight $m^{-2}day^{-1}$	
Sargasso Sea	0.5 g dry weight $m^{-2}day^{-1}$	
Silver Spring	17.5-35 g dry weight $m^{-2}day^{-1}$	
Total biosphere	2000 kcal $m^{-2}y^{-1}$	
Temperate forest	8000 kcal $m^{-2}y^{-1}$	3600 kcal $m^{-2}y^{-1}$

TABLE 3.4
Estimated gross primary production (annual basis) of the biosphere and its distribution among major ecosystems

Ecosystem	Area (10⁶ km²)	Gross Primary Productivity (kcal/m²/year)	Total Gross Production (10¹⁶ kcal/year)
Marine *)			
Open ocean	326.0	1,000	32.6
Coastal zones	34.0	3,000	6.8
Upwelling zones	0.4	6,000	0.2
Estuaries and reefs	2.0	20,000	4.0
Subtotal	362.4	-	43.6
Terrestrial **)			
Deserts and tundras	40.0	200	0.8
Grasslands and pastures	42.0	2,500	10.5
Dry forests	9.4	2,500	2.4
Boreal coniferous forests	10.0	3,000	3.0
Cultivated lands with little or no energy subsidy	10.0	3,000	3.0
Moist temperate forest	4.9	8,000	3.9
Fuel subsidized (mechanized agriculture	4.0	12,000	4.8
Wet tropical and subtropical (broadleaved evergreen forests)	14.7	20,000	29.0
Subtotal	135.0	-	57.4
Total for biosphere (not included ice caps) (round figures)	500.0	2,000	100.0

*) Marine productivity estimated by multiplying Ryther's (1969) net carbon production figures by 10 to get kcal, then doublin these figures to estimate gross production and adding an estimate for estuaries (not included in his calculations).

**) Terrestrial productivity based on Lieth's (1963) net production figure doubled for low biomass systems and tripled for high biomass systems (which have high respiration) as estimates of gross productivity. Tropical forests have been upgraded in light of recent studies, and the industrialized (fuel subsidized) agriculture of Europe, North America, and Japan have been separated from the subsistence agriculture characteristic of most of the world's cultivated lands.

TABLE 3.5 A
Annual yields of edible food and estimated net primary production of major food crops at three levels: (1) Fuel subsidized agriculture (U.S.A., Canada, Europe or Japan); (2) Little or no fuel subsidy (India, Brazil, Indonesia or Cuba); (3) World average

	Edible portions		Estimated net primary production	
	Harvest weight *) (kg/ha)	Caloric content **) (kcal/m^2)	Dry matter production §) (kcal/m^2)	Rate growing season o) (kcal/m^2/day)
Wheat				
Netherland	4,400	1,450	4,400	24.4
India	900	300	900	5.0
World average	1,300	430	1,300	7.2
Corn				
USA	4,300	1,510	4,500	25.0
India	1,000	350	1,100	6.1
World average	2,300	810	2,400	13.3
Rice				
Japan	5,100	1,840	5,500	30.6
Brazil	1,600	580	1,700	9.4
World average	2,100	760	2,300	12.8
White potatoes				
USA	22,700	2,040	4,100	22.8
India	7,700	700	1,400	7.8
World average	12,100	1,090	2,200	12.2
Sweet potatoes and yams				
Japan	20,000	1,800	3,600	20.0
Indonesia	6,300	570	1,100	6.1
World average	8,300	750	1,500	8.3
Soybeans				
Canada	2,000	800	2,400	13.3
Indonesia	640	260	780	4.3
World average	1,200	480	1,400	7.8
Sugar				
Hawaii (cane)	11,000	4,070	12,200	67.8
Netherlands (beets)	6,600	2,440	7,300	40.6
Cuba (cane)	3,300	1,220	3,700	20.6
World average (all sugar: beets and cane)	3,300	1,220	-	-

*) Mean value 1962-1966 compiled from "Production Yearbook" vol. 21 (1966), Food and Agricultural Organization, United Nations.
**) Conversion, kcal/gm harvested weight as follows: Wheat 3.3; Corn 3.5; Rice 3.6; Soybeans 4.0; Potatoes 0.9; Crude sugar 3.7 (see USDA Agriculture Handbook No. 8, 1963).
§) Estimated on basis of 3X edible portion for grains, soybeans and sugar, 2X for potatoes.
o) Estimated to be six months (180 days) except sugar cane where sugar yields are calculated on 12 months growing season (365 days).

TABLE 3.5 B
Net primary production and edible portion at 3 levels: 1) Developed countries (fuel-subsidized agriculture), 2) Developing countries (little or no fuel subsidy), 3) World average

	kJ m^{-2} Net primary production			kJ m^{-2} Edible portion		
	1	2	3	1	2	3
Wheat	18,000	3800	5400	6000	1300	1900
Corn	18,500	4500	10,000	6500	1600	3500
Rice	22,600	7000	9600	7500	2500	3500
White potatoes	17,000	5700	9000	8800	2850	4380
Sweet potatoes	15,000	4500	6400	7000	2500	3500
Soybeans	10,600	3300	6300	3500	1100	2000
Sugar	52,000	15,000	28,000	16,500	5000	9000

Primary production in terms of food is summarized in Table 3.5 A+B, where the major food crops in developed and developing countries are compared. As can be seen, the developing countries have a lower per area unit producion compared with the developed countries, which use fuel-subsidized agriculture.

In 1975 there were an estimated $4*10^9$ people in the world each requiring $4.2*10^6$ kJoules per year, amounting to a total of $17*10^{15}$ kJoules per year, which represents only about 0,5% of the net primary production of the biosphere.

However, we must add to this the huge food consumption of all domestic animals. The standing crops of livestock in the world are at least 5 times that of humans in terms of equivalent food requirements. Thus man and his domestic animals consume at least 6.5% of net production of the whole biosphere or at least 12.5% of that produced on land. In addition, man also consumes huge quantities of primary production in the form of fibers, so his percentage consumption of net production on land is actually 15%. But what if the population doubles? Will there be enough food? Can we still continue to use and eat animals?   The situation is rapidly becoming critical, and our planning for the future, must include the following points:

1. No more than about 25% of the land is truly arable. Irrigation of huge areas of dry land will require large expenditure of money (energy) and will have some severe ecological side-effects.
2. We are able to reduce the number of domestic animals, but man's need for animal proteins must not be underestimated.
3. Theoretically crop yields in the developing countries can be increased significantly, but these countries do not have the money to invest in agriculture and have growing populations. Even now it is almost impossible to maintain the same per capita production of food, which is already

too small.

4. High productivity can only be maintained by the use of energy subsidies and will be accompanied byn environmental pollution due to the heavy use of machinery, herbicides, insecticides and fertilizers.

These points illustrate well how the environmental crisis, the energy crisis and the food crisis are linked together. Only the solution of all three problems simultaneously, can be considered a real solution.

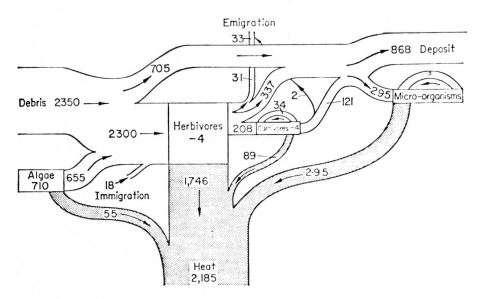

Fig. 3.4. Energy flow diagram for Root Spring, Concord, Mass., in 1953-54. Figures in kcal/m²/year, numbers inside boxes indicate changes in standing crops. (From Teal, 1957)

3.4. ENERGY IN ECOSYSTEMS.

The transfer of energy from its source in the plants through herbivorous animals (primary consumers) to carnivorous animals (secondary consumers) and further to the top carnivore (tertiary consumers) is called *the foodchain*. By using the first law of thermodynamics an energy diagram can be set up, as illustrated in Figs. 3.4 and 3.5. The components for such a model of ecological energy flow are shown in Fig. 3.6.

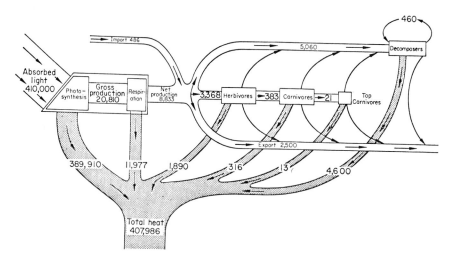

Fig. 3.5. Energy flow diagram for Silver Springs, Florida. Figures in kcal/m²/year. (Adapted from Odum, 1957)

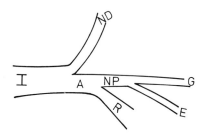

Fig. 3.6. Components for an ecological energy flow model. I = input of indigested food, ND = "nondigested" food or energy, A = assimilated energy, R = respiration, P = net production, E = excretion, G = growth (and reproduction). I = ND + A, A = NP + R, NP = G + E. Only G is available for next level in the foodchain at equilibrium.

In nature the food relationships are rarely as simple as a foodchain, but rather form a foodweb.

The relationships can also be illustrated by means of so-called ecological pyramids, which can either represent the number of individuals, the biomass (or energy content) or the energy flow on each level in the foodchain or foodweb (see Fig. 3.7). Only the energy flow forms a true pyramid, in accor-

dance with the first law of thermodynamics. The number pyramids are affected by variation in size and the biomass pyramids by the metabolic rates of
individuals.

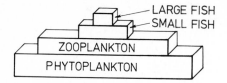

Fig. 3.7. An energy flow pyramid in the sea.

P.3.6. **Ecological energy flows are of considerable environmental interest as calculation of biological magnifications
are based on energy flows** (see also 2.8).

This concept is illustrated in Table 2.24 taken from Woodwell et al. (1967).
    Ecological efficiency should also be mentioned here, see Table 3.6, where
some useful definitions are listed. Some characteristic efficiencies are
listed in Tables 3.7 - 3.9.

TABLE 3.6
Ecological efficiency

Concept	Definition *)
Lindeman's efficiency	Ratio of energy intake level n to n-1: I_n/I_{n-1}
Trophic level assimilation efficiency	A_n/A_{n-1}
Trophic level production efficiency	NP_n/NP_{n-1}
Tissue growth efficiency	NP_n/A_n
Ecological growth efficiency	NP_n/I_n
Assimilation efficiency	A_n/I_n
Utilization efficiency	I_n/NP_{n-1}

*) For key to symbols used, see Fig 3.6

TABLE 3.7
Assimilation efficiency (A/I) for selected organisms
(after various authors)

Taxa	A/I value
Internal parasites	
Entomophagous Hymenoptera *Ichneumon* sp.	0.90
Carnivores	
Amphibian (*Nectophrynoides occidentalis*)	0.83
Lizard (*Mabuya buettneri*)	0.80
Praying mantis	0.80
Spiders	0.80 to 0.90
Warm- and cold-blooded herbivores	
Deer (*Odocoileus* sp.)	0.80
Vole (*Microtus* sp.)	0.70
Foraging termite (*Trinervitermes* sp.)	0.70
Impala antelope	0.60
Domestic cattle	0.44
Elephant (*Loxodonta*)	0.30
Pulmonate mollusc (*Cepaea* sp.)	0.33
Tropical cricket (*Orthochtha brachycnemis*)	0.20
Detritus eaters	
Termite (*Macrotermes* sp.)	0.30
Wood louse (*Philoscia muscorum*)	0.19
Soil-eating organisms	
Tropical earthworm (*Millsonia anomala*)	0.07

TABLE 3.8
Tissue Growth Efficiency (NP/A) for selected organisms
(after various authors)

Taxa	NP/A value	
Immobile, cold-blooded internal parasites		
Ichneumon sp.	0.65	
Cold-blooded, herbivorous and detritus-eating organisms		
Tropical cricket (*Orthochtha brachycnemis*)	0.42	
Other crickets	0.16	
Pulmonate mollusc (*Cepaea* sp.)	0.35	
Termite (*Macrotermes* sp.)	0.30	(?)
Termite (*Trinervitermes* sp.)	0.20	
Wood louse (*Philoscia muscorum*)	0.16	
Cold-blooded, carnivorous vertebrates and invertebrates		
Amphibian (*Nectophrynoides occidentalis*)	0.21	
Lizard (*Mabuya buettneri*)	0.14	
Spiders	0.40	
Warm-blooded birds and mammals		
Domestic cattle	0.057	
Impala antelope	0.039	
Vole (*Microtus* sp.)	0.028	
Elephant (*Loxodonta*)	0.015	
Deer (*Odocoileus* sp.)	0.014	
Savanna sparrow (*Passerculus* sp.)	0.011	
Shrews	Even lower values	

TABLE 3.9
Ecological Growth Efficiency (NP/I) for selected organisms
(after various authors)

Taxa	NP/I value
Herbivoruous mammals	
Domestic cattle	0.026 (0.44 x 0.057)
Impala antelope	0.022 (0.59 x 0.039)
Vole (*Microtus* sp.)	0.020 (0.70 x 0.285)
Deer (*Odocoileus* sp.)	0.012 (0.80 x 0.014)
Elephant (*Loxodonta*)	0.005 (0.30 x 0.015)
Birds	
Savanna sparrow (*Passerculus* sp.)	0.010 (0.90 x 0.011)
Herbivorous invertebrates	
Termite (*Trinervitermes* sp.)	0.140 (0.70 x 0.20)
Tropical cricket (*Orthochtha brachycnemis*)	0.085 (0.20 x 0.42)
Other crickets (New Zealand taxa)	0.050 (0.31 x 0.16)
Pulmonate mollusc (*Cepaea* sp.)	0.130 (0.33 x 0.30)
Detritus-eating and soil-eating invertebrates	
Termite (*Macrotermes* sp.)	0.090 (0.30 x 0.30)
Wood lause (*Philoscia muscorum*)	0.030 (0.19 x 0.16)
Tropical earthworm (*Millsonia anomala*)	0.005 (0.076 x 0.06)
Carnivorous vertebrates	
Lizard (*Mabuya* sp.)	0.100 (0.80 x 0.14)
Amphibian (*Nectophrynoides occidentalis*)	0.180 (0.83 x 0.21)
Carnivorous invertebrate	
Spiders	0.350 (0.85 x 0.42)
Internal parasites	
Ichneumon sp.	0.580 (0.90 x 0.65)

TABLE 3.10
Density, biomass, and energy flow of five primary consumer populations differing in size of individuals comprising the population

	Approximate density (m^2)	Biomass (g/m^2)	Energy flow ($kcal/m^2/day$)
Soil bacteria	10^{12}	0.001	1.0
Marine copepods (Acartia)	10^5	2.0	2.5
Intertidal snails (Littorina)	200	10.0	1.0
Salt marsh grasshoppers (Orchelimum)	10	1.0	0.4
Meadow mice (Microtus)	10^{-2}	0.6	0.7
Deer (Odicoileus)	10^{-5}	1.1	0.5

After E.P. Odum, 1968

P.3.7. There is a close relationship between energy flow rates and organism size.

Some of the most useful of these relationships are illustrated in Figs. 3.8-3.11 and in Table 3.10.

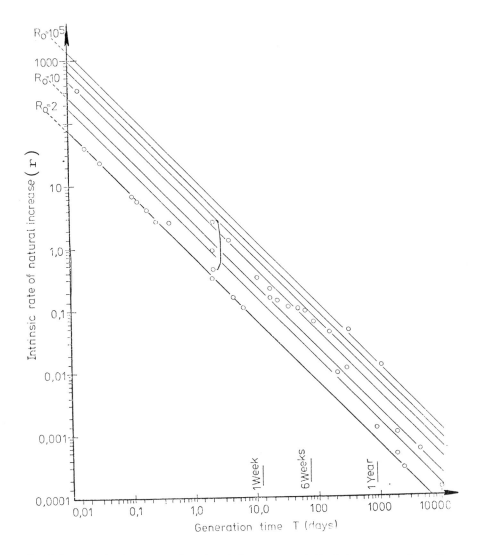

Fig 3.8. Intrinsic rate of natural increase plotted to generation time with diagonal lines representing net reproduction rate from 2 to 10^5 for a variety of organisms. (For definition of r, see 4.2)

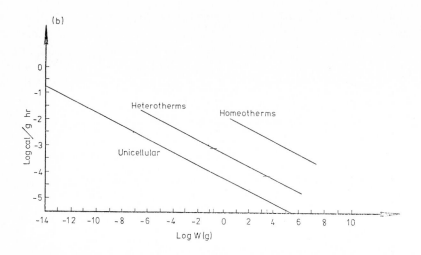

Fig. 3.9. The relationship of metabolic rate to weight for various animals.

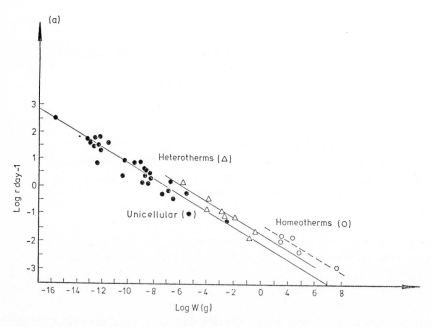

Fig. 3.10. Intrinsic rate of natural increase, r, to weight for various animals (for definition of r, see 4.2)

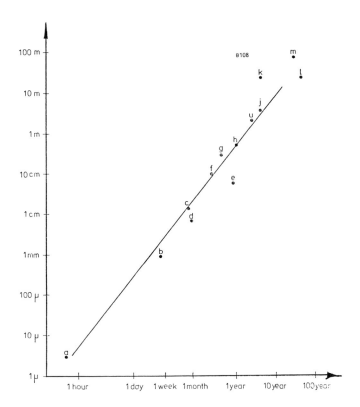

Fig. 3.11. Length and generation time plotted on log-log scale.

All these examples illustrate the fundamental relationship between size (surface) and the biochemical activity in the organisms. The surface determines the contact with the environment quantitatively, and thereby the possibility to take up food and excrete waste substances.

The same relationship explains the graphs shown in Figs. 3.12-3.15, where biochemical processes of toxic substances are involved.

These figures are constructed from literature data and, as seen, the excretion and uptake rates (for aquatic organisms) follow the same trends as the metabolic rate (Fig 3.9). This is of course not surprising, as the excretion is strongly dependent on the metabolism and the direct uptake is dependent on the surface.

The concentration factor, which indicates the ratio: concentration in the organism to the concentration in the medium, also follows the same lines, see Fig. 3.14. By equilibrium, the concentration factor can be expressed as the ratio between the uptake and excretion rates, as shown in Jørgensen (1979). As most concentration factors are determined by equilibrium, the relationship found in Fig. 3.14 seems reasonable. Intervals for concentration factors are here indicated for some species, in accordance with the literature (see Jørgensen et al., 1980).

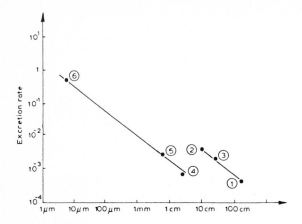

Fig. 3.12. Excretion of Cd ($24h^{-1}$) plotted to the length of various organisms: (1) Homo Sapiens (2) mice (3) dogs (4) oysters (5) clams (6) phytoplankton.

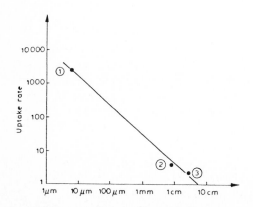

Fig 3.13. Uptake rate for Cd (μg/g 24H) plotted to the length of various organisms: (1) phytoplankton (2) clams (3) oyster.

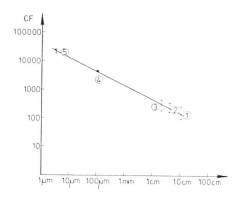

Fig. 3.14. CF for Cd versus size: (1) goldfish (2) mussels (3) shrimps (4) zooplankton (5) algae.

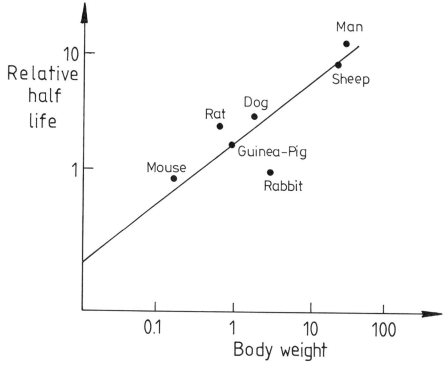

Fig. 3.15. Excretion of PCB and DDT versus organism size.

The principles illustrated in Figs. 3.12-3.16 can be applied generally. In other words, it is possible to find the uptake - and excretion rates and the concentration factors, provided that these parameters are available for the considered element or compound for one - but preferably several species. When a plot similar to Figs 3.12-3.16 is constructed, it is possible to read off the parameters when the size of the organism is known.

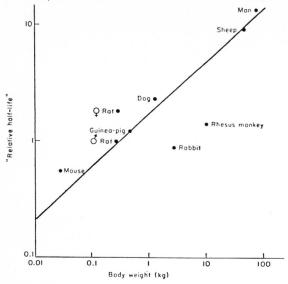

Fig. 3.16. The half life for various toxic substances relative to the half life time for rats.

There is a fixed upper limit to the total biomass of any species that can be supported in a given habitat. This fixed limit is referred to as the *carrying capacity* of that particular habitat for a particular species.

P.3.8. **The carrying capacity is related to the organism size in a similar manner to energy flow rates:**

$$C = w * p^{3/2} \qquad (3.7)$$

where C is the carrying capacity, w is the mean weight of individuals and p is the population density (White and Herper, 1970).

This means, since C is fixed, that we can increase the mean weight of the plants in a population or the population density, but not both. Space itself can be considered an important resource, independently of any other resources. Increasing the space per individual may have beneficial effects, even where other resources remain constant. The influence of space on the number of species is discussed in section 4.5 and a relationship is presented in P.4.24.

Any self-sustaining ecosystem will contain a wide spectrum of organisms ranging in size from tiny microbes to large animals and plants. The small organisms account in most cases for most of the respiration (energy turnover), whereas the larger organisms comprise most of the biomass.

Different species have very different types of energy use to maintain their biomass. For example the blue whale uses most (97%) of the energy available for increasing the biomass (G in Fig. 3.6) for growth and only 3% for reproduction. Many fishes, insects and other invertebrates use another partition of this energy. The adult female reproduces every season she is alive and the proportion going into reproduction can be over 50%.

Ecological energy flows are very sensitive to man's impact on ecosystems. A detailed picture of the energy flow in an ecosystem like that shown in Fig. 3.4 is a very useful tool for understanding this influence. The function of the ecosystem is closely related to the energy flow, and any change in the flow will mean a change in ecological function.

Fig. 3.17 shows the energy flow in a man-made ecosystem. By comparison with Fig. 3.4 it can be seen that the energy flow is simpler than in the natural system. Man-made ecosystems (chiefly agriculture systems) often have little ecological diversity (see also sections 4.5 and 4.8), in other words the number of species is small. In many cases, however, this renders the ecosystem very sensitive or vulnerable to any change.

P.3.9.   Man's influence on an ecosystem quite often means a simplification or decrease in its complexity, although the relationship between an ecosystem's stability and complexity is rather complicated (see sections 4.6 and 4.8).

Such an influence may have disatrous consequences for the stability of the ecosystem (see sections 4.6 and 4.8).

Generalizations such as these must, however, be used *very carefully*, as ecological stability is a very abused concept. See also sections 4.5 and 4.8.

A total energy balance for a terrestrial ecosystem can be set up by use of the first law of thermodynamics (P.3.1.):

$$R + C + H + ¥ * E = O \qquad\qquad (3.8)$$

where R is the net radiation, C is the conduction in solids, H is the convection in moving fluids and ¥ * E is the latent heat exchange associated with evaporation. C, H and ¥ * E are negative.
Net radiation is the dominant term in the equation. It is the balance between incoming and outgoing fluxes of shortwave and longwave radiation. R can be

found from:

$$R = (1-r) G + é(L-\partial T^4) \qquad (3.9)$$

where G and L are incoming fluxes of shortwave and longwave radiation, respectively; ∂ is a constant (= 81.7 * 10^{-12} for R in 1y min^{-1}), r is the reflectivity coefficient, é the emissivity coefficient and T the absolute temperature. r, é and T are ecosystem surface characteristics. Generally é≈ 1 for natural mineral and organic surfaces, while r, expressed as a percentage (= albedo), varies widely, see Table 3.11.
Table 3.12 shows the energy - and water budget for land areas of the earth.

TABLE 3.11
Typical albedoes for some natural surfaces

Surface	Description; conditions		Albedo
Water			
Liquid	Solar altitude:	60°	5
		30°	10
		20°	15
		10°	35
		5°	60
Solid	Fresh snow		75
	Old snow		50
	Glacier ice		30
Ground			
Soil	Dark organic		10
	Clay		20
	Light sandy		30
Sand	Gray:	wet	10
		dry	20
	White:	wet	25
		dry	35
Rock	Sandstone spoil, dry		20
	Black coal spoil, dry		5
Vegetation			
Grass	Typical field:	green	20
		dry	30
	Dry steppe		25
	Tundra, heather		15
Crops	Cereals, tobacco		25
	Cotton, potato		20
	Sugar cane		15
Trees	Rain forest		15
	Eucalyptus		20
	Red pine forest		10
	Hardwoods in leaf		18

TABLE 3.12

Mean values of the components of water and energy budgets for land areas of the earth.

Latitude (degrees)	Water budget (mm yr-1)			Energy budget (kly yr^{-1})		
	P	Q	E	IE	H	R
Polar (N)	176	-106	- 70	- 4	4	0
70-60	428	-227	- 201	-12	- 8	20
60-50	577	-259	- 318	-19	-11	30
50-40	535	-155	- 380	-22	-23	45
40-30	534	-122	- 412	-24	-36	60
30-20	611	-245	- 366	-21	-48	69
Tropics (N)	1,292	-436	- 856	-50	-22	72
Tropics (S)	1,576	-546	-1,030	-60	-13	73
20-30	564	- 88	- 476	-28	-42	70
30-40	660	-165	- 495	-29	-33	62
40-50	1,302	-914	- 388	-23	-18	41
50-60	993	-605	- 388	-23	- 8	31
60-70	429	-369	- 60	- 4	- 9	13
Polar (S)	148	-120	- 28	- 2	6	- 4

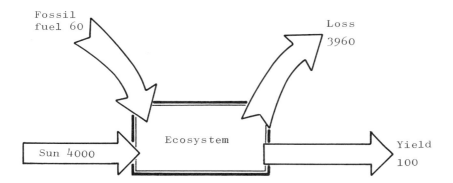

Fig. 3.17. Man-made (man-controlled) ecosystem (agriculture). Numbers in kcal m^{-2}day^{-1}.

3.5. ENERGY CONSUMPTION AND THE GLOBAL ENERGY BALANCE.

Recently the subjects of climate change and, particularly, the possible effects of CO_2 (carbon dioxide) on the climate are being widely discussed. Currently the question of interaction between atmospheric carbon dioxide and climate is a very critical one, that could greatly affect future energy strategies.

In 1975 the world primary consumption was about 7.6 TW, and the share of oil and gas 5.5 TW (Häfele, 1978). IIASA made an estimate of the future minimum energy demands based on the following considerations: The world is divided into seven regions roughly corresponding to North America, the Soviet Union and Eastern European countries, Western Europe, Australia and Japan, Central and South America, South East Asia and Africa, the Middle East and China.

Taking Keyfitz's population figures (1977), a total world population for 2030 not larger than 8 billion people is projected. The per capita energy demand growth rate in developed regions is assumed to decline steadily, coming close to saturation point 50 years from now, while higher growth rates and later saturation points are assumed for developing countries. A summary of this prediction is given Table 3.13.

TABLE 3.13
Demand scenario in the year 2030

Region	Population (10^6)	kW/capita	TW
North America	310	16.5	5.0
Soviet Union and Eastern Europe	460	11.6	5.3
Western Europe, Japan and Australia	780	10.5	8.2
Central and South America	720	3.3	2.4
South East Asia and Africa	3700	1.4	5.2
Middle East	300	4.8	1.4
China	1800	4.6	8.3
Total	8070	average: 4.4	35.8

This scenario gives a total global energy demand of 35 TW or almost 5 times present energy consumption. The average energy consumption is predicted at 4.4 kW per capita in 2030, while only 28 developed countries today have an energy consumption larger than 2 kW. The cumulated overall world energy demand by 2030 will be approximately 1000 TW for this reference scenario. The crucial questions are: can we meet the demand for such a high energy consumption? If so, how, and what will the environmental consequences be?

Another problem is of course the size of available fossil energy resources needed to supply this large energy demand. Fig. 3.18 lists the economically recoverable reserves of fossil energy and predict that in 2030 only a fraction of the 35 TW can be supplied in the form of conventional oil, gas and coal. The geological resources of fossil energy are much larger, but cannot be recovered under present and predicted technological and economic conditions.

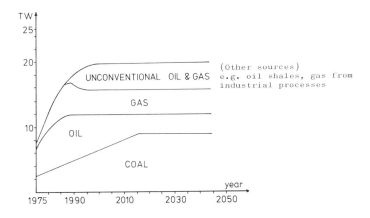

Fig. 3.18. Economically recoverable reserves of fossil energy.

A further constraint on the massive consumption of fossil fuel is represented by the CO_2 problem. And what we are considering here is: how such large scale use of fossil fuel would influence the global climate?

The use of fossil fuel means the oxidation of carbon or carbon hydrides, which produces a stoichiometric amount of carbon dioxide:

$$C + O_2 - CO_2 \tag{3.10}$$

$$C_xH_y + (x + y/4)O_2 - 1/2\, y\, H_2O + xCO_2 \tag{3.11}$$

Having determined that the burning of fossil fuels produces carbon dioxide, we can next consider how this carbon dioxide (and watervapour) produced by oxidation will influence the climate.

This is a very complex question and any attempt to an answer requires a comprehensive knowledge of the global cycle.

Another crucial issue related to the long-term prediction model, is the estimation of the ultimate recovery of coal, oil and natural gas. Zimen (1978) assumes a final input of carbon dioxide from fossil sources into the atmosphere to be 8 times the preindustrial amount of atmospheric carbon dioxide, taken as 292 ppm or $52.1 * 10^{15}$ moles. Baes et al. (1976 and 1977) estimate this figure to be 12 times as great, while Perry et al. (1977) take the figure to be 8.2 times. Table 3.14 gives the most recent estimates of global fossil fuel resources and the corresponding amounts of carbon dioxide that would be produced, calculated from the following information (Ziemen, 1977):

Coal 2.00 moles of carbon dioxide per MJ
Oil   1.68 moles of carbon dioxide per MJ
Gas 1.29 moles of carbon dioxide per MJ

TABLE 3.14
Recent estimates of global fossil fuel resources and the corresponding amount of carbon dioxide expressed as 10^{15} moles

	Estimates of global fossil fuel resources	
	10^{21} J	10^{15} mol CO_2
Solid fuel	150 - 300	300 - 600
Liquid fuel	10 - 80	15 - 130
Tars and shales	10 - 20	17 - 35
Natural gas	4 - 44	5 - 55

Rotty (1978) has suggested several possible patterns of future carbon dioxide production by use of the following expression:

$$\frac{1}{N} * \frac{dN}{dt} = 0.043 \left(1 - \frac{N}{A}\right)^x \qquad (3.12)$$

where
N is a function of time t and represents the total cumulative amount of carbon dioxide produced by use of fossil fuel up to the time t. A is the amount of carbon dioxide that would be produced from all of the fossil fuel ultimately recoverable (corresponds to $7.3 * 10^{12}$t C). x is a parameter applied to vary the emphasis on price, availability etc, as the fraction of

recoverable fossil fuel remaining is reduced.

Fig. 3.19 shows the annual production of carbon dioxide expressed as ppm per annum plotted against time for 3 different values of x.

Fig 3.20 gives the corresponding cumulative carbon dioxide production.

This model is, of course, very simple and does not consider the many interacting processes of the global carbon dioxide cycle (see 2.5).

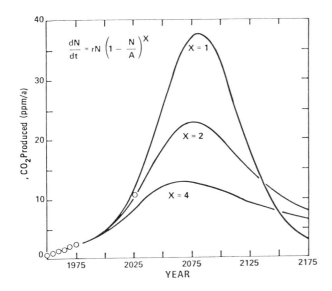

Fig. 3.19. Global production of CO_2 from fossil fuels.

One part of the problem is the problem is the quantitative determination of the carbon pools and the fluxes between these pools, another is the relationship between the atmospheric carbon dioxide concentration and the climate. This issue is also very complex as different subsystems interact with each other.

The global climatic system consists of 5 subsystems: the atmosphere, the oceans, the cryosphere (ice and snow), the land surface and the biosphere. These subsystems interact with each other through a wide variety of processes. The climatic state (Garp, 1975) is defined as the average and the variability of the complete set of atmospheric, hydrospheric and cryospheric variables over a specified time interval and in an specified domain of the earth-atmosphere system. The climatic states are subject to fluctuations of a statistical origin in addition to those of a physical nature. These statistical fluctuations in the weather are unpredictable over the time scales of climatological interest and are therefore referred to as "noise" or the "inherent variability of the climatic system".

With reference to the carbon dioxide problem, we are not concerned with the inherent variability of the climate, but with the natural variability caused by changes in external and/or internal conditions. However, any estimate of a change in climate caused by an increase in carbon dioxide concentration must take into consideration the fact that climate fluctuates naturally with time, and therefore that it takes a long time to detect a significant change in climate. It should also be pointed out that problems could arise because of the time scales of natural variability, e.g. the time constant for heat storage in the oceans.

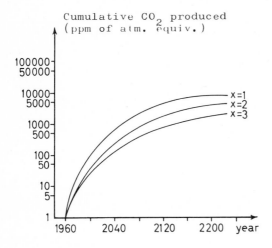

Fig. 3.20. Projected cumulative CO_2 production for x=1, 2 and 3 (for formula, see text).

A global average model can be used to compute a change in average climate for a given change in atmospheric carbon dioxide concentration. Such models are available today, although results are uncertain due to lack of knowledge of feedback mechanisms.

The question of regional climatic changes is, however, of more practical importance. In this context it should be stressed that

P.3.10. **it is possible to have drastic changes in the regional climate without any substantial change in the global ave-
rage situation.**

This is illustrated in Fig. 3.21, where changes in seasonal temperatures are plotted against latitude.

The global energy balance (see Fig. 3.22), is influenced by the carbon dioxide concentration in the atmosphere. The atmosphere is transparent for solar radiation, but prevents, to a certain extent the loss of heat from the earth's surface (by longwave radiation). The radiation balance of the atmosphere is essentially determined by the presence of optically active minor constituents, such as water vapor, carbon dioxide, ozone, aerosols and others. An increased carbon dioxide concentration means increased absorption of longwave radiation and consequently an increased global average temperature. Carbon dioxide has what is called a greenhouse effect, but the atmosphere also contains a number of other minor constituents capable of causing changes in the global average temperature. Table 3.15 lists the calculated surface temperature change for a 1% increase in each of the parameters given (the table is based upon the following sources: Reck, 1978; Wang et al., 1976; Reck et al., 1977; Ramathan, 1975; Hummel et al., 1978). However, from this table it can be seen that the concentration of carbon dioxide seems to be the most important agent in this respect, although the role of trace gases has not yet been investigated in detail.

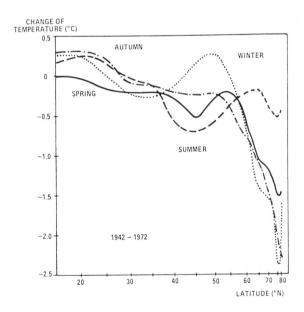

Fig. 3.21. Meridional profile of the change in each season of zonal mean temperature (°C) during the period 1942-1972. (Source: Williams 1976)

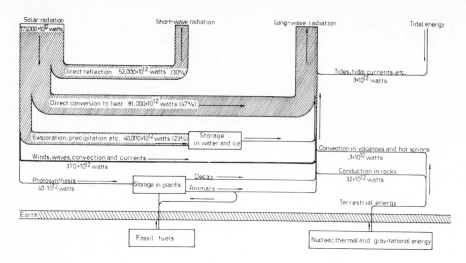

Fig. 3.22. The global energy balance. More than 99% of the input is solar radiation.

TABLE 3.15
Calculated surface temperature change for a 1% increase in each of the parameters given

	Present value	Units	ΔT_s	Relative rank
Albedo	14.01	%	-0.14	1
Relative humidity	74	%	0.065	0.46
CO_2	330	ppmv	0.020	0.14
Airborne paticles extraction coefficient	0.1	km^{-1}	-0.0063	0.057
N_2O	637	ppbv	0.0044	0.031
O_3	0.37	atm-cm	-0.0032	0.023
CH_4	1.6	ppmv	0.002	0.014
H_2S	3.5	ppmv	0.0012	0.0086
$CHCL_3$	$<2 *10^{-4}$	ppmv	0.0010	0.0071
NH_3	$6 *10^{-3}$	ppmv	0.0009	0.006
HNO_3	$10^{-3} - 10^{-2}$	ppmv	0.00060	0.004
$CFCl_3$	$2.3 *10^{-4}$	ppmv	0.00035	0.0025
SO_2	2	ppmv	0.00020	0.0014
CF_2Cl_2	$1.3 *10^{-4}$	ppmv	0.00019	0.0014
C_2H_4	$2 *10^{-4}$	ppmv	0.0001	0.0007
CH_3Cl	$5 *10^{-4}$	ppmv	0.0001	0.0007
CCl_4	$1 *10^{-4}$	ppmv	0.0001	0.0007

ΔT_s = surface temperature change (K) for 1% increase in the present average value

Augustsson et al. (1977) have calculated the relationship between the atmospheric carbon dioxide concentration and the surface temperature of the earth, see Fig. 3.23. Changes in cloud cover can be taken into account by

assuming either a fixed cloud top altitude (CTA) or a fixed cloud top temperature (CTT). Both assumptions are taken into consideration in Fig. 3.23, but
even if the concentration of carbon dioxide could be accurately predicted,
vast uncertainties about its effect on the climate would remain. However,
there is strong evidence to suggest that additional atmospheric carbon
dioxide will cause global warming in the range of an average 2 to 4 °C rise in
temperature for a doubling in the carbon dioxide concentration.

As well as the global average surface temperature the atmospheric circulation pattern is also affected. The general atmospheric circulation is
characterized by a poleward transport of heat energy. The general temperature increase caused by a higher carbon dioxide concentration enhances the
poleward transport of latent heat in middle latitudes, resulting in a general
movement of agroclimatic zones. For example, summer temperatures may become too high for corn and soybean, so that the corn belt will be shifted
north.
Agricultural productivity is very much dependent upon climate. Each crop
has its unique response to climatic variations (Thompson, 1975), as can be
seen in Tables 3.16-3.17

TABLE 3.16
Estimated percent change in corn yield due to changes in
temperature and precipitation. Source: Benci et al. (1975)

Temperature	Change in precipitation (% of normal)				
change (°C)	- 20%	- 10%	0	+ 10%	+ 20%
- 2.0	19.8	21.2	22.7	24.2	25.6
- 1.0	8.4	9.8	11.3	12.8	14.2
0	- 2.9	- 1.5	0	1.5	2.9
+ 1.0	-14.2	-12.8	-11.3	- 9.8	- 8.4
+ 2.0	-25.6	-24.2	-22.7	-21.2	-19.8

The response of snow and ice cover to climatic factors varies greatly in
time. Typical extents of cover are 0.1-1.0 years for seasonal snow, 1.0-10
years for sea ice and 1000-100,000 years for ground ice and ice sheets. Of
primary interest, of course, are the times of seasonal temperature
transition across the 0°C threshold (-1.8 °C for seawater). Another important
effect is the large albedo differences between snow cover (appr. 0.8) and
snow-free ground (0.1-0.25) or between ice (0.65) and water (0.05-0.1). At
present perennial ice covers 7% of the world's oceans and 11% of the land
surface (see Table 3.18), with a further percentage of land under seasonal
snow cover (see Table 3.18). Barry (1978) has attempted to predict the
effect on the cryosphere of a doubling in the atmospheric carbon dioxide

concentration, assuming that it will result in a global warming of 1.5°C with a rise of 5-6°C in polar latitudes (Manabe et al., 1975).

TABLE 3.17
Percent deviation from world rice production base as influenced by changes in temperature and precipitation

Precipitation change (%)	Temperature change (°C)						Total pre-cipitation change (%)
	- 2	- 1	-0.5	+0.5	+1	+2	
-15	-19	-13	- 8	- 4	0	3	- 8
-10	-17	-11	- 6	- 2	2	5	- 6
- 5	-13	- 7	- 2	2	6	9	- 2
+ 5	- 9	- 3	2	6	10	13	2
+10	- 5	1	6	10	14	17	6
+15	- 3	3	8	12	16	19	8
Total tempe-rature change	- 11	- 5	0	4	8	11	

Source: Stansel and Huke (1975)

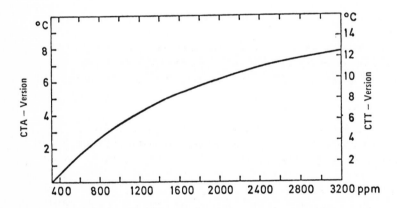

Fig. 3.23. CO$_2$ concentration and equivalent surface temperature.   Augustsson and Ramanathan model.

TABLE 3.18
Components of the cryosphere (Untersteiner (1975);   Kukla et al. (1974))

	Area (10^8 km^2)	Volume (10^8 km^3 as water)

Land ice	0.16	0.31
Antarctica	0.138	0.28
Greenland	0.018	0.027
Other areas	0.004	0.003
Ground ice (excl. Antarctica)	0.248	<0.025
Cont. permafrost	0.075	<0.02
Discont. permafrost	0.173	<0.005
Sea ice		
Antarctica - September	0.20	
Antarctica - March	0.025	
Arctic - September	0.054	
Arctic - March	0.15	
Total snow, land ice, sea ice		
N. hemisphere - January	0.58	
N. hemisphere - July	0.14	
S. hemisphere - January	0.18	
S. hemisphere - July	0.25	
Global		
mean annual	0.59	

It is concluded that:
1. there will be negligible changes for major ice sheets and ground ice on a time scale of centuries,
2. changes in seasonal snow cover may also be minimal. Any increase in winter snowfall would probably be offset by increased summer melt,
3. it is possible that greater melt of pack ice leading to more open water, would reduce the solar radiation input.

However, it is also concluded that major deficiencies exist in terms of data on snow cover extent and depth. Many crucial research problems remain, especially because our knowledge about the effects of a rise in temperature on the melting glaciers and ice sheets is rather limited.

Although much more research is needed before we can map the relationship between the consumption of fossil fuel and the climatic conditions, it can be concluded from this brief review that we have quantitative knowledge of the interactions between the subsystems to justify constructing a model of this relationship. It is only through the use of such a model that we can assess in which direction we should conduct our further research to discover ways of setting up more reliable models.

The problem discussed in this chapter is a very complex one, because its solution requires knowledge of the interactions between a large number of subsystems. A break-down of the problem is attempted in Fig. 3.24: the energy policy determines the energy demand and both determine the consumption of fossil fuel, which in turn is related to carbon dioxide production.

The concentration of carbon dioxide in the atmosphere is a function of both the global carbon cycle including all its processes, and the carbon dioxide production. The carbon dioxide concentration and the global average surface temperature are related in a very complex way, and the latter will cause several other global changes, such as higher sea level, shifts in crop belts etc - changes, which it is very difficult to predict with our present knowledge. However, such predictions of global changes are needed before we can set up a realistic energy policy.

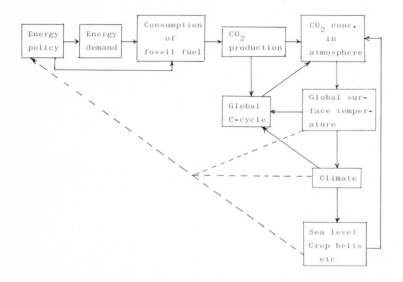

Fig. 3.24. A break-down of the CO_2-problem into subsystems.   Interactions between the subsystems are shown.

3.6.   ENERGY BUDGETS OF ANIMALS AND PLANTS.

P.3.11. An animal is coupled to the climate surrounding it by energy flow.

Animals are often in transient states for limited periods of time, but over a sufficiently long time interval they must be in an energy balance. Warm blooded animals (homeothermic) require a more or less constant body temperature, while cold blooded (poikilothermic) animals can allow their body temperature to vary over a considerable range. However, the principles of energy balance in physical terms are essentially the same for warm blooded and cold blooded animals.

The principle of an energy balance for food is shown in Fig. 3.6 but it might also be considered as an illustration of a mass balance, as mass can be converted to energy by multiplying it by the content of chemical energy per g of dry matter (expressed e.g. in the unit kJ g^{-1}). Within the animal metabolic heat is generated from food consumed and additional energy is acquired by absorption of incident radiation, in the form of direct sunlight, scattered skylight, reflected light from clouds and other surfaces, and thermal radiant heat from the environment.

The energy balance for the surface of an animal can be expressed as follows:

$$M - E + R = é\partial T_s^4 + k * V^{1/3} * D^{-2/3} (T_s - T_a) + C \qquad (3.13)$$

where

M = metabolic heat
E = evaporation
R = radiation absorbed
é = emissivity of the surface
∂ = Stefan-Boltzmann's constant
T_s = surface temperature in K
k = a constant
V = wind speed
D = diameter of body (cm)
T_a = air temperature in K
C = energy exchange by conduction

The rate at which metabolic heat flows out from the body through the insulating layer of fat, fur or feathers depends on the temperature gradient between the body temperature and the surface temperature. Hence, the rate of net heat production M-E is given by:

$$M - E = \frac{T_b - T_s}{I_b + I_f} + \frac{T_b - T_s}{I} \qquad (3.14)$$

where

T_b = body temperature

I_b = insulation quality by fat

I_f = insulation quality by fure (feather)

$I = I_b + I_f$

By eliminating M-E from the two equations (3.13) and (3.14) we get:

$$T_b - T_s = I(é\partial * T_s^4 - R + k * V^{1/3} * D^{-2/3} (T_s - T_a)) \qquad (3.15)$$

If T_b is fixed, as in homeotherms, the surface temperature T_s will adjust, so the equation is valid. For poikilotherms the body temperature will adjust to the surface temperature through the animals' insulation mechanisms, see equation (3.14). If M-E is very small, as it often is for some reptiles, then T_b and T_s are almost equal. R is of course a very complicated function for the animals in the open, particularly during the day time. However, with some approximations we have:

$$R = é\partial * T_a^4 \qquad (3.16)$$

The climate also has a pronounced effect on plants. If the air is cooler than the leaf surface, then the flow of air across the leaf surface will pick up heat from the leaf and carry away a definite amount of energy by means of convection. It the air is warmer than the leaf then there is an energy transfer from the air to the leaf. The larger the leaf, the thicker the boundary layer of air which adheres to the leaf surface. Therefore, the rate of heat transfer is inversely proportional to leaf size. Furthermore, the convection is proportional to the difference in temperature between leaf and air and varies directly with the square root of the wind speed.

A leaf surface absorbs a total amount of radiation and get rid of energy by reradiation.

If a leaf loses energy by radiation and convection, we can set up an energy balance as follows:

$$R = \partial T^4 + 6 * 10^{-3} \sqrt{V} * D^{-1} (T - T_a) \qquad (3.17)$$

where

V = the wind speed

D = the width of leaf (cm)

T_a = the plant temperature

If we add the plant's ability to cool by transpiration, the leaf temperature is reduced further as demonstrated in Table 3.19. The energy budget equation involving transpiration in addition to radiation and convection is expressed:

$$R = \partial T^4 + 6 * 10^{-3} \sqrt{V} * D^{-1} (T - T_a) + 580 \frac{d - rh * d_a}{r_e + r_a} \qquad (3.18)$$

where

d   =   saturation concentration of water (g cm^{-3}) in the leaf
d_a =   saturation concentration of water (g cm^{-3}) in the air
rh   =   relative humidity
r_e =   resistance of the leaf to water loss
r_a =   resistance of the leaf surface (boundary layer) to water loss

TABLE 3.19
Characteristic temperatures in degrees centrigrade for the conditions listed (°C)

R (cal cm^{-2} min^{-1})	Radiation only	Radiation and convection V= cm sec^{-1}			Radiation, convection and transpiration V= cm sec^{-1}		
		10	100	500	10	100	500
0.6	20	28	29	30	25	27	29
1.0	60	41	34	32	33	31	30
1.4	89	53	39	34	40	35	32

A modification of the normal energy balance for plants is widely used in agriculture. The heat load can be increased by use of heat trapping, as demonstrated in Fig. 3.25. This is the most widely used form of control in plant environments. Other possibilities are summarized in Table 3.20.

TABLE 3.20
Artificial control of plant environment

Control practice	Major beneficial results	Major adverse result
Heat trapping	Higher plant temperature	Too high plant temperature
Shading	Lower plant temperature in hot weather	Expensive
Flooding	Frost protection	Retarded warming after cold spell
Spraying cold water	Frost protection	Expensive
Burning fuels	Frost protection	Expensive
Wind machines	Frost protection	Expensive
Soil heating cables	Frost protection	Very expensive
Hot capping	Higher plant temperature	Very expensive
Mulching and plowing	Reduced water loss	Increased frost hazard
Irrigation	Reduced moisture stress	May benefit soilborne pests
Soil packing	Frost protection	Retard soil warming
Shelter belts and nurse crops	Reduce heat loss and erosion in spring	Soil moisture losses from deeper layers

Daily temperature patterns are also modified by topography. A slope facing the sun intercepts light beams more perpendicularly than does a slope facing away from it; as a result a South-facing slope in the Northern hemisphere receives more solar energy than a North-facing slope and heats up faster and gets hotter during the day (see Fig. 3.26). A South-facing slope is typically drier than a North-facing one, because it receives more solar energy, and therefore more water is evaporated.

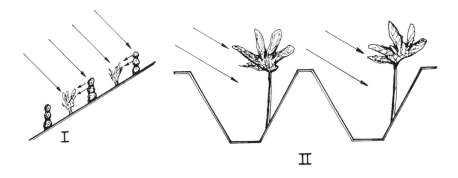

Fig. 3.25.. Heat trapping by stone walls between crop rows (I) and by furrow planting (II).

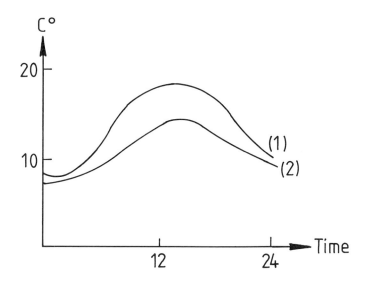

Fig. 3.26. Daily variations of temperature on an exposed South-facing slope (1) and a North-facing slope (2) during late summer in the Northern hemisphere.

QUESTIONS AND PROBLEMS.

1. Draw all possible parallels between a foodchain and an energy chain.

2. List at least 6 environmental factors, which influence the primary productivity.

3. Use Fig. 3.4 to answer the following question: If algae have a concentration factor of 1,000 relative to water containing 1 μg l^{-1} of a toxic and completely non-biodegradable compound, what will be the concentration of the toxic compound in carnivores and microorganisms, provided that the uptake efficiency from food is 90% and that no excretion takes place?

4. Find the approximate generation time, metabolic rate and intrinsic rate for a homeothermic animal with a length of 20 cm.

5. How would an increasing humidity influence the global average temperature?   What would be the temperature change for an increase form the present humidity of 74% to 80%, and how would such a change influence the corn yield and cryosphere?

6. What will be the sign of entropy change for denaturation of proteins?

7. Discuss the connection between the second law of thermodynamics and the ecological energy pyramid.

8. A chlorphenol has a solubility in water of 1.2 mg l^{-1}. Find the approximate concentration factor for mussels by use of the methods presented in sections 2.13 and 3.4.

9. What would be the expected increase in surface temperature if 10% of the earth's surface area now covered by snow and ice becomes tundra? Use Tables 3.11 and 3.15.

CHAPTER 4

ECOLOGICAL PRINCIPLES AND CONCEPTS

4.1. ECOSYSTEM CHARACTERISTICS.

In Chapter 2 we mentioned how man can influence his environment or even the entire ecosphere by changing the global cycles of elements (see 2.5) or by discharging chemical compounds e.g. organic matter (see 2.6), nutrients (see 2.7) and pesticides (see 2.8) into the environment. The effect of such activity on the natural mass flows presents a threat to man's own survival on earth.

Man is facing three crises - an environmental crisis, a food crisis and an energy crisis, all of which are closely linked together. It is important to realize that he cannot solve these crises by violating the first and second laws of thermodynamics. His impact on the environment may modify the global energy balance and thereby the climate (see 3.5), which might have disastrous consequences, but it is also important to understand the behaviour of energy at the ecosystem level, if mankind is to attempt to find a solution for all three crises. To this end chapter three was devoted to the energy flow in the ecosphere and in different ecosystems, and to fostering an understanding of how the first and second laws of thermodynamics could be applied to these problems.

In an ecosystem two major processes are in operation: transformation of high-quality energy to low-quality energy, which results in chemical cycling of important elements accompanied by energy cycling. The former is a condition of the latter and relates mass flows to energy flows.

We often use the term "balance of nature", which must *not* be interpreted as the ecosystems do not change with time. Ecosystems are not static, but dynamic. The biotic communities that make up an ecosystem are continually changing in response to environmental changes caused either by the communities themselves or by external stresses, among which is the impact of man's activities.

An ecosystem is able to maintain its overall stability by three major mechanisms:
1. by controlling the rate of energy flow through the system,
2. by controlling the rate of element cycling within the system,
3. by maintaining a diversity of species and foodwebs - an ecological structure.

In Chapters 2 and 3 we discussed the first two mechanisms; this chapter is devoted to an explanation of the third.

P.4.1. The ecosystem is capable of self-maintenance and self-regulation.

But how does an ecosystem maintain its stability in spite of disruption? As we have illustrated all the components of an ecosystem are in constant interaction, and information from each component is continually feed back into the system. This interaction preserves the integrity of the system. Thus the science of control mechanisms, cybernetics, has an important role in ecology, especially since man tends to disrupt natural control.

P.4.2. Homoeostatis is the term generally applied to the tendency for biological systems to resist changes and to remain in a state of dynamic equilibrium,

and this chapter attempts to explain how this homoeostasis functions, in the context of the interplay of material cycles and energy flows already discussed in Chapters 2 and 3.

A good example of cybernetic control at the organism level is the method by which the temperature of the human body is kept within a few degrees of 37°C.

P.4.3. Any cybernetic system can, however, be overloaded. In spite of the feedback system there is a limited range of tolerance to variation, termed the homoeostatic plateau, which applies equally at all levels within an ecosystem.

Our knowledge about the cybernetic system at the ecosystem level is quite limited due to the complexity of ecosystems. However, some ecological feed back mechanisms are known and can be used to illustrate how ecosystems respond to stress. The following mechanisms will be discussed:

1. Adaptation
2. Self-regulated growth
3. Interaction between two or more species
4. Organization, development and evolution of the ecosystem
5. Diversity
6. Buffer capacity in the environment

4.2.   ADAPTATION.

Living organisms occur in diverse habitats from tropical forests to the polar ice caps, and from small ponds to deep oceans.

P.4.4.   Adaptation means both the genetic process, by which organisms become increasingly better able to exist under prevailing environmental conditions, and the specific genetic (phenotypic) trait that renders one organism more capable of existence than another.

A related but non-genetic process is acclimation, the modification of an organism's phenotypic trait by the environment, i.e. the environmental modification of gene expression.

An example of acclimation is given in Fig. 4.1. As seen there is an upper and a lower temperature which is lethal to fish. In between these extreme temperatures is a zone of tolerance where progressive rises in the acclimation temperature lead to a corresponding, but smaller rise in the upper temperature limit. The point (see Fig. 4.1, point A), where the acclimation temperature and the thermal death point are the same, is characteristic of each different species of fish (Erichson-Jones, 1964). The area of the polygon (in $^\circ C^2$) is a measure of the *thermal tolerance* of the organism (see Fig. 4.1). *The tolerance to elevated temperature* is related to the area between a 45° line intersecting the origin, the y-axis and the upper LC_{50} line.

TABLE 4.1
Tolerance to elevated temperature

Fish	Tolerance $(^\circ C)^2$
Mosquito Fish	676
Gold Fish	575
Brook Trout	306
Sockeye Salmon	282
Chum Salmon	260

Compiled from McErlean et al., 1969.

Previously we have referred to species as if they constituted ecological units. However, a species consistes of local populations that may differ conspicuously in their adaptive properties.

P.4.5.   Genetic variation in a population is the raw material of adaptive change.

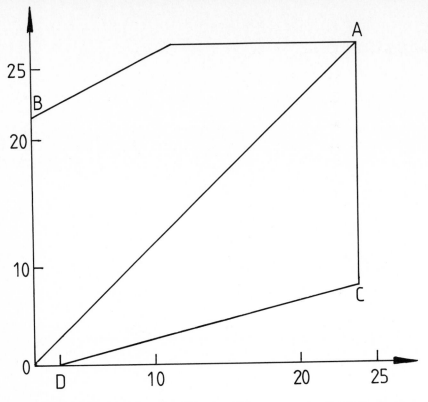

Fig. 4.1. Temperature of tolerance TT° versus acclimation temperature AT° for young sockeye Salmon (Brett, 1960). The area OBACDO is the thermal tolerance. OBA represents the tolerance to the elevated temperature.

Such variation may arise from **mutation**, a change in the basic composition of the genetic substance (DNA) - or from **recombination** - a change in the structural organization of the chromosomes.

At any instant in time a population will contain both old and new, or novel genes. Old genes are those which have been present in the population for a considerable period of time. During this period, the frequency of the occurrence of genes in the population depends on the mating patterns and natural selection occurring in the population. Novel genes may originate through mutation or by gene flow from other populations. The former is the most likely source for more or less isolated populations, while the latter is most likely to be responsible in a population in reproductive contact with several other populations.

We used to think of natural selection as a directional process, moving a population from one genetic constitution to another, but as pointed out by Mather (1953) this is only one of three possibilities (see Fig. 4.2).

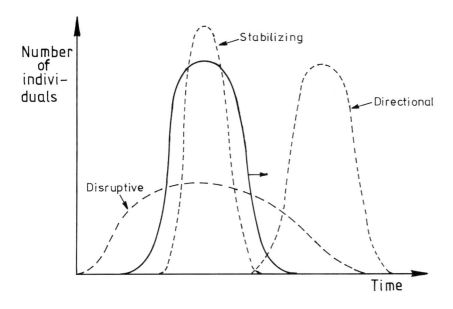

Fig. 4.2. Effect of different types of selection on the distribution of phenotypes in a population (Mather, 1953).

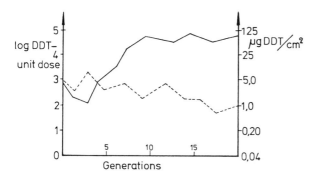

Fig. 4.3. Change in DDT resistance of fruit fly populations exposed to directional selection for high (——) and low resistance (———). The left hand scale = log of dose tolerated. The right hand scale = the actual dose, which could kill half a population sample in standard exposure time.

An example studied by Bennett (1960) and of environmental interest is illustrated in Fig. 4.3. Two populations of fruit fly were studied, one of which was exposed to DDT-impregnated paper. The LD_{50}-value was found (after 18 hours exposure) for each generation of the two populations. Fig. 4.3 shows that after only five generations the two populations exhibited marked differences in DDT resistance. After 18 generations the LD_{50}-value was 125 times higher for the exposed population than for the non-exposed population.

It is also due to adaptation that the synthetic pesticides (aromatic chlorinated hydrocarbons) introduced shortly after the Second World War, were not as successful as they were expected to be after the preliminary results.

P.4.6. **The application of this new generation of pesticides gave very promising results in the first couple of years, but then the insects became resistant to these chemicals.**

When we also take into account the other properties of this group of compounds, such as low biodegradability (see 2.13) and an ability to be concentrated through the foodchain (see 2.8), it is easy to understand that the application of these pesticides has been questioned. In many countries today it is no longer legal to use DDT and some related compounds.

One of the best examples of natural selection under field conditions has been provided by Kettewell (1956). He has studied the British moth, Biston betularia. This species has a series of different ecotypic populations occupying different areas, and varying in colour from almost white to very dark grey. One of the consequences of man's increased consumption of fuel has been the heavy deposition of soot in industrial areas. The dark type moth is inconspicuous in polluted areas and conspicuous in unpolluted areas, whereas the converse is true for the light type. The frequency of the types in the two habitats is distinctly different with substantially more dark grey moths in the polluted habitats and many more light forms in unpolluted habitats. The results of an examination of the phenotypic patterns are given in Table 4.2.

Direct observations by Kettlewell indicate that there was substantially more predation by birds upon the maladapted forms, which goes some way to explain the distribution of the two forms. It has recently been found (Cook et al., 1970) that pollution abatement programs are being accompanied by an increase in the frequency of light coloured forms in previously polluted areas, indicating that selection is now being reversed. (The original moths were mostly pale coloured).

Many more examples could illustrate the concept of adaptation but both examples demonstrate that

P.4.7. the ecosystem responds to external stress by selecting the form that can meet the stress by the smallest change in the ecosystem.

TABLE 4.2
Phenotypic patterns in a moth occupying woodlands with dark coloured tree trunks resulting from the death of light-coloured lichens on tree trunks from air pollution and woodlands with unpolluted, lichen-covered, and light-coloured tree trunks

Moth phenotype	Polluted area	Unpolluted area
	Percent of phenotypes in native population	
White	10	95
Grey	85	0
	Percent of phenotypes recaptured	
White	13	13
Grey	28	6
	Observed predation, number taken	
White	43	26
Grey	15	164

4.3. GROWTH AND SELF-REGULATION.

A population is defined as a collective group of organisms of the same species. Each population has several characteristic properties, such as population density (population size relative to available space), natality (birth rate), mortality (death rate), age distribution, dispersion growth forms and others.

A population is a changing entity, and we are therefore interested in its size and growth. If N represents the number of organisms and t the time, then dN/dt = the rate of change in the number of organisms per unit time at a particular instant (t) and $dN/(Ndt)$ = the rate of change in the number of organisms per unit time per individual at a particular instant (t). If the population is plotted against time a straight line tangential to the curve at any point represents the growth rate.

Natality is the number of new individuals appearing per unit of time and per unit of population.

We have to distinguish between absolute natality and relative natality, denoted B_a and B_s respectively:

$$B_a = \frac{\Delta N_n}{\Delta t} \tag{4.1}$$

$$B_s = \frac{\Delta N_n}{N \Delta t} \tag{4.2}$$

where ΔN_n = production of new individuals in the population.

Mortality refers to the death of individuals in the population. The absolute mortality rate, M_a, is defined as:

$$M_a = \frac{\Delta N_m}{\Delta t} \tag{4.3}$$

where ΔN_m = number of organisms in the population, that died during the time interval Δt, and the relative mortality, M_s, is defined as:

$$M_s = \frac{\Delta N_m}{\Delta t \ast N} \tag{4.4}$$

P.4.8. **When the environment has unlimited space and food and no other organisms exert a limiting effect, the specific growth rate, i.e. the growth rate of the population per individual, becomes constant and reaches a maximum permitted by the climatic conditions.**

Thus:

$$\frac{dN}{dt} = B_a - M_a = (B_s - M_s) \ast N = r \ast N \tag{4.5}$$

where
r = growth rate coefficient = $B_s - M_s$ (4.6)

or $\quad N_t = N_o \ast e^{r \ast t}$ (4.7)

or $\quad \ln \frac{N_t}{N_o} = r \ast t$ (4.8)

Fig. 4.4 shows a population growth curve based on equation (4.5). Here growth is **exponential, characterized by a fixed doubling time**. As seen in Fig. 4.5 this exponential growth curve is represented by a *straight* line in a *semologaritmic* plot.

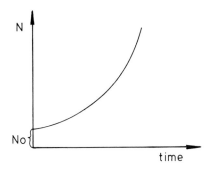

Fig. 4.4. Growth in accordance with $dN/dt = r * N$ $(r > 0)$ corresponding to exponential growth.

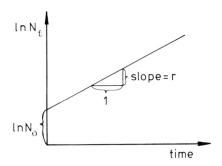

Fig. 4.5. $\ln N_t$ versus the time t. $\ln N_t = \ln N_0 + r * t$.

The growth coefficient, also denoted as *the intrinsic rate of natural increase, r,* should not be confused with *the net reproductive rate, R,* which is related to the *generation time, T* :

$$R = e^{r \cdot T} \qquad T = \frac{\ln R}{r} \qquad\qquad (4.9)$$

Some typical growth coefficients can be seen in Table 4.3.

TABLE 4.3

Typical growth coefficients (r) at optimal conditions

Escherichia coli	60 day $^{-1}$
Protozoa	0.8 - 1.3 day $^{-1}$
Phytoplankton	1 - 3 day $^{-1}$
Fish	0.02 - 0.04 day $^{-1}$
Octupus	0.006 - 0.02 day $^{-1}$
Ptinus tectus (insect)	0.05 day $^{-1}$
Daphnia pulex	0.1 - 0.4 day $^{-1}$
Flour beetle	0.71 week $^{-1}$
Human louse	0.78 week $^{-1}$
Brown rat	0.104 week $^{-1}$
Dog	0.06 week $^{-1}$
Man (world average)	0.02 year $^{-1}$

The growth rate represented by equation (4.5) assumes, as previously mentioned, *unlimited space and food* , but this is of course an unrealistic assumption. In reality the growth rate will slow down gradually as the environmental resistance increases, until a more or less equilibrium level is reached and maintained. These observations can be described by use of the following equation:

$$\frac{dN}{dt} = rN \frac{(K - N)}{K} \qquad (4.10)$$

As seen, the upper limit beyond which no increase can occur is represented by the constant K, denoted as

P.4.9. the carrying capacity. If N = K, the growth rate is zero in accordance with this equation.

A growth curve which follows this expression is called **a logistic growth curve** and has a characteristic s=shape, (see Fig. 4.5). On solving equation (4.10) we find:

$$N = \frac{K}{1 + e^{a-r^*t}} \qquad (4.11)$$

where

$$a = \ln \frac{K - N}{N} \quad \text{when } t = 0; \text{ it means } a = \ln \frac{K - N_o}{N_o} \tag{4.12}$$

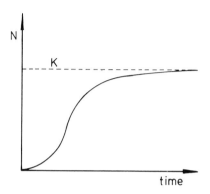

Fig. 4.6. Logistic growth.

This simple situation in which there is a linear increase in the environmental resistance with density seems to hold good only for organisms that have a very simple life history.

P.4.10. In populations of higher plants and animals, that have more complicated life histories, there is likely to be a delayed response.

Wangersky and Cunningham (1956 and 1957) have suggested a modification of the logistic equation to include two kinds of time lags: 1) the time needed for an organism to start increasing, when conditions are favourable, and 2) the time required for organisms to react to unfavourable crowding by altering birth- and death rates. If these time lags are $t - t_1$ and $t - t_2$ respectively we get:

$$\frac{dN(t)}{dt} = r N_{(t-t_1)} * \frac{K - N_{(t-t_2)}}{K} \tag{4.13}$$

Population density tends to fluctuate as a result of seasonal changes in environmental factors or due to factors within the populations themselves (so-called intrinsic factors). We shall not go into details here but just mention that

P.4.11. **the growth coefficient is often temperature dependent and since temperature shows seasonal fluctuations, it is possible to explain some of the seasonal population fluctuations in density in that way.**

Another modified logistic equation was introduced by Smith (1963):

$$\frac{dN}{dt} = r * N \frac{K - N}{K + \frac{r}{C} * N} \qquad (4.14)$$

where
C = the rate of replacement of biomass in the population at saturation density.

P.4.12. **The r value in equation (4.10) is strongly influenced by any environmental factor. Such physical factors as temperature, light, humidity, etc., or such chemical factors as the nutrient concentration and trace element concentration are examples of environmental factors affecting the growth of plants or animals. Toxic pollutants affect the growth of the individuals as well as the mortality rate,**

as demonstrated in Appendix 5. The data here are only a minor selection of the data available in the literature, see Jørgensen et al. (1979a).

The energy available to an organism capable of reproduction may be directed either toward the survival and growth of that organism, toward the production of offspring or apportioned between the two.

All the equations (4.1) - (4.14) consider growth in terms of the numbers in a population, but to understand fully the concepts of self-regulation, it is necessary to discuss in detail two possible strategies, called K-strategy and r-strategy.

P.4.13. **The K-strategists have a stable habitat with a very small ratio between generation time and the length of time, habitat remains favourable.**

They evolve toward maintaining their population at its equilibrium level, which is the carrying capacity in the logistic growth equation.

They will often be selected for *large size*, and *high levels of fecundity are not essential if the reduction in birthrate can be matched by increased survival* . K-strategists make a significant investment in defence mechanisms. Parental care is facilitated by low fecundity, longevity and size. This reduction in mortality may be considered to lead to a more efficient use of energy resources (Cody, 1966).

If K-strategists suffer perturbations, their populations need to return quickly to equilibrium levels or competitors may seize the resources. Because there is little mortality, this will tend to be accomplished through the birth rate, which will be very sensitive to the population density and will rise rapidly if density falls.

K-strategists are recognized by **large size, longevity, low recruitment and mortality rates, high competitive ability and a large in- vestment in each offspring.**

K-strategists will tend to have a stable equilibrium point E, to which the system returns after moderate disturbances. They are unlikely to be well adapted to recover from population densities significantly below their equilibrium level, and depressed to such low levels they may become extinct.

P.4.14. **They, rather than the r-strategists, therefore need the concern of the conservationist. They follow what is called Cope's rule: they will evolve toward increased size until extinction,**

a pattern often found in the fossil record.

The K-strategists become more and more adapted to a specialized and hitherto stable habitat (Bretsky and Lorenz, 1970). Thus the extinction of the extreme K-strategists dinosaurs was probably due to their inability to respond to the changes in climate at the end of the Cretaceous (Axelrod and Bailey, 1968; Southwood et al., 1974).

Other examples of extreme K-strategists are the Andean condor, the albatros and large tropical butterflies.

P.4.15. **The r-strategists are basically opportunistic "boom and burst" species. They are exposed to selection at all population densities and are continually colonizing habitats of a temporary nature. The r-strategist's population grows rapidly at low densities, has an equilibrium point about, which it is able to oscillate and crashes down from high densities.**

The ratio between generation time and the length of the time the habitat remains favourable is not small. *Migration* will be a major component of their population process and may even occur every generation (Dingle, 1974;

Southwood, 1962; Kennedy, 1975; Southwood et al., 1974). Selection will favour a high r arrived at by a **large fecundity** and **short generation time**. High competitive ability is not required and individuals will typically be **small in size**. Very high values of r lead to instability (May, 1974 and 1975) and their **high mortality, wide mobility** and **continuous exposure to new situations by migration** are likely to make them *fertile sources of speciation* . Perhaps the best way to see the difference between the two strategies is to look at the relationship between the population size at time t + 1 and the population size at time t (see Fig. 4.6).

Temporal heterogeneity in the habitat will increase instability, but animals with K-strategy will average out over the variations, while animals with r-strategy will closely follow the variations in the environment. This difference is illustrated in Fig. 4.7, where tawny owls (K-strategists) and voles and mice (r-strategists) have different scales and amplitudes of fluctuation.

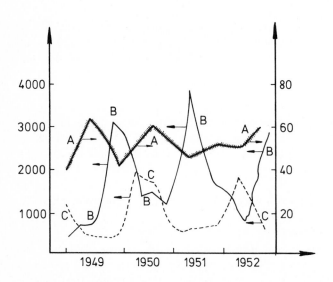

Fig. 4.7. Fluctuation in the number of animals in the same habitat: A) Tawny Owls (K-strategists), B) Bank Voles and C) Woodmice in Wythan Wood (both r-strategists). Data from Southern (1970). A) right hand scale, B) and C) left hand scale.

4.4.   INTERACTION BETWEEN TWO OR MORE SPECIES.

There are 8 possible observable interactions between two species (see Table 4.4).

P.4.16.   In terms of population growth these interactions involve adding positive, negative or zero terms to the basic equations presented in 4.3.

All these population interactions are likely to occur in the average community. For a given pair of species, the type of interaction may change under different environmental conditions.

Various terms have been suggested for a mathematical description of two-species interactions. The effects, beneficial or detrimental, of all the other species can be taken into account by adding $\pm C * N * N_1$ to the equation (4.10).

Some of the most widely discussed aspects of competition revolve around the so-called Lotka-Volterra equation, which consists of the following two expressions:

$$\frac{dN}{dt} = r * N \left(\frac{(K - N)}{K} - b * N_1 \right) \qquad (4.15)$$

$$\frac{dN_1}{dt} = r_1 * N_1 \left(\frac{K_1 - N_1}{K_1} - c * N \right) \qquad (4.16)$$

where
N = the first population
N_1 = the second population
r, r_1, K, K_1, b and c = constants

It is difficult to approach the subject of parasitism and predation (see Table 4.4) objectively, since we have a natural aversion to both types of interaction, although man is in fact the greatest predator the world has ever known. Predation and parasitism should be considered from point of view of the population and not from that of the individual.

P.4.17.   Predators and parasites play a major role as regulators, e.g. by keeping herbivorous insects at a low density.

TABLE 4.4
Two-species population interactions

0 : No significant interaction
+ : Positive term (effect) added to growth equation
- : Negative term (effect) added to growth equation
Population 1 is larger than population 2

Effect population 1	Effect population 2	0	+	-
0		Neutralism	Commensalism	Amensalism
+			Protocoopera-tion Mutalism	Predation
-		Amensalism	Parasitism	Competition

Deer populations are often cited as examples of populations that tend to erupt when predation pressure is reduced. The Kaibab deer herd, which was originally described by Leopold (1943) based on estimations by Rasmussen (1941), allegedly increased from 4000 (on less than 300,000 ha on the north side of the Grand Canyon in Arizona) in 1907 to 100,000 in 1924 coincident with an organized governmental predator removal campaign. Caughley (1970) has re-examined the case and concludes, that there is no doubt that the population did increase, overgraze and then decline drastically. There is, however, doubt about the extent of over-population and no real evidence has been presented to support the belief that it was due solely the removal of predators. Cattle and fire may have played a part, as well.

However, this example shows that human interaction can easily disturb the delicate balance of an ecosystem.

P.4.18. If an ecosystem is stressed by man it will, of course, find a new equilibrium after some time, but a disaster might occur before that new equilibrium is reached. Sudden changes in an ecosystem might cause such catastrophies.

An understanding of the influence, that all species in an ecosystem have on the overall balance is most important.

P.4.19. This is also the case when pesticides are used for pest control, but unfortunately many case studies illustrate that pesticides have been misused by ignoring ecological

principles. In such cases, far from controlling them, pesticides often cause pests to increase.

Table 4.5 shows results obtained from tests on the effects of DDT on red scale in a variety of citrus groves in southern California over a period of years. As can be seen the red scale population has risen considerably under the use of DDT (Debach, 1974).

TABLE 4.5
Adverse effects of DDT, southern California
Relative California red scale population densities in various biological control plots (untreated) and in plots in which natural enemy activity had been suppressed by prior application of a light DDT spray

| Property and location | No. of DDT applications | Final scale population density # | | Fold increase, DDT-sprayed/ untreated |
		Un-treated	DDT sprayed	
Bothin, Santa Barbara	11 +	3	425	142
Sullivan, Santa Barbara	11 +	16	575	36
Beemer, Pauma Valley	9 *	1	580	580
Irvine Company, Irvine	29 +	6	1336	223
Rancho Sespe, Fillmore	54 +	1	1250	1250
Ehrler, Riverside	10 *	4	390	98
Sinaloa Ranch, Simi	55 +	3	1015	299
Stow Ranch, Goleta	37 +	8	850	112
Hugh Walker Grove, Orange County	47 +	1	463	463

+ Monthly applications
* Quarterly applications
# Initial scale population densities were similar in all plots

There was a period following the widespread adoption of chlorinated hydrocarbons and organophosphorus pesticides when the first documented reports of striking upsets in crops generally met with disbelief or dis-approval from pesticide-oriented entomologists and industries. Today few, if any, knowledgeable professionals deny such effects to be rather common-place. They are in fact extremely commonplace, as can readily be ascertained from the entomological literature.

Table 4.6, taken from Huffaker et al. (1962), provides another example. Light dosages of DDT were used on olive trees to fight olive scale. The pro-cedure was repeated over a period of several years in many olive groves. Some typical results are summarized in Table 4.6, and as can be seen the relative population increase of olive scale in sprayed over unsprayed trees is truly amazing, ranging from 75 fold to nearly 1000 fold within a period of two years.

TABLE 4.6
Relative population density increase of the olive scale on unsprayed trees and on DDT-sprayed trees (Adapted from Huffaker, Kennett and Finney, 1962)

| | 1958 | | 1959 | | | | 1960 | | | | Relative fold |
| | Fall *) | | Spring | | Fall | | Spring | | Fall | | increase DDT- |
Location	DDT spr.	Un-spr.	DDT spr.	Un-spr.	DDT spr.	Un-spr.	DDT spr.	Un-spr.	DDT spr.	Un-spr.	spr. to un-spr.
Lindsay	0.0	0.0	0.3	0.3	5.5	1.2	25.5	1.9	67.6	0.9	75:1
Seville	2.0	1.6	4.2	9.0	2.7	0.0	12.6	0.0	29.8	0.03	993:1
Hills Valley	0.8	1.3	10.6	0.1	12.1	0.4	25.6	0.2	90.7	0.1	907:1
Clovis	1.2	3.0	12.4	0.2	43.9	0.5	55.6	1.9	287.8	2.5	140:1
Herndon	2.0	3.8	38.1	3.2	134.0	0.7	60.4	0.0	169.8	1.5	113:1
Madera	16.9	11.6	23.9	3.4	43.5	0.4	120.3	0.6	204.2	0.5	408:1

*) Pretreatment counts

Such results, of course, raise the question of how we can solve the pest problem. The answer is not an easy one, and will be dependent on the *ecosystem, the species, the climate and a number of other factors*. However, the general opinion among professionals today is that

P.4.20. **pesticides alone can never solve the problem and that they should be used only to a very limited extent and always selectively. Furthermore, only biodegradable pesticides should be used.**

New, far more selective, pesticides are under development - the so-called **third generation of pesticides**. These are based on hormones which affect the biochemical balance of the pest organism.

Another powerful method, that has developed rapidly during the last two decades, is the use of **biological control**. This can be defined as the regulation by natural enemies of another organism's occur. In other words the method uses the ecological possibility of interaction between two species. A comparison of biological and chemical control is given in Table 4.7, where the disadvantages of the two methods are listed.

Biological control has recently been successfully applied in control of the California red scale, although it had been regarded as a failure for nearly 60 years. Completely satisfactory control is established in some groves year after year and the pest is kept to reasonable levels in the entire district. Four species of parasites are responsible for the degree of biological control now achieved, all established between 1941 and 1957. Each of these can keep the red scale population at an extremely low level. These four parasites are the most recent enemies to become established and are by far the most effective of all the species of predators and parasites imported and liberated since 1890. This illustrates the value of continued research and indicates that it may be unwise ever to condemn a biological control project as a failure: The best weapon may as yet be undiscovered.

TABLE 4.7
A comparison of biological and chemical control

Category	Biololgical control	Chemical control
Environmental pollution; danger to man wildlife, other non-target organisms, soil etc.	None	Considerable
Upsets in natural balance and other ecological disruptions	None	Common
Permanency of control	Permanent	Temporary - must repeat one to many times annually
Development of resistance to the mortality factor	Extremely rare, if ever	Common
General applicability to broad-spectrum pest control	Theoretically unlimited but not expected to apply to all pests. Still underdeveloped. Initial control may take 1-2 years but then pest remains reduced	Applies empirically to nearly all insects but not satisfactory with some. Can rapidly reduce outbreaks but they rebound. Psychologically satisfying to the user at first

Other non-chemical approaches to pest management include three methods of proven practical use in the field as well as several still in the theoretical stage or in pilot tests. The three methods now in practical use are:

1. **the development of plant resistance to insects and plant diseases,**
2. **cultural techniques or habitat modification designed to control pests,**
3. **the genetic technique of sterile male release in the field** which mate with wild females and prevent progeny production.

Two useful general references covering this subject are Kilgore and Doutt (1967) and the National Academy of Sciences (1969).

In order to be succesful, a biological control agent must possess certain attributes:

1. The species must be able to find whichever species it is supposed to contol. This is termed *searching ability.*
2. It must have a high *enough reproductive rate* to build up in numbers fast enough to prevent build-up in numbers of the pest.
3. It must *not interfere unduly with other members* of its own species.
4. It must *perform well in the climate* in which it is released.
5. It must be *intensitive to competition* from other enemies of the pest that might be released or already present in nature.
6. It must be certain that it will *confine its diet to the pest species* and avoid interference with other species of economic value.
7. It should have *no special requirements for resources* that the environment cannot supply in sufficient quantity to maintain the agent's density.

Several equations have been proposed to account for the numbers of the pest that are attacked over a given time interval.

A simple preliminary model, which is very useful but which has now been superseded by slightly better models, is:

$$N_A = P * K(1 - e^{-a*N_0*p} (1 - b)) \qquad (4.17)$$

where N_A is the number of pests attacked by parasites, N_0 the numbers vulnerable to attack and P the parasite population density. a, b and K are constants, namely the searching rate, the parasite intraspecific competition pressure and the maximum egg-laying rate, respectively.

4.5.   ORGANIZATION OF ECOSYSTEMS.

Out of thousands of species that might be present in an ecosystem relatively few exert on major controlling influence by virtue of their numbers, size, production or other activities. Intracommunity classification therefore goes beyond the taxonomic listing and attempts to evaluate the actual importance of organisms in the community. Three types of indices are used to describe the organization in an ecosystem:

1. **indices of dominance,**
2. **indices of similarity** and
3. **indices of diversity.**

Some of the most commonly used indices in these three groups are listed in Table 4.8.

P.4.21. **Species diversity tends to be low in physically controlled ecosystems and high in biologically controlled ecosystems, but this rule is not always valid, and takes only the tendency into account.**
There is a relationship between the stability of the ecosystem and its diversity, but the detailed relationship has still to be discovered.

The general relationship between species and numbers of individuals takes the form of a curve, see Fig. 4.8. Stress will tend to flatten the curve as illustrated in Fig. 4.8 by the dotted line.

Two broad approaches are used to analyze species diversity in different situations, namely:
1. a comparison based on **diversity indices** (see Table 4.8),
2. a comparison based on the **shapes, patterns and equations of species abundance curves.**

P.4.22. **Two stability concepts are generally used. Stability is either understood as maintenance of constant biological structure or as resistance to changes in external factors.**

However, to avoid misunderstandings this last stability concept will be defined as an **ecological buffer capacity**, see section 4.8, while "stability" will be reserved for the former definition. The relationship between this stability concept and diversity has been widely discussed in the literature, see e.g. Jacobs (1974). One of the crucial questions is whether *stable physicochemical conditions will cause high diversity or whether high diversity will produce a stable environment.* For those who want further information about this complex relationship, see Margalef (1969); Pielou (1966); Slobod- kin and Sanders (1969); Paine (1966); Sanders (1968); Boesch (1974); Goodmann (1974); May (1974); Orians (1974); Preston (1969).

It is essential to recognize that species diversity has a number of components which may respond differently to external and internal factors of importance to environmental conditions.

TABLE 4.8
Some useful indices of species structure in communities

A. Index of dominance (c) [+]

$$c = \Sigma(ni/N)^2 \quad \text{where} \quad ni = \text{importance value for each species (number of individual, biomass, production, and so forth)}$$
$$N = \text{total of importance values}$$

B. Index of similarity (S) between two samples [o]

$$S = \frac{2C}{A+B}$$

where A = number of species in sample A
B = number of species in sample B
C = number of species common to both samples

Note: Index of dissimilarity = 1 - S.

C. Indices of species diversity
(1) Three species richness or variety indices (d)[++]

$$d_1 = \frac{S-1}{\log N} \qquad d_2 = \frac{S}{\sqrt{N}} \qquad d_3 = S \text{ per 1000 ind.}$$

where S = number of species
N = number of individuals, etc.

(2) Evenness index (e)[§]

$$e = \frac{H}{\log S}$$

where H = Shannon index (see below)
S = number of species

(3) Shannon index of general diversity (H)[*]

$$H = -\Sigma \left(\frac{ni}{N}\right) \log \left(\frac{ni}{N}\right)$$

or

$$-\Sigma P_i \log P_i \quad \text{where} \quad ni = \text{importance value for each species}$$
$$N = \text{total of importance values}$$
$$P_i = \text{importance probability for each species} = ni/N$$

[+] See Simpson (1949)
[o] See Sørensen (1948), for a related index of "% difference" see E.P. Odum (1950)
[++] d_1 see Margalef (1958). d_2 see Menhinick (1964). d_3 see H.T. Odum, Cantlon and Kornicker (1960).
[§] See Pielou (1966); for another type of "equitability" index, see Lloyd and Ghelardi (1964).
[*] See Shannon and Weaver (1963); Margalef (1968).

Note: In d_1, e and H natural logarithms ($\log_e$) are usually employed, but $\log_2$ is often used to calculate H so as to obtain "bits per individual".

Areas with very predictable and stable climates *tend to support* fewer different plant life forms than regions with more erratic climates, as is demonstrated in Fig. 4.9.

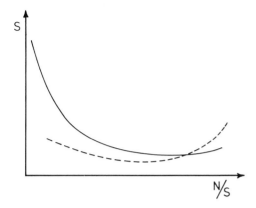

Fig. 4.8. Relationship between the number of species (S) and the number of individuals (N) per species. Environmental stress tends to flatten the curve as indicated with the dotted line.

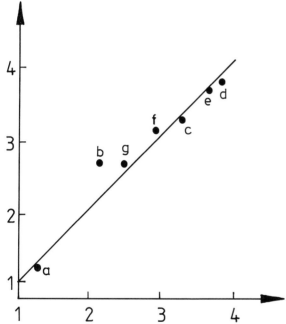

Fig. 4.9. Rainfall unpredictability or climate diversity is plotted versus plant life form diversity (from May (1975) and other sources). Both diversities are computed as Shannon indices. a: Tropical rain forest, b: Subtropical forest, c: Deciduous forest, d: Mediterranean scrub, e: Desert, f: Steppe, g: Arctic tundra.

The most widely used index of diversity is **the Shannon index**. It is *normally distributed* (Bowman et al., 1970; Hutcheson, 1970), so the *routine statistical methods* can be used to test for magnificance of differences between means. It is also reasonably independent of sample size as illustrated in Fig. 4.10.

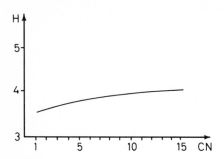

Fig. 4.10. The effect of increasing sample size on the Shannon index. H is plotted against cumulative number of samples (CN).

Higher diversity means longer foodchains, more cases of symbioses and greater possibilities for negative feedback control, which will reduce oscillations and hence increase stability. This expresses the general interest for the ecologist in the concept diversity, as he sees in any measure of diversity an expression of the possibilities for constructing feedback systems (Margalef, 1960).

P.4.23. Diversity indices can be used to evaluate man-made stresses on ecosystems,

as clearly demonstrated in Figs. 4.11 - 4.14.

The use of grafic methods in this context is illustrated in Fig. 4.15.
Before diversity indices are used to compare one situation with another, the effect of sample size should be determined (compare with Fig. 4.8).
The relationship between diversity and buffering capacity, and the development of diversity in an ecosystem will be discussed in the following sections.
Ecosystems are usually very difficult to manipulate experimentally. For this reason, ecologists have long been especially interested in islands, which constitute some of the finest natural ecological experiments. Through such studies it has been found that larger islands generally support more species of plants and animals than smaller ones. In many cases, a tenfold increase in

area corresponds to an approximate doubling of the number of species.

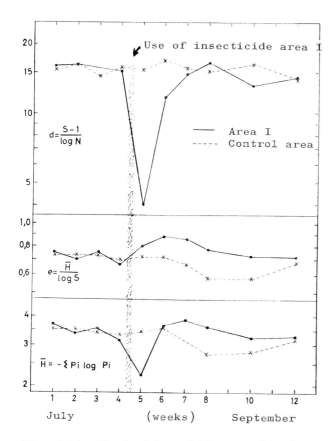

Fig. 4.11. Effect of insecticides on diversity.

P.4.24. **The number of species, S is related to the area of the islands concerned as follows**

$$S = C * A^Z \qquad (4.18)$$

where C is a constant that varies between taxa and from place to place, and
z generally ranges from about 0.24 to about 0.34.
Estimated z-values are shown in Table 4.9.

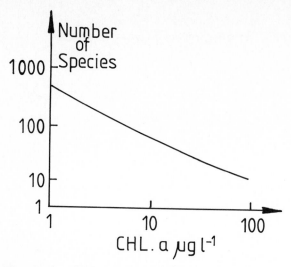

Fig. 4.12. Weiderholm (1980) obtained the shown relationship in a number of Swedish lakes between number og species and eutrophication, measured as chlorophyll a (μg l^{-1})

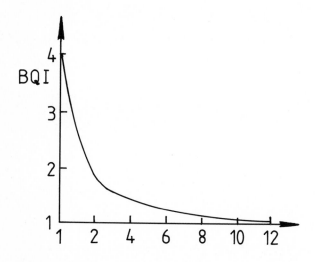

Fig. 4.13. Ahl & Weiderholm (1977) found the shown relationship between BQI and total phosphorus concentration divided by depth. BQI is a diversity index for the benthic fauna (benthic quality index). It is defined as

$$BQI = \sum_{i=0}^{i=5} \frac{k_i n_i}{N}$$

where k_i represents a value for each species, n_i the number of individuals in the various groups and N the total number of indicator species.

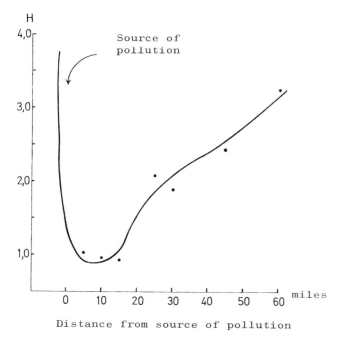

Fig. 4.14. Shannon index (H) plotted versus distance (0 = discharge of waste water).

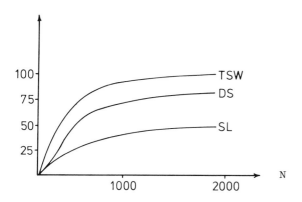

Fig. 4.15. Number of species plotted against number of individuals. Typical plots are shown for TSW (Tropical Shallow Water), DS (Deep Sea) and SL (Shallow Lake).

TABLE 4.9
Estimated z-values for various terrestrial plants and animals on different island groups. (MacArthur and Wilson, 1967)

Flora or fauna	Island group	z
Carabid beetles	West Indies	0.34
Amphibians and reptiles	West Indies	0.30
Birds	West Indies	0.24
Birds	East Indies	0.28
Birds	Islands of Gulf of Guinea	0.49
Birds	East Central Pacific	0.30
Land plants	Galapagos Islands	0.33
Land Vertebrates	Islands of Lake Michigan	0.24
Ants	Melanesia	0.30

Variety of species and their relative abundance are by no means the only factors involved in community diversity. Arrangement patterns also contribute to community function and stability. Many different kinds of arrangements in the standing crop of organisms contribute to pattern diversity in the community, following for example: 1. *horizontal zonation pattern*, 2. *stratification pattern*, 3. *reproductive pattern* (parent/offspring), 4. *social pattern* (flocks, herds etc), 5. *periodicity pattern*, 6. *stochastic pattern* (results of random forces).

There seems no doubt about the existence of some relationship between diversity in space and species and the ecosystem stability in its broadest sense, but a quantitative ecosystem theory which includes these relationships is still not available. However, an examination of the diversity is very useful for providing information about ecological conditions, although such information is qualitative and cannot be used to give quantitative assessments of environmental impact.

To understand diversity fully we must consider the related concepts of **the ecotone** and **the ecological niche**.

An ecotone is a transition zone between two or more diverse communities, for example between forest and grassland, between a fjord and the open sea, etc. The ecotonal community commonly contains many of the organisms of both overlapping communities, as well as organisms, characteristic of and often restricted to the ecotone. Often, the number of species and the population density of some of the species are greater in the ecotone than in the communities flanking it. The tendency for increased variety and density at community junctions is known as the "*edge effect* ".

The ecological niche is primarily the physical space occupied by an organism (the so-called spatial niche), but the term also encompasses its functional role in the community and its position in environmental gradients of temperature, humidity, pH, soil composition, etc. In this context such expressions as *trophic niche* or *multidimensional niche* are used.

A complete description of the ecological niche for species would require an infinite set of biological characteristics and physiocochemical parameters. The concept is most often used in terms of differences between species. Organisms that occupy the same or a similar ecological niche in different geographical regions are called ecological equivalents. Species that are ecologically equivalent tend to be closely related taxonomically in regions, which are contiguous, but are not necessary closely related in regions widely separated or isolated from each other. There are several exceptions for this rule, however.

The role of diversity is very well illustrated by presentation of the advantages of mixed forests: these can be listed as follows:

1. A greater density of crop is attainable in mixed forests. A complete utilization of the soil can only be attained when each portion is stocked with the species best suited for growth there. Furthermore, they maintain a closed canopy for the longest period of time, thereby fully utilizing the light resource as well as the soil. The different growth rhythms can be utilized with suitable thinning methods in order to enable the maximum volume and dry-matter production to be achieved.

2. Different light requirements of the species participating in a mixture may produce an increased assimilation efficiency of mixed stands compared with pure stands. The intermediate and lower stories of shade- and semishade-tolerant species are capable of utilizing the light which has been transmitted through the crowns of light-demanding species in the upper canopy, thus producing an additive increment.

3. If the mixed species occupy different root horizons, the sites can be utilized more fully. The species with strong roots are able to open soil layers in which rooting is otherwise difficult, enabling the whole to benefit from this and from an improved nutrient supply by way of the litter from such mixed species.

4. The character of the litter from mixed species, especially litter which is rich in nitrogen and easily decomposed, improves productive capacity by stimulating the soil fauna. A recent result of planting nitrogen-fixing species as an admixture to more classical tree crops appears most encouraging.

5. Plants have favourable as well as unfavourable effects on each other. It has been demonstrated that some chemical substances might have a strong effect on germination.

6. Natural reproduction of mixed woods is easier than for pure stands. Mixed stands have characteristics which are more favourable for the evolution of abundant fruit production than those usually encountered in pure stands.

7. Mixed forests are less exposed than pure forests to external disruptions. Shallow-rooting species, for instance, are much less exposed to damage from storms when mixed with deeper-rooting kinds of trees than when grown alone.

4.6   DEVELOPMENT AND EVOLUTION OF THE ECOSYSTEM.

Ecosystem development or succession, as it is often called, is a process, which can be defined by the following three steps:
1. The changes in the **physicochemical environment** initiate the process. These changes can be either natural or a result of human interference.
2. The rate of change in the community is **community controlled** and it is an orderly process involving changes in species structure and community processes with time.
3. The process is a development toward **a stabilized ecosystem** (here defined as an ecosystem which has the highest possible ability to maintain its ecological structure independent of external changes).

The whole sequence of communities that replace each other in a given area is called the sere; the relatively transitory communities are called seral stages, while the final stabilized system is known as *the climax* . Species replacement in the sere occurs because populations tend to modify the physical environment making conditions favourable for other populations until an equilibrium between biotic and abiotic factors is achieved.

The ecological succession process includes a series of changes, of which changes in adaptation, diversity and other mechanisms play a part. Ecological investigations, and more recently functional considerations, have led to the results presented in Table 4.10, which is based on works by Odum and Pinkerton (1955), Lotka (1925), Margalef (1963 and 1968) and Jørgensen et al. (1977 and 1979). The concept of ecological buffering capacity is included in the table, but will be discussed in the next paragraphs. All the other concepts presented in this table have been defined already. Trends are emphasized by contrasting the situation in early and late developmental stages of the ecosystem. Early and late are understood here to represent the time after a substantial change in the ecological condition has occured. The time required for the transition form one stage to another may vary with climatic, physiographic and ecosystem factors. However, the changes will usually occur more rapidly when the ecosystem is far from stabilization.

P.4.24. One of the characteristic trends to be expected in the development of an ecosystem is that the P/R ratio (P = gross production, R = respiration) approaches 1 and that P/B (B = biomass) decreases.

This is illustrated in Fig. 4.16 based on data of Cooke (1967) and Fig. 4.17 based on results presented by Kira and Shidei (1967).

TABLE 4.10
Ecological succession

Properties	Early stages	Late or mature stage
A Energetics		
P/R	>>1 <<1	Close to 1
P/B	High	Low
Yield	High	Low
Entropy	High	Low
Exergy	Low	High
Information	Low	High
B Structure		
Total biomass	Small	Large
Inorganic nutrients	Extrabiotic	Intrabiotic
Diversity, ecological	Low	High
Diversity, biological	Low	High
Patterns	Poorly organized	Well organized
Niche specialization	Broad	Narrow
Size of organisms	Small	Large
Life cycles	Simple	Complex
Mineral cycles	Open	Closed
Nutrient exchange rate	Rapid	Slow
C Selection and homoeostatis		
Internal symbiosis	Undeveloped	Developed
Stability (resistance to external perturbations)	Poor	Good
Ecological buffer capacity	Low	High
Feedback control	Poor	Good
Growth form	Rapid growth	Feedback controlled growth

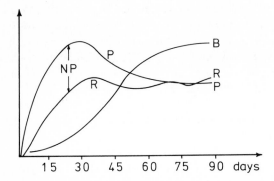

Fig. 4.16. Microcosm succession. P = gross production, NP = net production, R = respiration and B = biomass plotted against time.

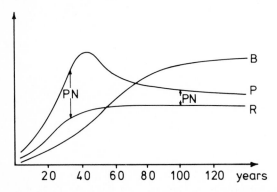

Fig. 4.17. Forest succession. P = gross production, PN = net production, R = respiration and B = biomass are plotted against time.

Communities can be *classified in accordance with the P/R ratio,* see Fig. 4.18. The diagram consists of three areas. Ecosystems with a high P/R ratio are represented in upper left of the diagram, while those with a low P/R ratio are situated in the lower right of the diagram. The direction of succession is indicated by the arrows pointing toward what is called the balanced ecosystem.

Balanced ecosystems represent the mature stage, and although there is a pronounced difference in production and respiration in the ecosystems in this part of the diagram all have a P/R ratio close to 1. The production in a coral reef is almost 1000 times greater than that in a desert. The diagram is based on the classification of Odum (1956).

Figs. 4.19 - 4.24 show some characteristic observations of succession patterns in ecosystems. As seen, the figures demonstrate the expected trends, compare with Table 4.10.

The changes referred to in these diagrams and in Table 4.11 are all brought about by biological processes within the ecosystem. Geological and human forces acting on the system can reverse the trends, in which case the ecosystem has to start again from the very beginning.

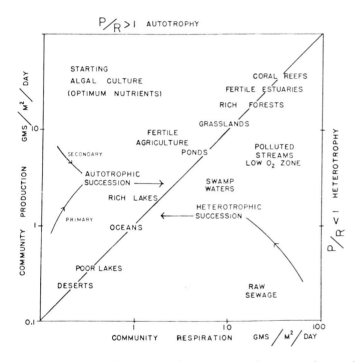

Fig. 4.18. Position of various community types in a classification based on community metabolism. Gross production (P) exceeds community respiration (R) on the left side of the diagonal line (P/R) greater than 1 (= autotrophy), while the reverse situation holds on the right (P/R) less than 1 (= heterotrophy). The latter communities import organic matter or live on previous storage or accumulation. The direction of autotrophic and heterotrophic succession is shown by the arrows. Over a year's average, communities along the diagonal line tend to consume about what they make and can be considered to be metabolic climaxes. (Redrawn from H.T. Odum, 1956).

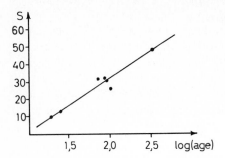

Fig. 4.19. Number of plant species, S, against log(age) in years in a pond. (Godwin, 1923).

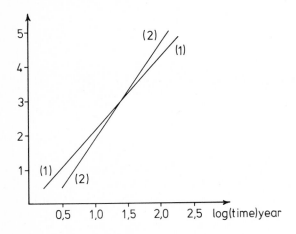

Fig. 4.20. Succession of bird community. (1) Number of pairs per ha or (2) number of species per 10 ha follow the same pattern, as shown on the figure. (Johnston and Odum, 1956).

Eutrophication of lakes is a good example, where th high nutrient concentration pushes the system back to a simpler system with high P/R ratio and little diversity. Such a system is better able to cope with the high nutrient concentration by high production, because more dissolved nutrient is removed from the water, (see also next paragraph, where the concept ecological buffering capacity is discussed - the high production in a eutropic lake acutally gives a better buffering capacity for changes in nutrient concentrations).

P.4.25. The structure of the ecosystem is changed in accordance

with changing environmental factors. (In ecological modelling often named forcing functions).

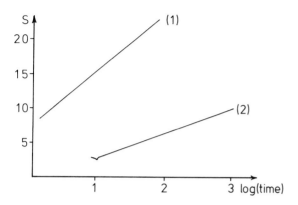

Fig. 4.21. Number of species (S) plotted against log(time) in years for (1) a plant community in Illinois, (2) in Idaho. In the midler climate (1) species are added more rapidly. (Bazzas, 1968 and Chadwick and Dalke, 1965, respectively).

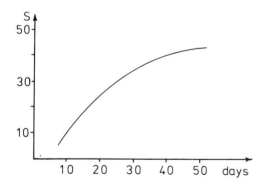

Fig 4.22. Number of protozoan species on spongy blocks suspended in a lake plotted against time (days).

In the eutrophic lake the nutrient concentration is high and consequently the ecosystem will change in structure to become one, which in the long term, can reduce the nutrient concentration. Recent studies on land sediment as well as theoretical considerations have indicated that lakes can and do progress to more oligotrophic conditions when the nutrient input ceases, see

Mackers (1065); Cowgill and Hutchinson (1964); Harrison (1962); Jørgensen (1976); Jørgensen and Mejer (1977), see also section 4.8.

Coevolution is a type of community evolution which occurs in ecosystem development.

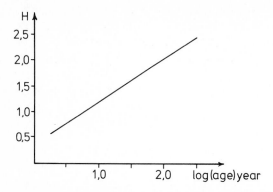

Fig. 4.23. H of bird community plotted against time. (Johnston and Odum, 1956).

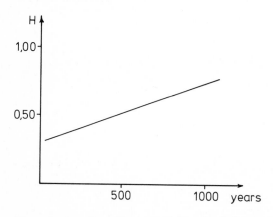

Fig. 4.24. Density based H of cladoceran community in Lago di Monterosi, Italy, against age in years. (Goulden, 1969).

P.4.26. Coevolution involves reciprocal selective interaction between two major groups or organisms with a close ecological relationship, such as plants and herbivores, or parasites and their hosts.

Ehrlich and Raven (1965) have outlined the theory of coevolution, which can be described by an example: Suppose a plant produces, through mutation or recombination, a chemical compound not directly related to those involved in its basic metabolism, by chance one or these compounds serves to reduce the palattability of the plants to herbivores. Such a plant is now better protected and will in a sense enter a new adaptive zone. Selection will now carry the line of the herbivores into a new adaptive zone, allowing them to diversify in the absence of competition with other herbivores. Thus, the diversity of plants may not only tend to augment the diversity of herbivores, but the converse may also be true. In other words the plant and the herbivores evolve together in the sense that the evolution of one is dependent on the evolution of the other. This is named coevolution.

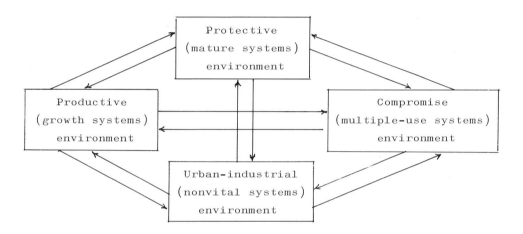

Fig. 4.25. Compartment model of the basic kinds of environment required by man, partitioned according to ecosystem development and life-cycle resource criteria. (After E.P. Odum, 1969)

There is apparently a *conflict between the strategies of man and nature* . The goal of agriculture and forestry is highest possible production, while nature attempts to achieve maximum support of a complex ecosystem structure, which implies thet P/R must be close to one and that net production is around zero. However, the problem is not that simple, since man is dependent not only on food, fibers and wood, but also on the oxygen concentration in the atmosphere, the climate buffer, and clean air and clean water for many uses. Until recently man has taken it for granted that nutrient cycling, water purification and other protective functions of self-maintaining ecosystems will work independently of environmental manipulaitons such as discharge of pollutants, or changes to ecological structures (forest to agriculture land). The most pleasant landscape to live in from an aesthetic point of view and the safest ecologically is one containing a variety of crops, forest, lakes etc. i.e. a **large pattern diversity.**

In other words it seems essential for man to plan the landscape with a large pattern of diversity, and which allows for exchange between different types of ecosystem. Odum (1969) has conceptualized these considerations in the construction of a compartment model of the basic kinds of environment required by man, see Fig. 4.25. Thus *the preservation of natural areas is not a luxury for society* but a capital investment of crucial importance for our civilization and our life on earth.

4.7.   pH-BUFFERING CAPACITY IN ECOSYSTEMS.

The concept of buffering capacity is generally used in chemistry to express the ability of a solution to maintain its pH value.
Buffering capacity in this context is defined as:

$$\beta = \frac{dC}{dpH} \tag{4.19}$$

β = buffering capacity
C = added acid or base in moles H^+ resp. OH^- per l

As pH is one of the important factors determining life conditions directly or indirectly, it is *crucial for ecosystems to have a high pH-buffering capacity,* see Figs. 4.26 and 4.27, and Tables 4.11 - 4.12.

TABLE 4.11
Fish status for 1679 lakes in Southern Norway grouped according to pH

pH	No. of lakes in pH range	% of lakes with no fish	% of lakes with sparce populations	% of lakes with good populations
<4.5	111	73	25	2
4.5-4.7	245	53	41	6
4.7-5.0	375	38	41	21
5.0-5.5	353	25	40	35
5.5-6.0	164	8	36	56
>6.0	431	1	13	86

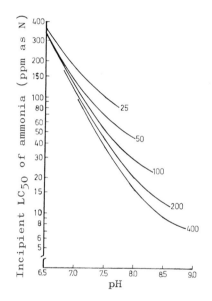

Fig. 4.26. LC_{50} of ammonia for trouts versus pH at different HCO_3^- alkalinities.

P.4.27. A low pH implies that hydrogen carbonate is coverted into free CO_2, which is toxic to fish and other animals as respiration is controlled by the difference between CO_2-concentration in the blood and the environment.

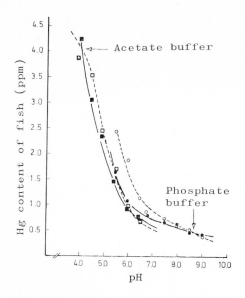

Fig. 4.27. Effect of pH on the mercury content of fish.
Emerald shinner ¤; Fathead minnows ¤; Emerald shinner o;
Fathead minnows o. Exposure to 1.5 ppm Hg from $HgCl_2$ solutions.

TABLE 4.12
Effects of pH values on fish (Alabaster and Lloyd, 1980)

pH-range	Effect
3.0 - 3.5	Unlikely that any fish can survive more than a few hours
3.5 - 4.0	Lethal to salmonoids. Some other fish species might survive in this range, presumably after a period of acclimation to slightly higher pH
4.0 - 4.5	Harmful to salmonoids, bream, goldfish and carp, although the resistance to this pH increases with the size and age
4.5 - 5.0	Likely to be harmful to eggs and fry of salmonoids. Harmful also to adult salmonoids and carp at low calcium, sodium and/or chloride concentration
5.0 - 6.0	Unlikely to be harmful, unless concentration of free CO_2 is greater than 20 mg l^{-1} or the water contains freshly precipitated $Fe(OH)_3$
6.0 - 6.5	Harmless unless concentration of free $CO_2 > 100$ mg l^{-1}

The processes involved can be described by the following chemical equations:

$$CO_3^{2-} \xrightarrow{+H^+} HCO_3^- \xrightarrow{+H^+} H_2O + CO_2 \qquad (4.20)$$

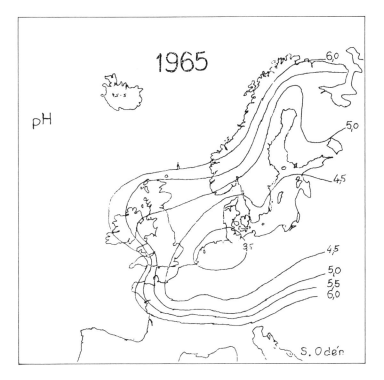

Fig. 4.28. pH is lowest close to industrial areas.

A lowered pH will affect the entire ecosystem and the ecological balance. Three serious environmental problems will be mentioned below to illustrate the importance of pH as an environmental factor, and consequently of a high pH-buffering capacity in ecosystems:

1. The acidification of lakes caused by SO_2-pollution.
2. The effect of decreased pH on soil and plants.
3. The possibility of maintaining the pH in the oceans in spite of the increased uptake of CO_2.

1. Continuous acidification has been observed in many lakes in parts of North America and Scandinavia. The geographical position (see Fig. 4.28) and the inflow conditions of these lakes suggest that the increase in acidity is the result of deposition of an air-borne substance that lead to the formation of acid, such as SO_2. This deposition is, of course, not restricted to lakes, but also affects soil, forests, etc. (see point **2** below).

Affected areas are characterized by a low pH-buffering capacity. Surface water in these areas is soft and has a low conductivity. The buffering capacity of surface water is mainly related to hydrogen carbonate ions, which are present in small concentrations in lakes. These ions are able to take up hydrogen ions in accordance with the following process:

$$HCO_3^- + H^+ \; — \; H_2O + CO_2 \uparrow \qquad\qquad (4.21)$$

During the combustion of sulphurous fuels, sulphur is primarily converted to SO_2 (97-98%), but 2-3% is oxidized to SO_3, which reacts with water to form sulphuric acid:

$$SO_3 + H_2O \; — \; H_2SO_4 \qquad\qquad (4.22)$$

The SO_2 on the other hand when released into the atmosphere comes into contact with very small particles covered by an aqueous film or with water droplets, and forms H_2SO_3, a medium strong acid:

$$H_2SO_3 \; — \; H^+ + HSO_3^- \qquad\qquad (4.23)$$

Under the influence of iron compounds acting as catalysts, HSO_3^- and SO_3^{2-} are rapidly oxidized to sulphuric acid (Brosset, 1973).

The geographical distribution of acids in precipitation over Europe as a consequence of these processes is shown in Fig. 4.28. Values are expressed as annual mean pH. Together with the increased combustion of fossil fuels, the observed pH values in precipitation have decreased with the time, see Fig. 4.29. Fig 4.30 shows the total SO_2-emission in Europe plotted against the time.

Similar trends have been observed in many North American and Scandinavian lakes and even some of the great lakes of Europe have been affected by this acidification process, as shown in Table 4.13. In smaller lakes, where precipitation forms a greater portion of lake volume, pH values as low as 4.0 have been observed. The effect of low pH on the fish population can be seen in Table 4.14, which is based on an examination carried out by Jensen and Snekvik (1972).

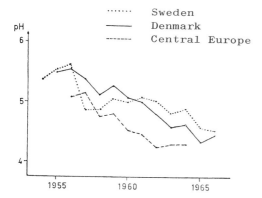

Fig. 4.29. pH in rainwater against year.

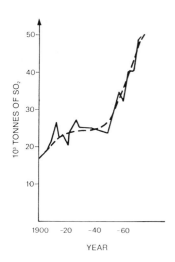

Fig. 4.30. Emissions of sulphur dioxide from the combustion of fossil fuels in Europe.

Two other observed effects are illustrated in Table 4.15 and in Fig. 4.31, where it is shown that

P.4.28. the number of fish and zooplankton species is decreasing in acidified lakes.

TABLE 4.13
Acidification of soft-water lakes in Scandinavia and North America

Region	No. of lakes	pH Early measurements	pH Recent measurements	Average change ΔpH/year
Scandinavia				
Central Norway	10	7.3 ± 0.8 (1941)	5.8 ± 0.7 (1975)	-0.05
Westcoast of Sweden	6	6.6 ± 0.2 (1933-35)	5.4 ± 0.8 (1971)	-0.03
	8	6.8 ± 0.2 (1942-49)	5.6 ± 0.9 (1971)	-0.04
West central Sweden	5	6.3 ± 0.3 (1937-48)	4.7 ± 0.2 (1973)	-0.06
South central Sweden	5	6.2 ± 0.2 (1933-48)	5.5 ± 0.7 (1973)	-0.03
Southernmost Sweden	51	6.76 ± 0.14 (1935)	6.23 ± 0.44 (1971)	-0.015
North America				
La Cloche Mtns.	7	6.3 ± 0.7 (1961)	4.9 ± 0.5 (1972-73)	-0.06
Ontario	8	5.0 ± 0.7 (1969)	4.8 ± 0.5 (1972-73)	-0.05
North of La Cloche	7	6.6 ± 0.8 (1961)	5.9 ± 0.7 (1971)	-0.07
Mtns. Ontario	19	6.7 ± 0.8 (1968)	6.4 ± 0.8 (1971)	-0.10
Adirondack Mtns.				
New York	8	6.5 ± 0.6 (1930-38)	4.8 ± 0.2 (1969-75)	-0.05

TABLE 4.14
Trout population and pH in 260 lakes

No. of lakes	Population	pH 4.00-4.50 No.	pH 4.00-4.50 %	pH 4.51-5.00 No.	pH 4.51-5.00 %	pH 5.01-5.50 No.	pH 5.01-5.50 %	pH ≥5.51 No.	pH ≥5.51 %
33	Empty	3	9.1	17	51.5	7	21.2	6	18.2
87	Sparse population	2	2.3	15	17.2	21.2	24.1	49	56.3
82	Good population	-	-	9	11.0	14	17.1	59	72.0
58	Over-populated	-	-	3	5.2	13	22.4	42	72.3
260	total	5	1.9	44	16.9	55	21.2	156	60.0

The decreasing pH has a striking effect on the fish population. At extremely low pH values all yong fish disappear completely (Almer, 1972). Nevertheless, spawning and fertilized eggs have been observed even at low pH values, so **it seems that the development of eggs may be disturbed by high activity of hydrogen ions in aquatic environments.**

This is illustrated in Fig. 4.32, where percentage of hatched eggs is plotted against pH value. The period from fertilization to hatching also tends to be prolonged at low pH values, as shown in Fig 4.33.

TABLE 4.15
Occurrence of fish species before acidification and species found during the 1973

Lake	Species earlier forming permanent stocks	Species found 1973	Species reproducing 1973
Bredvatten	Pe Pi E	(E)	
Lysevatten	Pe Pi R E	Pe (E)	Pe
Gårdsjön	Pe Pi R T C E	Pe Pi E	Pe
Örvattnet	Pe St M	Pe	Pe
Stensjön	Pe Pi R St M	Pe Pi R	Pe Pi
Skitjärn	Pe Pi R L	Pe Pi R L	Pe Pi

Pe=perch (Perca fluviatilis, Pi=pike (Esox lucius), E=eel (Anguilla vulgaris), R=roach (Leuciscus rutilus), T=tench (Tinca tinca), C=Crucian carp (Carassius carassius), St=Brown trout (Salmo trutta), L=Lake whitefish (Coregonus albula), M=Minnow (Phoxinus phoxinus).

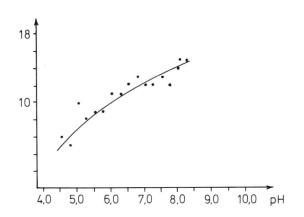

Fig. 4.31. Number of zooplankton species in Swedish Lakes according to pH. From a comprehensive examination of 84 lakes in Sweden.

Further to the primary biological effects of a continuous supply of acid substances on individuals and populations in a lake, more profound long-term changes, that force the lake into an increasingly more oligotrophic state, also take place, as suggested by Grahn et al. (1974). pH generally increases with eutrophication due to uptake (removal) of hydrogen carbonate ions and carbon dioxide by photosynthesis. This is demonstrated in Fig. 2.11, which demonstrates the seasonal pH-variations in an hypereutrophic lake.

P.4.29. **In an acidified lake the phytoplankton concentration will decrease, and so will the uptake of CO_2 and HCO_3^-, and the transparency of the water will increase.** (see Fig. 4.34)

P.4.30. **The biological pH-buffering capacity is reduced by this process and by means of this feedback mechanism the process of acidification is further accelerated.**

Finally, it should be mentioned that

P.4.31. **the concentration of free metal ions will increase with decreasing pH due to release of methal ions from sediment and their higher solubility and lower tendency to form complexes at lower pH.**

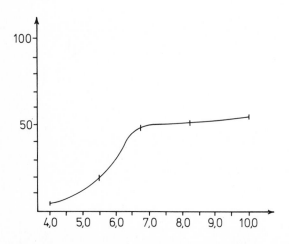

Fig. 4.32. % eggs hatched according to pH. Total number of eggs reared 253-274 by Brachydanio rerio Ham.-Buch.

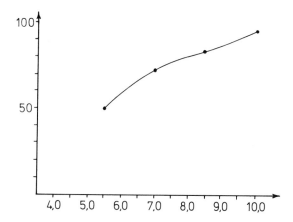

Fig. 4.33. Eggs hatched during the first 96 h after fertilization as % of total number of hatched eggs according to pH. Conditions, see Fig. 4.28.

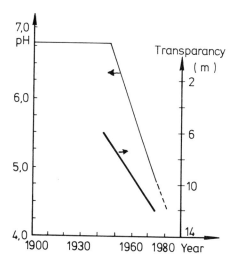

Fig. 4.34. pH and transparency of Lake Stora Skarsjön plotted against time.

In Sweden the addition of calcium hydroxide is widely used to rivers and lakes to reduce the damage of low pH. Hundred of millions of Swedish Krona are spent every year on increasing the pH of natural waters. This amount should be compared with the billions of dollars it would cost to reduce the sulphur content of fossil fuel to an acceptable level in Nortwest Europe.

TABLE 4.16.
[HA] and [A⁻] at various pH-values

pH	log [HA]	log [A⁻]
<< pK_a	log C	pH - pK_a + log C
>> pK_a	-pH + pK_a + log C	log C
= pK_a	log C/2 = log C-0.3	log C/2 = log C-0.3

2. Decreased pH values in the lithosphere also cause an overall deterioration in the environment.

P.4.32. The leaching of nutrients from soil is increased at lower pH.
as demonstrated for calcium ions in Fig. 4.35.

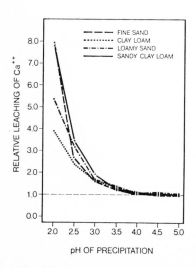

Fig. 4.35. Leaching of calcium in forest soil exposed to precipitation adjusted to pH values from 2.0 to 5.0 during a period of 40 days. The Ca leaching in the control (distilled water) is set at 1.0. Precipitation: 500 mm/month.

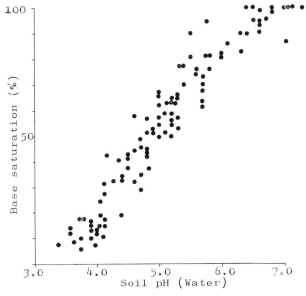

Fig. 4.36. The relationship between pH and base saturation illustrated in samples from the organic surface layer of Swedish soils. Base saturation determined through extraction with 1-N acetic acid and 1-N ammonium acetate (Brown's method), pH in water extracts (1:1). In both cases dried samples have been used.

P.4.33. The ability of the soil to bind ions is decreased at lower pH.

as illustrated in Fig. 4.36. Here the so-called CEC (Cation Exchange Capacity) is plotted against the pH value.

The leaching of ions from soil produces a change in the chemical composition of surface waters, which is illustrated in Fig. 4.37.

Further to these observations:

P.4.34. SO_2 has a direct effect on the degradation of plant pigments, and detritus is decomposed at reduced rate at lower pH,

as seen from Fig. 4.38, where the enzymatic decomposition of celulose is plotted as a function of pH. Plant growth is furthermore dependent on pH, as illustrated in Fig. 4.39. What effect the ever-decreasing pH in the soil of Scandinavia will have on the forestry, on which so many people are dependent is a worrying problem.

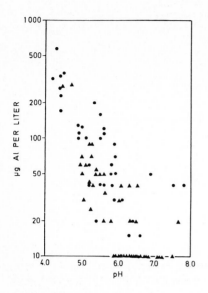

Fig. 4.37. Aluminium concentration in clear water lakes in Sweden (circles) and in Norway (triangles, unpublished data). Values plotted as 10 µg/l are below the analytical detection limit.

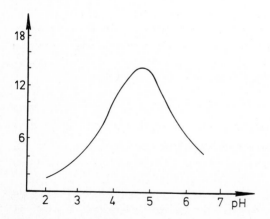

Fig. 4.38. Relative decomposition rate of cellulose plotted against pH.

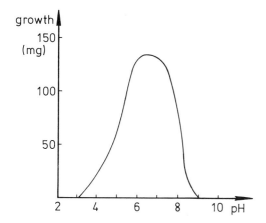

growth (mg)

Fig. 4.39. Growth of a fungus Merasmius graminum as a function of pH.

3. Seawater contains several proteolytic species, including hydrogen carbonate (about 2.4 mM), borate (about 0.43 mM), phosphate (about 0.0023 mM), silicate and fluoride in various stage of protonation (=uptake of hydrogen ions).

The concentrations of proteolytic species are characterized by the total alkalinity A, and pH. The total alkalinity is determined by adding an excess of a standard acid (e.g. 0.1 M), boiling off the carbon dioxide formed and titrating back to a pH of 6. During this process all the carbonate and hydrogen carbonate are converted to carbon dioxide and expelled and all the borate is converted to boric acid. The amount of acid used (i.e. the acid added minus the base used for back titration) then corresponds to the alkalinity, Al, and the following equation is valid:

$$Al = C_{H_2BO_3^-} + 2C_{CO_3^{2-}} + C_{BO_3^-} + (C_{OH^-} C_{H^+}) \qquad (4.24)$$

where C = the concentration in moles per litre for the indicated species.

In other words the alkalinity is the concentration of hydrogen ions that can be taken up by proteolytic species present in the sample examined.

Obviously, the higher the alkalinity, the better the solution is able to maintain a given pH value if acid is added. The buffering capacity and the alkalinity are proportional (see e.g. Stumm and Morgan, 1970).

Each of the proteolytic species in an aquatic system has an equilibrium constant. If we conseder the acid HA and the dissociation process:

$$HA \rightarrow H^+ + A^- \qquad (4.25)$$

we have

$$K_a = \frac{[H^+]\,[A^-]}{[HA]} \qquad (4.26)$$

where K_a = the equilibrium constant.

It is possible, when the composition of the aquatic system is known, to calculate both the alkalinity and the buffering capacity, using the expression for the quilibrium constants. However, these expressions are more conveniently used in logarithmic form. If we consider the expression for K_a for a weak acid, the general expression (4.24), may be used in a logarithmic form:

$$pH = pK_a + \log \frac{[A^-]}{[HA]} = pK_a + \log [A^-] - \log [HA] \qquad (4.27)$$

multiplying both sides of the equation with -1 and using the symbol p for -log and pH for -log H^+.

It is often convenient to plot concentrations of HA and A^- versus pH in a logarithmic diagram. If C denotes the total concentrations C=[HA]+[A$^-$], we have at low pH:

$$[HA] \approx C \qquad (4.28)$$

$$\log[A^-] = pH - pK_a + \log C \qquad (4.29)$$

This means that log[A$^-$] increases linearly with increasing pH, the slope being +1. The line goes through (log C, pK_a) as pH=pK_a gives log[A$^-$] = log C, see equation (4.29).
Correspondingly, at high pH, [A$^-$] = C and

$$\log [HA] = pK_a - pH + \log C \qquad (4.30)$$

which implies that log[HA] decreases with increasing pH, the slope being -1. This line also goes through (logC, pK_a).

At pH = pK_a, [A$^-$] = [HA] = C/2 or log [A$^-$] = log [HA] = log C - 0.3

Table 4.16 and Fig. 4.40 show the result of these considerations for a single acid-base system.

Note that for H_2A the slope will be -2 at $pH>pK_2$, corresponding to the dissociation of $2H^+$:

$$H_2A \quad - \quad 2H^+ + A^{2-}$$ and for A^{2-} the slope will be +2 at $pH<pK$. This is demonstrated in Fig. 4.41.

ß, the buffer capacity, is defined (4.19).
It can now be shown that

$$\log \left(\frac{ß}{2.3} \right) = \log \left([H_3O^+] + [OH^-] + \Sigma \frac{[A^-] \, [HA]}{c} \right) \qquad (4.31)$$

At log pH [HA] = C, and only $[H_3O^+]$ plays a role.

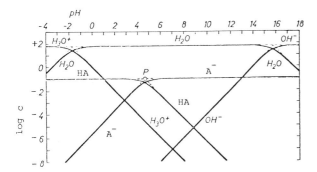

Fig. 4.40. H_3O^+, OH^- and water + HA/A⁻. $pK_a = 4.64$ and C = 0.1 M.

At higher pH, also $\dfrac{[A^-] \, [HA]}{c} = [A^-]$ contributes to $\dfrac{ß}{2.3}$

where $[H_3O] = [A^-]$, $\log \left(\dfrac{ß}{2.3} \right) = \log \left(2 \, [H_3O^+] \right) = -pH + 0.3 = \log \left(2 \, [A^-] \right)$.

At still higher pH, but with values of $pH<pK_a$, $\log [A^-]$ dominates.

- 239 -

At pH = pK$_a$ $\log \dfrac{\beta}{2.3}$ = $\log \dfrac{\frac{C}{2} \cdot \frac{C}{2}}{C}$ = $\log \dfrac{C}{4}$ = log C - 0.6

At pH>pK$_a$, [A$^-$] = C and log [HA] contributes the most to $\dfrac{\beta}{2.3}$

At very high pH, log [OH$^-$] will dominate. These considerations are used in the construction of Fig. 4.43.

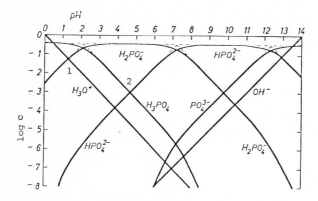

Fig. 4.41. pH-log c diagram for phosphoric acid.

Fig. 4.42 is a double logarithmic diagram for seawater. The proteolytic species mentioned above are represented in their appropriate concentrations. The important species are hydrogen and hydroxide ions, boric acid (HB) and carbonate ions (C^{2-}). The arrow in the diagram indicates the pH value of seawater - about 8.1.

Based on such a diagram it is possible to set up another diagram, representing the buffering capacity as a function of pH, see Fig. 4.43. For those who are interested in the relationship between the two diagrams, see Hägg (1979).

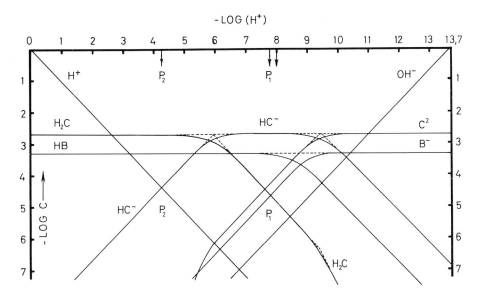

Fig. 4.42. pH-diagram. $H_2C = H_2CO_3$, $HC^- = HCO_3^-$, $C^2 = CO_3^{2-}$, $B^- =$ borate. pH of the sea is indicated by an arrow.

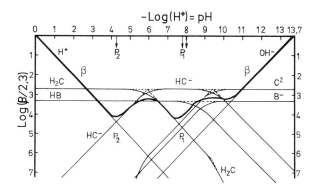

Fig. 4.43. Buffering capacity of the sea as a function of pH (ß).

From Fig. 4.43 we can conclude *that the buffering capacity of seawater is very limited. About 3 mM of strong acid will change the pH from 8 to 3.*

However, the pH of the entire ocean is remarkably resistant to change in pH caused by the addition of naturally occurring acids and bases, while the limited buffering capacity of an isolated litre of seawater accords with diagram 4.37. Sillen (1961) has suggested an explanation for the observed buffering capacity of the sea. A pH-dependent ion exchange equilibrium between solution and aluminosilicates (clay minerals), suspended in the sea is the main buffering system in oceans.

This buffering system may be represented by the following simplified equation:

$$3Al_2Si_2O_5(OH)_2 + 4SiO_2 + 2K^+ + 2Ca^{2+} + 12H_2O \longrightarrow$$
$$2KCaAl_3Si_5O_{16}(H_2O)_6 + 6H^+ \qquad (4.32)$$

The pH-dependence is indicated by the corresponding equilibrium expression in logarithmic form:

$$\log K = 6 \log(H^+) - 2 \log K^+ - 2 \log Ca^{2+} \qquad (4.33)$$

Sillen (1961) **estimated the buffering capacity of these silicates to be about 1 mole per litre or approximately 2000 times the buffering capacity of carbonates.** However, as pointed out by Pytkowitcz (1967), the buffering capacity of aluminosilicates has a much larger time scale than the buffering capacity based on the carbonate system.

P.4.36.   **In conclusion it seems that radical changes in the pH value of the oceans should not be expected as a result of increased combustion of fossil fuel, although the effect is cumulative.**

However, the regional effects of combustion of fossil fuel on pH, due to deposition of sulphuric acid, is a very severe environmental problem. Its solution lies in either drastic reduction in the combustion of fossil fuel or a corresponding reduction in the sulphurous content of the fossil fuel.

At present the pH values of lakes situated in districts where water has little buffering capacity are steadily decreasing and the only available remedy is the addition of calcium hydroxide to surface water. This will, however, change the chemical composition of the aquatic ecosystem in the entire region, which again will involve ecological changes. So, **the only real remedy is to reduce the emission of sulphurous compounds to an environmentally acceptable level.**

4.8.   OTHER BUFFERING EFFECTS IN ECOSYSTEMS.

The ecologist working with environmental probelms is concerned with the response of the system to changed external factors. Fig. 4.44 shows the response of Lake Fure to increased phosphorus loading. During the period 1945-1973 increased loading was almost proportional to time, but as can be seen the concentration of phosphorus in the water remained almost unchanged for the first two decades, mainly because the added phosphorus was stored in the sediment by the following chain of processes:

Soluble P in water — uptake by algae — settling — P in sediment

Although some of the phosphorus in sediment was released to the water, a substantial part was stored in the sediment.

P.4.37.   An ecosystem is able to minimize changes caused by external sources.
However, this buffering capacity is finite and once the capacity is used, the changes become more pronounced.

The course of the response then parallels the change in pH, caused by the addition of acids or alkalies (see 4.7).

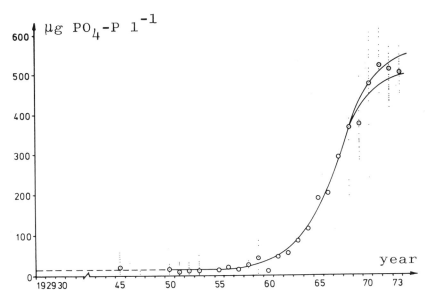

Fig. 4.44.   The phosphorus concentration in Lake Fure plotted against time (years). During the period considered the phosphorus loading is increased.

It is of course extremely important to recognize the response of ecosystems to changes in external factors, such as increased discharge of pollution. As seen, throughout this book

P.4.38. the response is very rarely proportional to the external factor.

If such an assumption is used the prediction of ecosystem responses would be completely wrong in most cases and too high an environmental impact might be tolerated. If, for example, the observed response to increased phosphorus loading was used for Lake Fure in the period 1950-1956 (see Fig. 4.44) removal of phosphorus from the waste water would not have been required, as it is today, Of course the entire problem of lake management is much more complicated than that expressed in Fig. 4.44 (see Jørgensen, 1980), as many different processes are involved. But these observations demonstrate the non-linear response of ecosystems, which can be explained by the concept of buffering capacity.

Generally, the concept of **ecological buffering capacity** can be defined as:

$$\text{ß} = \frac{\text{d(external variable)}}{\text{d(internal variable)}} \tag{4.34}$$

where
external variable ˷ external factors ˷ forcing functions
internal variable ˷ state variable

The ecological buffering capacity is the reciprocal sensitivity of the considered state variable (see, e.g. Halfon, 1976):

$$\text{sensitivity of state variable} = \frac{\text{d(state variable)}}{\text{d(external factor)}} \tag{4.35}$$

As it is not possible to measure differential coefficients, the ecological buffering capacity is found as (Jørgensen et al., 1977):

$$\text{ß} = \frac{\Delta\text{external variable}}{\Delta\text{internal variable}} \tag{4.36}$$

An ecosystem has almost an infinite number of buffering capacities. The buffering capacity related to the changes in soluble phosphorus caused by increased discharge of phosphorus can be defined as:

$$\beta_P = \frac{\Delta P\text{-loading}}{\Delta P\text{-soluble}} = \frac{\Delta P\text{-total}}{\Delta P\text{-soluble}} \qquad (4.37)$$

The change in the total phosphorus concentration is measured by the increased discharge of phosphorus, since the law of mass conservation states that phosphorus (mainly soluble phosphorus) put into the lake must be present in one form or another. But due to the buffering capacity, or rather the many processes in which phosphorus takes part, the soluble phosphorus in the lake will increase less than would be expected from its input, because the system will minimize the direct effect of an increased input of soluble phosphorus.

A similarly definition can be applied to the buffering capacity in response to chaged temperature. As it is difficult to compare changes in temperature with changes in concentrations of phytoplankton and zooplankton (or with other words the ecological buffer capacity should have no unit) it seems more reasonable in this case to use a related expression:

$$\beta_{T\text{-rel}} = \frac{\Delta Temp * Phyt}{Temp * \Delta Phyt} \qquad (4.38)$$

where
Temp = temperature
Phyt = phytoplankton concentration

The concept of ecological buffering capacity has been examined by Jørgensen et al. (1979) using ecological models. They set up the following hypothesis:

The ecosystem will respond to changes in external factors, such as nutrient input, climatic changes of temperature and irradiance, etc, by developing a structure with the best possible ability to meet perturbations. It can be shown that an ecosystem with little diversity (only a lake has been considered) has a better buffering capacity at a high nutrient input that a more complex ecosystem, which explains why many eutrophic lakes have low diversity. They also found that complex ecosystems have a very high buffering capacity for temperature changes and that species with a temperature dependence better suited to a new temperature pattern contributed more to the buffering capacity than species "ill" suited to the temperature pattern.

It seems, from these consideration, that ecosystems not only have a buffering capacity, which minimizes changes in the system caused by external factors, but also have the ability to change their structure in such a way that the highest possible buffering capacity is available for actual changes in external factors.

However, like buffering capacity, the ecosystem's ability to adapt its structure to a new situation is also finite. Drastic changes in an ecosystem cannot be tolerated under any circumstances, as we consider a simplification of an ecosystem (in terms of reduced diversity) to be an ecological deterioration. The buffering capacity may be increased for nutrient input or input of other pollutants by decreasing the complexity of the ecosystem, but the buffering capacity for temperature or other climatic variations is decreased simultaneously, which makes the ecosystem more susceptible to natural variations in external factors.

In Table 4.17 is listed some ecosystem reactions to changes in external factors, which can be interpreted by use of the ecological buffer capacity.

TABLE 4.17
Ecosystem reactions and ecological buffer capacities

Change in external variable	Ecosystem reactions	Increased buffer capacity
P-loading increased	Lower diversity. Species with lower specific surface, and thereby slower nutrient uptake	β_P
Temperature	Species with other temperature. Optimum take over	$\beta_{T\text{-rel}}$
Increased conc. of toxic substances	Some species become extinguished. Less suspectible species become more dominant	$\beta_{TOX} = \dfrac{\Delta \text{toxic subst.}}{\Delta \text{biomass}}$
Increased BOD$_5$	Some species become extinguished. Species less suspectible to low oxygen conc. become more dominant	$\beta_{BOD} = \dfrac{\Delta \text{BOD}}{\Delta \text{biomass}}$

4.9.   THE ECOSPHERE.

The ecosystems, which constitute the ecosphere, can be classified according to their physical-chemical-biological characteristics.

Each type of ecosystem has its own characteristic habitat, which to a certain extent defines the environmental problems it experiences.

TABLE 4.18
Ecosystems

A. Aquatic Ecosystems
   FRESHWATER
   Lotic   Rivers and streams: flow characteristic,
   (BOD_5/O_2-balance of importance)
   Rapids: $\geq$ 0.5 msec^{-1}, bottom particles $\geq$ 5 mm
   $\leq$ 0.5 msec^{-1}, bottom particles $\leq$ 5 mm
   Pools: very slow streams

   Lentic   Lakes and ponds: slow water renewal,
   (eutrophication ond acidification)
   Deep lakes
   Shallow lakes
   Ponds: shallow enough for sunlight to reach the
   bottom everywhere

   MARINE
   Littoral   Shoreline: reduced circulation (all types of
   aquatic pollution)
   Rocky
   Sand
   Neritic   Continental shelt: oil-pollution
   Upwellings (Eutrophication)
   Coral reef
   Pelagic   Open sea (oil pollution)
   Epipelagic
   Mesopelagic
   Bathypelagic
   Abyssal

B. Terrestrial Ecosystems
   Desert   Annual rainfall $\leq$ 25 cm. Water shortage.
   Hot
   Cold

   Tundra   Long, cold harsh winter and short, cool summer.
   Treeless from forest limit to the ice caps or
   glaciers. Low diversity.
   Arctic
   Alpine

   Prairie   Grassland, now widely cultivated.
   Rainfall 40-100 cm year^{-1}
   Moist
   Dry

   Savannah   Trees scattered within a grassland matrix.
   Rainfall 50-150 cm year^{-1}. 5-20°N or S.
   High diversity. (Maintenance of ecological
   balance and a high natural diversity)

   Forest   Coniferous: Dominated by needle-leaved trees.
   Man's source of wood. (Deforestration)
   Deciduous: Warm summers, cold to cool winters.
   Broadleaves species. 60-150 cm yr^{-1} preci-
   pitation

Tropical: 20°S-20°N. High rainfall.
Very high diversity. Mean annual temperature
~ 25°C

Bogs Have cushionlike growths of small plants, accumulation of peats.
North America, Northern Europe. Include raised bogs or high moors

Marshes Dominated by grasses, sedges, bulrushes, cattails etc.
Warmer climate

Swamps Are wooded. Higher diversity than marshes

TABLE 4.19

A. Primary productivity, net production and plant biomass of large biomes (expressed in tonnes of dry organic matter) (from Lieth & Whittaker, 1975)

Ecosystem type	Area (10^6 km^2)	Net primary productivity ($g\,m^{-2}\,yr^{-1}$) Normal range	Mean	Net production worldwide ($10^9 t\,yr^{-1}$)	Biomass per unit area ($t\,ha^{-1}$) Normal range	Mean	Total biomass worldwide (10^9t)
Tropical rain forest	17.0	1000-3500	2200	37.4	60-800	450	765
Tropical seasonal forest	7.5	1000-2500	1600	12.0	60-600	350	260
Temperate evergreen forest	5.0	600-2500	1300	6.5	60-2000	350	175
Temperate deciduous forest	7.0	600-2500	1200	8.4	60-600	300	210
Boreal forest (taiga)	12.0	400-2000	800	9.6	60-400	200	240
Woodland and shrubland	8.5	250-1200	700	6.0	20-200	60	50
Savanna	15.0	200-2000	900	13.5	2-150	40	60
Temperate grassland	9.0	200-1500	600	5.4	2-50	16	14
Tundra	8.0	10-400	140	1.1	1-30	6	5
Desert and semi-desert scrub	18.0	10-250	90	1.6	1-40	7	13
Extreme desert (sand), polar regions	24.0	0-10	3	0.07	0-2	0.2	0.5
Cultivated land	14.0	100-3500	650	9.1	4-120	10	14
Swamp and marsh	2.0	800-3500	2000	4.1	30-500	150	30
Lake and stream	2.0	100-1500	250	0.5	0-1	0.2	0.05
Total continental	*149.0*		*773*	*115.0*		*123*	*1837*
Open ocean	332.0	2-400	125	41.5	0.01-0.05	0.03	1.0
Upwelling zones	0.4	400-1000	500	0.2	0.5-10	2.0	0.008
Continental shelf	26.6	200-600	360	9.6	0.1-4.0	1.0	0.27
Algal bed and corel reef	0.6	500-4000	2500	1.6	0.4-40	20	1.2
Estuaries	1.4	200-3500	1500	2.1	0.1-60	10	1.4
Total marine	*361.0*		*152*	*55.0*		*0.1*	*3.9*
Full total	*510.0*		*333*	*170.0*		*36*	*1841*

TABLE 4.19

B. Secondary production and productivity in the biosphere (from Lieth & Whittaker, 1975)

Ecosystem type	Leaf-surface area (10^6 km^2)	Biomass of litter (10^9 t)	Animal consumption (10^9 t/yr)	Secondary production (10^9 t/yr)	Secondary productivity of animal matter (kg/ha/yr)	Animal biomass (10^9 t)
Tropical rain forest	136	3.4	2600	260	152.9	330
Tropical seasonal forest	38	3.8	720	72	96.0	90
Temperate evergreen forest	60	15.0	260	26	52	50
Temperate deciduous forest	35	14.0	420	42	60	110
Boreal forest (taiga)	144	48.0	380	38	31.7	57
Woodland and shrubland	34	5.1	300	30	35.3	40
Savanna	60	3.0	2000	300	200	220
Temperate grassland	32	3.6	540	80	88.9	60
Tundra	16	8.0	33	3	3.8	3.5
Desert and semi-desert scrub	18	0.36	48	7	3.9	8
Extreme desert (sand), polar regions	1.2	0.03	0.2	0.02	0.008	0.02
Cultivated land	56	1.4	90	9	6.4	6
Swamp and marsh	14	5.0	320	32	160	20
Lake and stream			100	10	50	10
Total continental	*644*	*111*	*7811*	*909*	*61*	*1005*
Open ocean			16600	2500	75.3	800
Upwelling zones			70	11	275.0	4
Continental shelf			3000	430	161.7	160
Algal bed and corel reef			240	36	600	12
Estuaries			320	48	342.9	21
Total marine			*20230*	*3025*	*83.8*	*997*
Full total			*28041*	*3934*		*2002*

Ecosystems can be divided into two main groups: aquatic and terrestrial. The classification of ecosystems is presented in Table 4.18. Source of the major environmental problems is indicated in brackets for some ecosystems.

Table 4.19 gives area, primary productivity, net production, plant biomass, secondary production, secondary productivity and animal biomass of various types of ecosystems or large biomes (see also Table 2.16). From this table it is noticeable that no proportional relationship exists between total biomass and production.

4.10.   APPLICATION OF ECOLOGICAL ENGINEERING.

Ecological engineering is the discipline that deals with methods of assisting and modifying ecosystems to overcome the impact of pollution. These methods are alternatives to the use of environmental technology, which is covered in part B of this book. However, combinations of the two types of methods might give an optimum solution in many cases.

It has been pointed out throughout this book that a close to optimum solution of an environmental problem can only by found by a quantification of the problem combined with a right selection among the wide spectrum of applicable methods. This principle is even more important to use when the solution requires interference with the ecosystem. Without a quantification of the problem, we are unable to measure and compare the effects of the various alternatives.

Ecological engineering is obviously more applicable when the pollutants are threshold, rather than non-threshold agents, although it has also been applied on non-threshold pollutants.

In this section an overview of ecological engineering methods will be given. Those who seek a more comprehensive introduction into the application of these methods are referred to Jørgensen and Mitsch (1988). The most important methods are listed below and a brief description of the application, advantages and disadvantages is given.

Ecological Engineering Methods.

1. *Diversion*   of waste water has been extensively used, often to replace waste-water treatment. Discharge of effluents into an ecosystem which is less susceptible than the one used at present is, as such, a sound principle, which under all circumstances should be considered, but a quantification of all the consequences has often been omitted. Diversion might reduce the number of steps in the treatment, but **cannot replace** waste-water treatment totally, as discharge of effluents, even to the sea, always should require at least mechanical treatment to eliminate suspended matter. Diversion has often been used with a positive effect when eutrophication of a lake has been the dominant problem. Canalization, either to the sea or to the lake outlet, has been used as solution in many cases of eutrophication. However, effluents must be considered as a fresh-water resource. If it is discharged into the sea, effluent cannot be recovered; if it is stored in a lake, after sufficient treatment of course, it is still a potential water resource, and it is far cheaper to purify eutrophic lake-water to an acceptable drinking-water standard than to desalinate seawater. Diversion is often the only possibility when a *massive* discharge of effluents goes into a suscep-

tible aquatic ecosystem (a lake, a river, a fjord or a bay). The general trend has been towards the construction of larger and larger waste-water plants, but this is quite often an ecologically unsound solution. Even though the waste-water has received multistep treatment, it will still have a high amount of pollutants relative to the ecosystem, and the more massive the discharge is at one point, the greater the environmental impact will be. It it is considered that the canalization is often a significant part of the overall cost of handling waste-water, it might often turn out to be a both better and cheaper solution to have smaller treatment units with individual discharge points.

2. *Removal of superficial sediment* can be used to support the recovery process of very eutrophic lakes and of areas contaminated by toxic substances (for instance, harbours). This method can only be applied with great care in small ecosystems. Sediments have a high concentration of nutrients and many toxic substances, including trace metals. If a waste-water treatment scheme is initiated, the storage of nutrients and toxic substances in the sediment might prevent recovery of the ecosystem due to exchange processes between sediment and water. Anaerobic conditions might even accelerate these exchange processes; this is often observed for phosphorus, as iron(III) phosphate reacts with sulphide and forms iron(II)sulphide by release of phosphate. The amount of pollutants stored in the sediment is often very significant, as it reflects the discharge of untreated waste-water for the period prior to the introduction of a treatment scheme. Thus, even though the retention time of the water is moderate, it might still take a very long time for the ecosystem to recover.

   The method is, however, costly to implement, and has therefore been limited in use to smaller systems. Maybe the best known case of removal of superficial sediment is Lake Trummen in Sweden. The transparency of the lake was improved considerably, but decreased again due to the phosphorus in overflows from rainwater basins. Probably, a treatment of the overflow after the removal of superficial sediment would have given a better result.

3. *Uprooting and removal of macrophytes* has been widely used in streams and also to a certain extent in reservoirs, where macrophytes have caused problems in the turbines. The method can, in principle, be used whereever macrophytes are a significant result of eutrophication. A mass balance should always be set up to evaluate the significance of the method compared with the total nutrient input. A simultaneous removal of nutrients from the effluent should also be considered.

4. *Coverage of sediment by an inert material* is an alternative to removal of superficial sediment. The idea is to prevent the exchange of nutrients (or maybe toxic substances) between sediment and water. The general applicability of the method is limited due to the high costs, even though it might be more moderate in cost than removal of superficial sediment. It has only been used in a few cases and a more general evaluation of the method is still lacking.

5. *Siphoning of hypolimnic water* is more moderate in cost than methods 2 and 4. It can be used over a longer period and thereby gives a pronounced overall effect. However, the effect is dependent on a significant diffe- rence between the nutrient concentrations in the epilimnion and the hypolimnion, which, however, often is the case if the lake or the reser- voir has a pronounced formation of a thermocline. This implies, on the other hand, that the method will only have an effect during the period of the year when a thermocline is present (in many temperate lakes from May to October/November), but as the hypolimnic water might have a concentration 5-fold or higher than the epilimnic water, it might have a significant influence on the nutrients budtget to apply the method anyhow.

   As the hypolimnic water is colder and poorer in oxygen, the ther- mocline will move downwards and the possibilities of anaerobic zones will be reduced. This might have an indirect effect on the release of nutrient from the sediment.

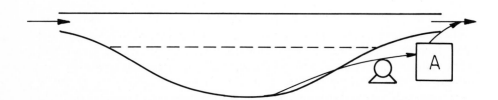

Fig. 4.45. Application of siphoning and ion exchange of hypolimnic water. The dotted line indicates the thermocline. 1 indicates the ion exchanger.

   If there are lakes or reservoirs downstream, the method cannot be used, as it only removes, but does not solve the problem. A possibility in such cases would be to remove phosphorus from the hypolimnic water before it is discharged downstream. The low concentration of phosphorus in hypolimnic water (maybe 0.5 - 1.0 mg l^{-1}) compared with waste-water makes it impossible to apply chemical precipitation, see example 6.6 in section 6.3. However, it will be feasible to use ion exchange, because the capacity of an ion exchanger is more dependent on the total amount of phosphorus removed and the flow than on the total volume of water treated (see section 6.3). Figure 4.45 illustrates the use of siphoning and ion exchange of hypolimnic water.

6. *Flocculation* of phosphorus in a lake or reservoir is another alternative. Either alumsulphate or iron(III)-chloride can be used. Calcium hydroxide cannot be used, even though it is an excellent precipitant for waste-water, as its effect is pH-dependent and a pH of 9.5 or higher is required.

    The method is not generally recommendable as 1) it is not certain that all flocs will settle and thereby incorporate the phosphorus in the sediment, 2) the phosphorus might be released from the sediment again at a later stage.

7. *Circulation* of water can be used to break down the thermocline. This might prevent the formation of anaerobic zones, and thereby the release of phosphorus from sediment.

8. *Aeration* of lakes and reservoirs is a more direct method to prevent anaerobic conditions from occuring. Aeration of highly polluted rivers and streams has also been used to avoid anaerobic conditions.

9. *Regulation of hydrology* has been extensively used to prevent floods. Lately, it has also been considered as a workable method to change the ecology of lakes, reservoirs and wetlands. If the retention time in a lake or a reservoir is reduced with the same annual input of nutrients, eutrophication will decrease due to decreased nutrient concentrations. The role of the depth, which can be regulated by use of a dam, is somehow more complex. Increased depth has a positive effect on the reduction of eutrophication, but if the retention time is increased simultaneously, the overall effect cannot generally be quantified without the use of a model. The productivity of wetlands is highly dependent on the water level, which makes it highly feasible to control a wetland ecosystem by this method.

10. *Fertilizer control* can be used in agriculture and forestry to reduce the nutrient loss to the environment. Utilization of nutrients by plants is dependent on a number of factors (temperature, humidity of soil, soil composition, growth rate of plant (which again is dependent on a number of factors), chemical speciation of nutrients,. etc.). Models of all these

processes are available today on computers, and in the nearest future it must be foreseen that the fertilization scheme will be worked out by a computer on the basis of all the above-mentioned information. This will make it feasible to come closer to the optimum fertilization from an economic-ecological point of view.

11. *Insecticide control* by use of computer models has now been in use for several years. The idea is to use the pesticides when they have an effect on the harmful organisms - not when they might harm their predators. Observations on the appearance of the relevant species are used as input to a computer model, which gives the scheme for the use of insecticides.

12. *Application of wetlands or impoundments as nutrient traps* could be considered as an applicable method, whereever the non-point sources are significant. The use of wetlands has also been applied as a direct waste-water treatment method, for instance in Florida, but it will probably be most effective in dealing with nutrient losses from agricultural areas. Inputs of nutrients into a wetland will be denitrified, adsorbed on the sediment or used for growth of algae and macrophytes (phragmites, etc.), if it is not found in the outflows. Management of a wetland or an impoundment as a nutrient trap obviously requires that a major part of the nutrient input is removed by denitrification (nitrogen only) and stored in sediment and plants. The storage capacity is often large, but of course limited. This implies that nutrients must be removed by harvest of macrophytes, which is quite feasibel mechanically. However, if the retention time of a lake or reservoir in the temperate zone is short, it might be favorable to let the inflowing water pass through a wetland or an impoundment. In the wintertime, when almost no growth takes place in the wetland, the nutrients are washed out, but due to the short retention time, the water with high nutrient concentration will have passed the lake or reservoir before the spring bloom starts. The nutrients of the water which passes the wetland during the spring and summer will, on the other hand, be used to a large extent in the wetland or the impoundment for growth of algae and macrophytes. The water flowing to the lake or reservoir will therefore have a significantly lower nutrient concentration at the time of the year when eutrophication may appear.

The role of wetlands in maintaining an ecological balance is still not fully understood, although our knowledge of the topic is far better today than 10 or 20 years ago. Lately, many large land-reclamation projects have been questioned due to increased experience in the field. Indications show that wetlands are important not only as nutrient traps but also for maintenance of species diversity and for an ecologically sound hydrology in the region.

13. *Calcium hydroxide* is widely used to neutralize low pH-values in streams and lakes in those areas where acidic rain has a significant impact.

14. *Biomanipulation* covers a wide range of possibilities. The role of diversity in a forest has been touched on in 4.5; it can be considered as biomanipulation when additional tree species are actually planted in a forest. Introduction of herbivorous fish (carp, etc.) into streams and lakes which suffer from eutrophication or uncontrolled growth of macrophytes is another example of biomanipulation.

It has also been considered to reduce eutrophication by the introduction of top carnivorous fish species into lakes and reservoirs. They will reduce the populations of zooplankton-eating fish species and cause an increase in zooplankton populations, which again will imply reduced phytoplankton concentrations. The observed effects are clear, but it is indeed questionable whether the effect is permanent, if nutrient inputs continue. Some observations indicate that the ecosystem in this case will go back to its old equilibrium with reduced populations of top carnivorous fish species (observations by M. Straskraba and others).

QUESTIONS AND PROBLEMS

1. A fish has a growth coefficient of 0.02 day^{-1}. The carrying capacity in a lake is 120,000. At a given time 100 fish are introduced.
When will the carrying capacity -2% relative be achieved? A logistic growth is assumed.
What would the result be if the time lags t_1 and t_2 in (4.13) are 75 days?
Only approximate result is required in this case.

2. Indicate how the development of man made ecosystem consisting of a concrete basin full of municipal waste-water will occur.

3. Indicate the expected difference in diversity between a shallow and deep lake and between a eutrophic and an oligotrophic lake.

4. What is the difference between fresh and salt water in relation to a)

solubility of oxygen, b) LC_{50} value for total ammonia $(NH_3 + NH_4^+)$?

5. Explain using the concept of ecological buffering capacity and energy, why anaerobic conditions give lower diversity than aerobic conditions.

6. In an island with an area of 100 km^2 120 bird species were found. How many would one expect to find on an island in the same region, but with an area of 20 km^2?

7. Find the buffer-capacity of a lake with 100 mg Ca^{2+} l^{-1} and a lake with 10 mg Ca^{2+} l^{-1} at pH = 8.0 and pH = 5.0. It is presumed that the buffer capacity is entirely related to HCO_3^- and this ion is equivalent with the calcium concentration. How much would pH change in the 4 cases, if the lakes are 10 m deep and they receive sulphuric acid by precipitation corresponding to 0.5 g S m^{-2}?

CHAPTER 5

AN OVERVIEW OF THE MAJOR
ENVIRONMENTAL PROBLEMS OF TODAY

5.1. APPLICATION OF PRINCIPLES TO ENVIRONMENTAL PROBLEMS.

In chapters 2, 3 and 4 we outlined some of the principles applicable to
environmental problems, and used examples to illustrate how these principles can be applied. Throughout we have concentrated on creating an understanding of the problem, rather than determining all the different environmental problems of today. Environmental problems change with time. What is
crucial today might be solved during the next decade, during which time new
problems will have appeared.

The principles we might employ to solve these problems will also change
with time, but at a much slower rate, so we can probably use the same
principles to solve the different and novel problems of the next decade. This
chapter is devoted to a brief discussion of those environmental problems of
major concern to man today.

To understand the basis of an environmental problem a long list of
questions and problems, has to be answered. This may not always be possible
because of lack of data. However, when the data are available the principles
explained in chapters 2, 3 and 4 can be applied to formulate and answer the
relevant questions. Which questions to ask is sometimes difficult to decide
upon, but an attempt can be made as follows:

1. What is the source of pollution?
2. What is the distribution pattern of the pollutant in time and space in the
    ecosphere and in the ecosystem?
3. Is the pollutant harmful to plants and animals? What is its toxic effect?
    What is its level of toxicity?
4. Does the pollutant take part in chemicobiological processes in the
    environment? Will such pollution affect the ecological balance?
5. Will there be any chance of biomagnification? What is the ecological
    buffer capacity related to the pollution?
6. Will the pollutant accumulate in organs?
7. Will the pollutant accumulate somewhere in the ecosystem?
    (as for example heavy metals and nutrients accumulate in sediments)
8. What is the overall effect on man, bearing in mind all the pathways from
    the source to man?

Fully quantitative answers to such a list of questions often require the application of ecological models, as many interacting processes are involved and the system is very complex. However, without sufficient data it is impossible to solve any environmental problem, and an ecological model cannot give the answer, as it is only a tool to cope with complicated systems *provided that sufficient data are available.* But some data at least are available for environmental problems and in these cases it is possible partially to understand the core of the problem. Let us consider the DDT problem as an example of how far we can deal with most problems today. We know that the source of DDT is the application and the production of a particular pesticide. The problem is related to the toxicity, the low biodegradability and the biomagnification effects of this compound. Data focusing on these aspects of DDT are available and by applying the relevant principles (see sections 1-2.3) it can be concluded that only minor use of this pesticide can be accepted, and that a better solution would be to introduce alternative (biological) methods or compounds with better biode-gradability and less chance of biomagnification. These conclusions can be made without the use of ecological models; on the other hand, a quantitative relationship between a given application of DDT and its final concentration in top carnivores in an ecosystem would require the use of an ecological model.

Many environmental problems cannot be solved easily, but require *the imposition of legislative measures* to remove the source of pollution. This has been the case with DDT, which is banned in many countries. Other problems such as the use of nutrients cannot be solved in that way, but will require prudent planning of their use with ecological aspects taken into consideration. But in both cases it is crucial to understand the problem fully if the right decisions are to be made.

The following paragraphs list the major environmental problems of today. References to a more comprehensive and detailed treatment of these problems can be found in the reference list. The problems have been divided into the following categories:

1. Air pollution. Pollution problems mainly related to the atmosphere.
2. Pesticides and other toxic organic compounds which affect the hydrosphere, lithosphere and atmosphere.
3. Heavy metal pollution, which like (2) affects all three spheres.
4. Water pollution problems, mainly related to the hydrosphere.
5. Noise pollution.
6. Pollution from solid wastes.
7. Food additives.
8. Energy alternatives.

5.2. AIR POLLUTION.

Air pollution problems are concerned either with global effects on the climate or local (regional) effects due to toxicity of air pollutants.

There are several ways in which man's activities may affect global climatic patterns in the future. One of these has been mentioned in sections 2.5 and 3.5, i.e. the increasing carbon dioxide concentration in the atmosphere produced by the combustion of fossil fuel. Decreasing atmospheric transparency by the injection of particulate matter (dust, sulphates, liquid droplets) into the atmosphere from industry, vehicles, space heating, agriculture and land clearing activities might, however, have the reverse effect on the climate by increasing the reflection of solar radiation (Watt, 1972). Man is also changing the albedo (i.e. the percentage of incoming solar radiation that is directly reflected back into space) of the earth's surface through irrigation, urbanization, deforestation and agriculture. Furthermore, the rate of thermal energy transfer between the oceans and the atmosphere is altered by oil pollution in the hydrosphere (see also 5.5) and finally there is the problem of the direct emission of heat to the atmosphere by the burning of fossil and nuclear fuels.

Regionally, the emission of toxic gases is of great importance. The problem is related to the following compounds: carbon monoxide, nitrogenous gases with the composition NO_x, where x is between one and two, sulphur dioxide, particulate matter, chlorine, hydrogen chloride, hydrogen sulphide and hydrogen flouride and the formation of the so-called photochemical smog, which consists of ozone, PAN (peroxyacetyl nitrate), PPN (peroxypropanyl nitrate) and PBN (peroxybuturyl nitrate).

In this context we must also mention the possible effects of nitrogen oxide input to the stratosphere by supersonic aircrafts. Although we do not know enough to predict in detail the effect of these emissions, they could not only alter global climatic patterns, but also partially deplete the ozone layer, which protects us from harmful wave length of ultraviolet radiation. Johnston (1972) has projected that 500 supersonic aircrafts could halve the amount of ozone within as little as one year. The possible risk is easy to understand when it is added that a 5% decline in ozone would produce at least 8000 additional cases of skin cancer per year among the U.S. white population (National Academy of Sciences, 1972).

Table 5.1 gives a survey of gaseous pollutants, their sources, major effects and concentration levels.

Carbon monoxide is toxic because of its ability to supersede oxygen bound to hemoglobin (Hb) in blood. The relationship between the carbon monoxide concentration and the percentage COHb has been found by a statistical analysis.

TABLE 5.1
Survey of gaseous pollutants, their sources, major effects and concentration levels

Gaseous pollutants	Source	Effect	Concentration level
CO_2	Fossil fuel	Global heat balance	0.032%
CO	Vehicles	Toxic	10-75 ppm (motor ways, free ways)
SO_2, SO_3	Fossil fuel	Toxic Corrosive	≥ 0.15 mg m^{-3} in cities
NO_x	Vehicles	Toxic	Few ppm
Carbon hydrides	Fossil fuel Industry Vehicles	Toxic Photochemical smog	1-25 ppm
Particulates	Fossil fuel Industry Vehicles	Respiratory damage	≥ 200 μg m^{-3} in cities
Photochemical oxidants, ozone aldehydes and others	Fossil fuel Industry Vehicles	Photochemical smog Damage to rubber	Ozone 0.1 ppm PAN 0.01 ppm
Chlorine, hydrogen chloride	Industry Incineration of PVC, Organic chloride	Toxic	Local
Hydrogen fluoride	Industry	Toxic Damage to glass	Local
Iron dust	Industry	toxic	Local

5.3. PESTICIDES, AND OTHER TOXIC COMPOUNDS.

Pesticides are chemical compounds devised to kill insects (insecticides), weeds (herbicides), rodents (rodenticides), fish (piscicides), mites (miticides) and fungi (fungicides). Before the second world war, most pesticides were non-persistent naturally occurring compounds. For example, nicotine sulphate was widely used as pesticide.

The use of pesticides is vast and increasing. More than 1 billion tons of pesticides were produced in the U.S. during 1970.

The problems of using persistent pesticides have already been touchen upon in 2.8, 2.13, 4.2 and 4.4. They can be summarized as follows:

1. Low biodegradability, which implies accumulation in the environment.
2. Resistance is developed relatively quickly (see 4.2).
3. Biomagnification (see 2.8 and 3.4)
4. Toxicity for humans (see 2.3).
5. Destruction of non-target organisms, often including the target pests natural predators (see 4.4).

Pesticides are not the only compounds to be magnified in the foodchain. PCBs (polychlorinated biphenyls) were found in relatively high concentrations in fish, eagles, human tissue, etc. (Jensen, 1966). PCB is a mixture of many compounds derived from biphenyl by chlorination:

```
    x   x        x   x
     \ /          \ /
x --< O >---< O >-- x        (x possible Cl-substitutions)
     / \          / \
    x   x        x   x
```

It is used widely for a variety of purposes (as transformator oil, and as a softener in paint and plastics). Today the use of PCB in many countries is limited to such uses where its emission into the environment can be avoided because of its observed biomagnification.

Appendix 5 gives a list of LD_{50} values for some commonly used pesticides. Table 5.2 summarizes the properies of environmental interest of some commonly used pesticides and PCB.

Many organic compounds represent a threat to the environment. It is not possible here to list all these chemicals, as more than 30,000 compounds are commercially applied in such quantities as to be of environmental concern.

TABLE 5.2
Some commonly used synthetic pesticides

Class	Examples	Major use
A. Chlorinated hydrocarbons	DDT, DDE, DDD, aldrin, dieldrin, endrin, heptachlor, toxaphene, lindane, chlordane	Broad spectrum insecticides (kill a wide variety of target and nontarget organisms)
B. Organic Phosphates	Malathion, parathion, Azodrin, Phosdrin, Diazinon, TEPP	Broad and narrow spectrum insecticides and a few fungicides and herbicides
C. Carbamates	Carbyl (Sevin), Zireb, Maneb	Broad and narrow spectrum insecticides, fungicides and herbicides
D. Phenoxy herbicides	2,4-D and 2,4,5-T	Herbicides

Class	Action	Persistence	Toxicity
A.	Attack central nervous system of insects causing convulsions, paralysis and death	High (2 to 15 years)	Relatively low for humans
B.	Nerve poisons that inactivate the enzyme that transmits nerve impulses	Low to moderate (normally 1 to 12 weeks but up to several years)	Very high for humans and other animals
C.	Nerve poisons	Usually low (days to 2 weeks)	Low to high for humans and other animals
D.	Cause metabolic changes in plants leading to leaf drop or death	Low to moderate (days to several weeks)	Low for humans and other animals

5.4. THE PROBLEM OF HEAVY METALS.

In recent years there have been a series of alarms about toxic metallic substances, such as lead, mercury and cadmium. Although the latter is not a heavy metal, i.e. one which has a specific gravity equal to or greater than

that of iron, it is often considered with them, as its environmental effects
are similar to those of lead, silver, chromium, nickel, mercury and others.
Accurate knowledge of the environmental effects of metals and their
compounds is very difficult to obtain, as some metals are essential to life in
small concentrations but toxic in higher concentrations. The situation is
further complicated because they are introduced into the environment as may
different compounds or become converted into other compounds in the
environment with different effects and toxicity.

Environmental problems related to the use of metallic compounds have
been illustrated for lead in 2.10, but to complete the picture we should take
a broader view.

Recently a number of pollution-producing industries have moved to
Thailand, where environmental legislation is relatively ineffective. One of
the results of this development is shown in Table 5.3 A + B.

Mercury contamination of fish in Thailand has been one of the lowest in
the world, with a mean of 0.07 ppm (range 0.002/0.30 ppm) in flesh. Recently
mercury values varying from 0.32 to 3.6 ppm in fish flesh have been observed
in the vicinity of an established caustic soda factory. In other words the
pollution problem has been exported from a developed country (in this case,
Japan) to a developing country (compare with Fig. 5.1), but the problem of
mercury pollution in connection with caustic soda production has not been
solved.

TABLE 5.3 A
Development of chemical industry in Thailand. The data indicates production in
thousands of metric tons

Type of industry	1967	1968	1969	1970	1971	1972	1973	1974	Average increase in annual rate of production (%)
Caustic soda	10	18	25	33	39	47	47	53	30
Sulphuric acid	12	14	18	15	14	47	47	47	39
Cement	1734	2168	2403	2627	2771	3378	3706	3923	13
Printing and writing paper	21	24	30	32	38	42	40	34	8
Hydrochloric acid	13	25	31	32	37	34	46	56	26
Washing powder and detergents	21	24	27	27	32	40	47	46	12
Non-cellusic continuous fibers	0.5	0.5	0.9	1.2	4.8	9.8	16.3	15.2	82

TABLE 5.3 B
Mercury content (on wet weight basis) in flesh of fish from some freshwater localities
in Thailand. The caustic soda factory (TACSCO) is to be found in locality 8.

Species lity	Loca- analyzed	No. of fish samples (g)	Weight range (cm)	Length range range	mg Hg/kg
Ophiocep-	1	12	280 - 660	29 - 41	0.002 - 0.19
halus	2-6	8	300 - 600	30 - 40	0.004 - 0.16
striatus	7	11	240 - 1120	31.5- 51	0.008 - 0.30
	8	10	120 - 360	21 - 35.5	0.32 - 3.6
Mystus	3	1	660	40	0.10
nemurus	5	1	900	41	0.08
Notopterus chitala	3	3	500 - 2080	38 - 63	0.07 - 0.12
Charias macroce-	1	1	280	30	0.04
phalus	6	3	300 - 400	32 - 33	0.003 - 0.004
Pangasius pangasius	1	1	4540	64	0.23

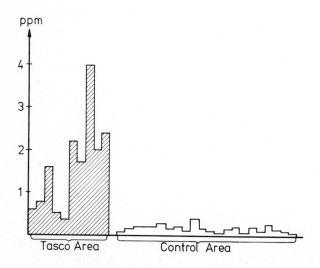

Fig. 5.1. Mercury contents in the flesh of the fish Ophiocephalus striatus
collected from the vicinity of the TACSCO caustic soda factory, compared
with fish from other parts of Thailand (control area).

That severe legislation is of great importance as illustrated in Fig. 5.2, where the decrease in mercury contamination was pronounced after the use of mercury in fungicides was banned in 1969 (compare also with Fig. 5.3).

How widely the use of all metallic compounds is spread is illustrated in Table 5.4. Even the sludge from biological treatment plants receiving only municipal waste water contains metals in minor concentrations, although sludge from plants treating mixed minicipal and industrial waste-water has a higher content of metals originating from specific industries (see Table 5.5). This indicates that industrial waste-water, which contains substantial amounts of heavy metals, should be treated on the spot.

Table 5.6 gives a survey of sources and effects on health of some widely used metals. Table 5.7 shows the electronegativity of carcinogenic metals.

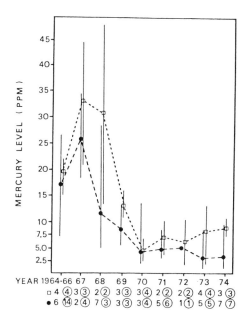

Fig. 5.2. Mercury levels (ppm) in feathers of adult eagle owls from southeast Sweden. ¤ annual means for the coastal population. • annual means for the inland population. The range in each year is also given. The figures below the year classes show the number of nests from which feathers were collected. The figures in circle represent the number of feathers analyzed.

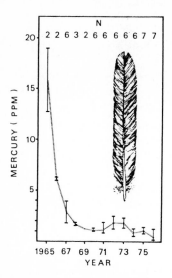

Fig. 5.3. Mercury levels in tail feathers of young marsh harrier in Kvismaren, in central Sweden 1965-1976. The levels are given in ppm on a dry weight basis. The number of nests (N) from which feathers from young were analyzed is given in the upper part of the graph. The intervals between the highest and the lowest levels are given annually.

Two factors of major importance for future trends of heavy metal pollution should be mentioned.

The current increase in the price of oil has accelerated the use of coal as an energy source. However, coal ash is rich in heavy metals as can be seen from Table 5.8. The concentration of metals in ash is not only dependent on the composition of coal, but is also influenced by the combustion temperature, the filter efficiency, etc. Table 5.9 shows the average concentration of several elements in coal with indications of their typical ranges. Their emission can be found as follows:

Emission = consumption of coal * concentration * emission factor.

Typical emission factors are given in Table 5.10.

TABLE 5.4
A.   Global heavy metal pollution

Element	Extracted per year (10^6 t y^{-1})	Transported to the sea (10^6 t y^{-1})	Washout by precipitation (10^6 t y^{-1})
Pb	3.5	0.74	0.3
Cu	6	0.25	0.2
V	0.02	0.03	0.02
Ni	0.5	0.01	0.03
Cr	2	0.04	0.02
Cd	0.01	0.0005	0.01
As	0.06	0.07	
Hg	0.009	0.003	0.08
Ca	5	0.7	
Se	0.02	0.007	
Ag	0.01	0.01	

B.   Heavy metal transportation in the River Rhine

Element	t/year	Ratio:	$\dfrac{\text{conc. in the Rhine}}{\text{conc. in the North Sea}}$
Cr	1000		20
Ni	2000		10
Zn	20,000		40
Cu	200		40
Hg	100		20
Pb	2000		700

TABLE 5.5
Concentration of heavy metals in municipal sludge from mechanical-biological treatment plants with more than 5,000 p.e. (Denmark) (g t^{-1} dry matter)

Locality	Cr	Ni	Co	Zn	Cd	Cu	Pb	Hg	Ag	Bi
Vejen	50	49	8	1386	10	123	218	5.2	13	<25
Græsted	46	17	4	1538	6	190	261	2.7	14	75
Farum	56	22	8	1487	9	284	305	6.8	27	<25
Ringe	42	20	6	1586	7	232	293	6.5	19	<25
Randers V	70	37	6	1545	10	186	332	3.6	12	<25
Sorø	43	19	7	2105	8	263	296	4.4	38	<25
Sønderborg N	40	18	4	2446	10	298	239	5.3	18	<25
Rungsted	51	18	4	2694	8	322	396	4.8	23	200
Slagelse	47	23	2	2896	7	309	222		32	<25
Hillerød	56	28	6	1778	9	219	261	12.7	97	<25
Odense, Ejby	268	50	3	3484	12	302	401	5.0	55	<25
Ringkøbing	37	22	3	1687	7	220	1164		39	<25

TABLE 5.5 - continued

Locality	Cr	Ni	Co	Zn	Cd	Cu	Pb	Hg	Ag	Bi
Kjellerup	85	37	3	1218	6	106	188	21.7	24	<25
Ringsted	76	21	4	2288	5	273	412	11.9	56	<25
Usserød	179	25	10	4235	7	336	229	6.0	45	75
Århus Viby	163	33	4	3665	10	514	317	32.5	100	<25
Odense NV	525	34	3	3941	13	771	768	5.1	18	<25
Roskilde	3575	218	3	4165	24	377	367	3.2	41	150
Herning	808	327	3	17414	30	217	322	5.0	21	<25
Ålebækken	1068	212	8	2302	11	2264	3898	27.8	87	50
Anlæg Y	201	100	3	4657	58	571	2517	9.3	55	<25
Lundtofte	825	251	59	2532	51	1568	2774	20.5	56	<25

TABLE 5.6
Sources and health effects of some widely used metals

Element	Sources	Health effects
Class 1: Pose serious threats now		
Cadmium	Burning of coal, zinc mining water mains & pipes, tobacco smoke	Cardiovascular disease and hypertension in humans
Lead	Auto exhaust (leaded gasoline) and paints (made before 1948)	Brain damage, convulsions, behavioral disorders, death
Mercury (as methyl mercury)	Burning of coal, electrical batteries, and many industrial uses	Nerve damage, death
Nickel (as nickel carbonyl)	Diesel oil, residual oil, burning of coal, tobacco smoke, chemicals and catalysts, steel and non-ferrous alkyls, gasoline additives	Lung cancer
Beryllium	Burning of coal and increasing industrial use (includes nuclear power industry and rocket fuel)	Acute and chronic respiratory diseases, lung cancer, beryllosis
Class 2: Potential hazards if levels increase		
Antimony	Industry, typesetting, enamel ware	Shortened life span in rats, heart disease
Arsenic	Burning of coal and oil, detergents, pesticides, mine tailings	Cumulative poison at high levels, may cause cancer in man
Selenium	Burning of coal, oil and sulphur, some paper products	Essential to man in trace amounts but rising use could increase levels of exposure, causes cancer in rats, may cause dental caries in man
Manganese	Metal alloys, smoke suppressant in power plants, may be used in place of lead as an anti-knock agent in gasoline	Essential to man in trace amounts but if it replaces lead in gasoline, levels could rise and imperil health from nerve damage

TABLE 5.7
Electronegativity of carcinogenic metals

Element	Chemical carcino- genicity	Electro- nega- tivity	Element	Chemical carcino- genicity	Electro- nega- tivity
Cs	none	0.7	Ni	positive	1.8
K	none	0.8	Sn	positive	1.8
Ru	none	0.8	Pb	positive	1.8
Na	none	0.9	Mn	suspected	1.8
Ba	none	0.9	Fe	suspected	1.8
Li	none	1	Si	suspected	1.8
Ca	none	1	Mo	none	1.8
Sr	none	1	Tl	none	1.8
La	none	1.1	Ge	none	1.8
Y	suspected	1.2	Cu	suspected	1.9
Ha	none	1.2	Te	none	1.9
Sc	suspected	1.3	Re	none	1.9
Zr	positive	1.4	Hg	none	1.9
Be	positive	1.5	Sb	none	1.9
Cr	positive	1.5	Bi	none	1.9
Al	suspected	1.5	As	suspected	2
Ti	suspected	1.5	Te	none	2.1
Zn	positive	1.6	Rh	suspected	2.2
Ga	suspected	1.6	Ru	none	2.2
V	none	1.6	Os	none	2.2
Cd	positive	1.7	Ir	none	2.2
W	none	1.7	Pt	none	2.2
In	none	1.7	Se	positive	2.4
Co	positive	1.8	Ag	none	2.4
			Au	none	2.4

As the consumption of coal is increased emission of harmful elements will also increase, unless the emission factor is reduced. A radical reduction in the emission factor is absolutely essential if a drastic increase in the global pollution of heavy metals is to be avoided. The second factor is the increasing cost of many metals due to speculation on price fluctuations and shortage of easily mined ores. This has caused a growing interest in the recovery of metals, which must be considered a positive development from an ecological point of view. Table 5.11 lists the percentage recovery for a number of widely used metals in the U.S. and Sweden.

TABLE 5.8
Concentration of metals in coal ash

Element	mg per kg
As	10 - 2000
Ag	1 - 100
B	20 - 2000
Be	1 - 200
Bi	5 - 100
Co	5 - 500
Cu	10 - 3000
Cd	1 - 200
Cr	3 - 2000
Mo	2 - 600
Hg	0.01 - 50
Pb	10 - 7000
Ni	3 - 1300
Mn	30 - 4000
Zn	30 - 10,000
Sb	1 - 500
Se	1 - 700
V	30 - 10,000
Tl	4 - 100
W	0.1 - 20
Te	15 - 60
U	10 - 60
Sr	10 - 10,000
Zr	10 - 2000

TABLE 5.9
Concentration of metals in coal (typical ranges are indicated in brackets)

Element	mg per kg
As	5 (1-15)
Ag	0.5 (0.1-1)
B	50 (10-200)
Ba	100 (20-400)
Be	1 (0.1-3)
Bi	1 (0.1-2)
Cd	1 (0.3-10)
Co	10 (1-20)
Cr	20 (5-30)
Cu	15 (5-40)
Ga	4 (1-10)
Hg	0.2 (0.05-1)
La	10 (0.1-40)
Mn	50 (10-100)
Mo	4 (1-10)
Ni	20 (5-50)
Pb	15 (2-70)
Sb	1 (0.2-2)
Se	2 (0.5-5)
Sn	3 (0.1-5)
Sr	100 (20-200)
Tl	2 (1-4)
U	1 (0.1-5)
V	30 (10-50)
Zn	30 (10-100)
Zr	100 (10-200)

TABLE 5.10
Typical emission factors (mg kg^{-1})

Element	value	Element	value
Ag	0.01	La	0.005
As	0.04	Mn	0.005
Ba	0.002	Mo	0.01 (0.001-0.1)
Be	0.003	Ni	0.02 (0.01-0.2)
B	0.05	Pb	0.03 (0.02-0.08
Cd	0.03	Sb	0.03
Co	0.01	Sc	0.05
Cr	0.01	Se	0.15 (0.1-0.4)
Cu	0.01	Sn	0.002
Hf	0.003	U	0.007
Hg	0.9 (0.5-0.98)	V	0.01
		Zn	0.03

TABLE 5.11
Percentage recovery of metals in U.S. and Sweden (relative to production)

Metal	%U.S.	% Sweden	Metal	%U.S.	% Sweden
A l	5	5	Mo		1 2
Au	9	3 2	Ni	2 5	1 4
Fe	2 5	2 7	Ag	3	3 9
Co	2		Sn	2 0	5 0
Cu	2 0	2 3	Pb	3 4	4 5
Cr	9	1 2	T i	3 3	
Mg	0.3		W	3	
Mn		9	Zn	6	6

References related to the tables:
Anderson and Smith (1977), Baby (1975), Bencko and Symons (1977), Billings and Matson (1972), Block and Dams (1976), Bolton et al. (1975), Bouilding (1976), Coles et al. (1979), Davidson et al. (1974), Eriksson and Jernelöw (1978, Friberg (1977), Gladney et al. (1978), Gutenmann et al. (1976), Joensuu (1971), Kalb (1975), Kantz et al. (1975), Klein et al. (1975a), Klein et al. (1975b), Klein and Russel (1973), Kaakinen et al. (1975), Lindberg et al. (1975) Linston et al. (1976), Magee et al. (1973), Natusch (1978), Natusch et al. (1974), Piperno (1975), Ragaini and Ondov (1975), Schwitzgebel et al. (1975), Smith et al. (1979a), Smith et al. (1979b), Swaine (1977), van Hook (1978), von Lehmden et al. (1974).

5.5 WATER POLLUTION PROBLEMS.

Water pollution problems can be divided into four groups:

1. The discharge of biodegradable organic waste, which consumes oxygen by the processes of mineralization. This problem has been discussed in detail in paragraph 2.6.
2. The discharge of nutrients from fertilizers, which might cause eutrophication problems, as discussed in paragraph 2.7.
3. The discharge of hydrocarbons into marine ecosystems. Massive oil pollution, as observed by a few catastrophies (for example, in the North Sea in 1976, in Brittany in 1977, and in the Gulf of Mexico in 1979) damages the entire ecosystem involved. Experience from the two former catastrophies shows, that it takes years for the ecosystem to recover and even though all the possible technical equipment and methods have

been applied to remove the oil. At this stage it is not possible to state whether the ecosystem will suffer any long-term effects. In addition to this problem is the effect of the continuous discharge of small amounts of oil into the oceans. Although facilities are available in most ports, all over the world, for oil deposition, many tankers and oil-industries discharge oil directly into the sea. The possible effects of this pollution problem include

a. An alteration in the rate of thermal energy transfer between the oceans and the atmosphere due to the formation of a thin oilfilm.
b. An alteration in the rate of oxygen transfer between the atmosphere and the oceans.
c. Increased cancer risk as some hydrocarbons are carcinogenic.
d. Damaged biochemical signals in some species of aquatic animals.
e. Possibel biomagnification of hydrocarbons.

4. The discharge of toxic compounds, such as pesticides and PCB, heavy metals and others (see 5.3 and 5.4).

5.6. NOISE POLLUTION.

Everyday noise or unwanted sound is assaulting citizens of all big towns. This is a very subjective form of pollution and one which runs into the problem of human value judgements. In general, however, sounds classified as noise include 1) loud sounds, 2) unpleasant sounds and 3) sudden sounds.

Noise is most commonly measured in decibels (db). This unit indicates the loudness, or sound intensity, which is related to the pressure on the ear. Since high frequency tones are more annoying, another scale has been introduced, which gives more weight to the high frequency tones and less weight to the low frequency tones. This scale is represented by dbA (A indicating the weighting used). The db and dbA scales are both logarithmic so that a rise of 10 db represents a tenfold increase in sound intensity, and a rise of 30 db represents a one thousandfold increase in loudness.

Table 5.12 lists some common noise levels to illustrate the scale.

TABLE 5.12
Common noise levels

Examples	Decibel (dbA)	Relative sound intensity	Effects with prolonged exposure
Jet takeoff (close range),	150	1,000,000,000,000,000	Eardrum raptures
Aircraft carrier deck	140	100,000,000,000,000	
Armoured personnel carrier	130	10,000,000,000,000	
Thunderclap, jet takeoff (200 feet)	120	1,000,000,000,000	Human pain threshold
Steel mill, live rock music, riveting, autohorn (3 feet)	110	100,000,000,000	
Jet at 1,000 feet, subway,outboard motor, power mower, motorcycle (25 feet, farm tractor, printing plant, jackhammer, blender	100	10,000,000,000	Serious hearing damage (8 hours)
Busy urban street, diesel truck, garbage disposal, clothes washer	90	1,000,000,000	Hearing damage (8 hours)
average factory, freight train (50 feet), noisy office, dishwasher	80	100,000,000	
Freeway traffic (50 feet), vacuum cleaner	70	10,000,000	Annoying
Conversation in restaurant, typical suburb	60	1,000,000	Intrusive
Quiet suburb (daytime), conversation in living room	50	100,000	Quiet
Library	40	10,000	
Quiet rural area (nighttime)	30	1,000	
Whisper, rustling leaves	20	100	Very quiet
Breathing	10	10	
	0	1	Threshold of audibility

5.7. SOLID WASTE POLLUTION.

The amount of waste per capita in the developed countries has increased rapidly during the last few decades. In the richer countries the average per capita amount has now reached approximately 20 t per year, including waste produced both directly and indirectly.

The direct production of solid wastes in Western Europe and North America is around 600 kg per capita per year:

Paper and paper products	260 kg/capital/year
Metals	130 kg/capital/year
Glass	110 kg/capital/year
Organic matter (excl. paper)	60 kg/capital/year
Ash	30 kg/capital/year
Other	20 kg/capital/year

In addition to direct production, waste is produced indirectly from agriculture and industry:

Agricultural wastes	8000 kg/capital/year
Mining and mineral wastes	6500 kg/capital/year
Industrial wastes	500 kg/capital/year

TABLE 5.13
Summary of advantages, disadvantages and cost for various methods of waste disposal now in use

Erratum:

Due to a technical error Table 5.13 on page 274 was ommitted. It is printed here below:

TABLE 5.13

Summary of advantages, disadvantages and cost for various methods of waste disposal now in use

Methods	Advantages	Disadvantages	Cost*
Littering	Easy (throwaway mentality)	Unsightly, expensive to clean up, wastes resources	No direct cost but removal and disposal cost from $40 to $4,000 per ton because it is spread out
Open dumps	Easy, the cheapest official method if external costs are not considered	Unsightly, breeds pests, causes air pollution when wastes are burned, smells bad, wastes resources, wastes land, can contaminate ground water and nearby streams from leaching and runoff	$1.50 to $3 per ton plus collection
Sanitary landfill	Can be attractive with no objectionable odors or pests, cheapest acceptable method, ground water and stream pollution minimized, can be set up in a short time, can receive all kinds of wastes, may enhance value of submarginal land (when fill is completed, land can be converted to other uses),¹ should be restricted to areas where precipitation is light, water tables low, and low cost land available	Wastes resources, land intensive (land may not be available near urban area), site must be carefully selected so that ground water is not polluted, ecologically valuable marshes and other wetlands may erroneously be considered "useless" and thus suitable for filling filled land may settle so periodic maintenance is required, difficult to locate suitable areas because of citizen opposition	$3.50 to $5 per ton plus collection
Incineration	Can handle about 80% of urban wastes, reduces volume of wastes by about 90%, requires little land, can produce some income from metal and glass salvage and use of waste heat	Large initial cost to build facilities, causes air pollution unless sophisticated pollution controls installed, some kinds of wastes (such as plastics)⁵ can damage equipment, frequent maintenance, costly repairs, skilled operators required	$6 to $12 per ton plus collection
Composting	Converts organic wastes to fertilizer for reuse on land	Relatively expensive at present, can be used only for organic wastes, more expensive than synthetic fertilizer, wastes have to be separated	$4 to $8 per ton plus collection and separation

* These costs vary widely with location and have been rising rapidly. They are shown for relative comparison only. Figures are for 1969 from U.S. Department of Health, Education, and Welfare and Environmental Protection Agency. Collection costs in urban areas are estimated at about $20 to $30 per ton.

Most of these wastes are dumped (more than 50% or left uncollected (24%). Sanitary landfills use 10-15% of the wastes, while 5-10% is incinerated. Only a relatively minor amount is composted, a method which from an ecological point of view must be considered the most attractive for organic wastes such as paper and sludge produced in biological plants treating municipal waste water. The ecological and economic advantages and disadvantages of these various disposal methods are summarized in Table 5.13. Notice, as is characteristic for many environmental problems, *that some of the methods solve one problem but simultaneously create another.* In environmental management it is therefore vital to attempt to find total solutions, and to take all aspects of an environmental problem into consideration.

5.8 FOOD ADDITIVES.

More than two thousand different chemicals are used as food additives in most developed countries, and it is vital to determine whether there are potentially harmful chemicals in both natural and synthetic foods. Unfortunately, the answer is not simple because 1) individuals vary widely in their suspectibility to different chemicals, and 2) otherwise harmless chemicals may interact synergistically. Extensive tests for a single food additive or drug are therefore complicated, can take up to 8 years to complete and are very expensive.

Table 5.14 summarizes some of the major classes of food additives. In most countries a positive list principle for food additives is used, meaning that only chemicals included in an authorized list can be used as food additives.

Although it is safer to use a positive than a negative list (i.e. one which lists chemicals that must not be used), knowledge about the effect of chemicals on man is very limited, because the problem is very complex.

In addition to the problems of food additives are several pollution problems related to food production, such as:

1. Accidental contamination, for example with residues of DDT, PCB, vinylchloride, heavy metals, etc. (see also 5.3 and 5.4).
2. Bacterial contamination that can cause Salmonella food poisoning and botulism.

TABLE 5.14

Some common food additives, processes and contaminants

Class	Function	Examples	Some typical uses
Preservatives	To retard spoilage from bacterial action and molds (fungi), especially in foods containing carbohydrates and proteins	Processes: drying, smoking, curing, canning process (heat and sealing), dehydration, freezing, pasteurization, refrigeration Chemicals: salt, sugar cure, sodium nitrate, sodium nitrite, calcium and sodium propionate, sorbic acid, potassium sorbate, benzoic acid, sodium benzoate, citric acid, sulfur dioxide	Breads, cheeses, cakes, jellies, chocolate syrups, fruits, vegetables, meats
Anti-oxidants (oxygen interceptors or freshness stabilizers)	To retard spoilage of fats by excluding oxygen or slowing down the rate of chemical breakdown of fats (rancidity)	Processes: sealed cans, wrapping, refrigeration Chemicals: lecithin, butylated hydroxyanisole (BHA), butylated hydroxytoluene (BHT), propyl gallate	Cooking oils, shortenings, cereals potato chips, crackers, salted nuts, soups, Pop Tarts, Dream Whip, Tang, and many other foods
Nutrition supplements*	To increase nutritive value of natural food or to replace nutrients lost in food processing †	Vitamins and essential amino acids	Bread and flour (vitamins and amino acids), milk (vitamin D), rice (vitamin B-1), corn meal, cereals
Flavors and flavor enhancers	To add or enhance flavor	Over 1,100 substances including saccharin, monosodium glutamate (MSG), essential oils such as cinnamon, banana, vanilla, bitter almond	Artificial flavors for ice cream, artificial fruit juices, toppings, soft drinks, candy, pickles, salad dressings, spicy meats, low-calorie foods and drinks (sweeteners), and most processed heat-and-serve foods
Coloring agents	To add color for esthetic or sales appeal or to hide colors that are either unappealing or show a lack of freshness	Natural color dyes and synthetic coal tar dyes	Soft drinks, butter, cheese, ice cream, breakfast cereals, candies cake mixes, sausages, puddings and many other foods
Acidulants	To provide tart taste or mask undesirable aftertastes	Phosphoric acid, citric acid, fumaric acid	Cola and fruit soft drinks, desserts fruit juices, cheeses, salad dressings, gravies, soups
Alkalis	To reduce natural acidity	Sodium carbonate, sodium bicarbonate	Canned peas, some wines, olives, coconut cream pie, and chocolate eclairs
Emulsifiers	To disperse droplets of one liquid (such as oil) in another liquid (such as water)	Lecithin, propylene glycol, mono- and diglycerides, polysorbates	Ice cream, candy, margarine, cake icings, nondairy creamers, dessert toppings, mayonnaise, salad dressings, shortening
Stabilizers and thickeners	To provide smooth texture and consistency, prevent separation of components, and provide body	Vegetable gums (gum arabic, gum ghatti, and others), sodium carboxymethyl cellulose, seaweed extracts (agar, algin), dextrin, gelatin	Cheese spreads, ice cream, sherbet, pie fillings, salad dressings, icings, dietetic canned fruits cake and dessert mixes, syrups, pressurized whipped creams, instant breakfasts, beer, soft drinks diet drinks
Sequesterants (chelating agents or metal scavengers)	To tie up traces of metal ions that catalyze oxidation and other spoilage reactions in food, to prevent clouding in soft drinks, and to add color, flavor, and texture	EDTA (ethylenediamine-tetraacetic acid), citric acid, sodium phosphate, chlorophyll	Soups, desserts, artificial fruit drinks (Tang, Awake), salad dressings, canned corn and shrimp, soft drinks, beer, cheese, canned frozen foods
Contaminants	Not deliberately added to foods	DDT, PCBs, compounds of mercury, lead, and other heavy metals, radioisotopes, bacteria from poor hygiene and improper storage and processing, insects	Can occur in a wide variety of foods depending on accidental exposure or improper food processing

5.9.   ALTERNATIVE ENERGY.

Most of the energy used today comes from finite sources, such as nuclear energy and fossil fuel. In addition, these energy sources are connected with serious pollution problems. The problems associated with the use of fossil fuels have already been mentioned in 2.5 and 3.5.

There is general agreement that nuclear fission is potentialy the most harzardous of all energy sources in use today. It is not possible here to enter into a debate on the nuclear energy dilemma, as it requires answers to several questions, about which the experts are not in agreement at present. These questions include:

1. what are the problems and the risks of storing nuclear wastes for long periods of time?
2. what are the short- and long-term risks and benefits of energy produced by nuclear fission as compared to other energy sources?
2. how safe are nuclear power plants?
4. what are the consequences of radioactive fuel escapes?

One major disadvantage of today's nuclear reactors is that they consume our limited supply of uranium-235 fuel. Most experts agree that the supply of relatively low cost uranium is very limited, while there are relatively large amounts of medium and high cost uranium. That is the background for current research program aimed at developing a so-called fast breeder reactor, which uses a mixture of abundant non-fissionable uranium-238 and fissionable plutonium-239 produced by present reactors. Under bombardment with fast neutrons, plutonium-239 undergoes fission and the uranium-238 is converted to plutonium-239. The net results is a hundredfold mulitplication of our usable uranium reserves.

Fast breeder reactors using only known uranium resources could provide all our electricity needs for at least 50,000 years. However, there are several technical problems to be solved before large-scale breeder reactors become commercially feasible. An even more serious drawback is the fact, that each typical commercial reactor will contain over 1 metric ton of plutonium, which is an extremely dangerous element. Only 30 g of plutonium would be needed to kill every human being on earth if it was widely dispersed.

Plutonium-239 emits alpha-particles, which have a low penetration power, but inhalation, ingestion or absorption of only a few particles can eventually cause cancer or death.The present maximum tolerance limits of 2 µg plutonium are 150,000 to 300,000 times too high because of its long half-life of 24,000 years. In addition plutonium must be stored for 200,000 years or more, before it decays sufficiently to be released with safety into the environment. Consequently it can be concluded that the breeder reactor is not a safe overall solution to the coming energy demand.

A long-term solution must rely on essentially infinite and unpolluting energy sources, such as nuclear fusion, solar energy, geothermal energy and wind power. These energy alternatives will be mentioned briefly below, and their advantages and disadvantages surveyed.

Advantages of fusion energy:
1. Its fuel sources are essentially infinite and cheap. The deuterium in the oceans could supply all the world's power for several billion years.
2. It is much less dangerous than fission energy. The amount and hazard of the radioactive materials produced would be less than with conventional fission reactors.
3. Fusion energy could be used for the cheap production of hydrogen gas, which could serve as a cleaner replacement for natural gas and gasoline (petrol).

Disadvantages of fusion energy:
1. Release of radioactive tritium either as gas or in water. This pollution problem requires a solution.
2. Thermal pollution equal to or slightly lower than that from fossil fuel plants.
3. It will require several decades of research before all the technical problems involved in the use of this energy source can be managed safely and effectively.

Advantages of solar energy:
1. It is an infinite and readily available energy source.
2. It is the cleanest and safest energy source of all.
3. No entropy is built up in the atmosphere.
4. It could be used to produce hydrogen gas as a replacement for natural gas and gasoline (petrol).

Disadvantages of solar energy:
1. Its economic feasibility on a large scale is a major problem today, but further development during the next two decades might improve the efficiency sufficient to solve this problem.
2. Solar energy is not directly available at night when the needs for electricity are higest. Thus we must have a method of storing energy received in the daytime to solve this problem in other ways.

Advantages of geothermal energy:
1. Almost infinite energy source.

2. Easily converted into electricity.

Disadvantages of geothermal energy:
1. Geothermal deposits are located at specific underground sites.
2. Mineral-laden water wastes are produced.
3. Present techniques limit the available resources.

Advantages of wind power:
1. Almost infinite source.
2. Relatively clean and safe.
3. Easily converted into electricity.

Disadvantages of wind power:
1. Relatively high cost at present technical level.
2. Causes landscape pollution on a large scale.

In this context, however, we should also mention a fifth energy alternative. Better insulation and improved efficiency in the conversion of energy from one form to another, in homes as well as in industry, could cut the energy consumption more than 50% of the present level. The interrelation between pollution problems and energy consumption underlines the possibilities. Increased consumption of energy will create more pollution, so to reduce pollution, energy exploitation must be reduced or eliminated.

PRINCIPLES OF ENVIRONMENTAL TECHNOLOGY

As mentioned in the introduction, this part consists of 4 chapters: Water Pollution Problems, including Water Resource Problems, Air Pollution Problems, Solid Waste Problems, and Examination of Pollution.

In connection with each environmental problem, the source of pollution will be given. All technological methods will be mentioned with special emphasis on their principles and an overall evaluation of their environmental consequences.

Principles and processes are marked in the text as in Part A.

Selection of methods will be discussed with reference to the principles mentioned in Part A. Environmental technology is often based upon the same processes as found in nature for elimination of pollutants; in this case, reference to the natural process already mentioned in Part A will be given.

After each chapter there is a list of questions or problems, which to a certain extent are linked with material presented in Part A.

Generally there are 4 principal methods for solving pollution problems:

A. To reduce the amount (energy and/or material) discharged by use of alternative technology. It may often be necessary to enforce the use of alternative technology by legislation. For example, reduction of sulphur in fossil fuel and lead in gasoline.

B. To recycle or reuse waste products. This method is very attractive from an environmental point of view, as a resource and an environmental problem are solved simultaneously. For example, recovery of chromium from waste-water and production of animal feed from slaughterhouse waste.

C. To decompose the waste to harmless components. For example, biological treatment of municipal waste-water.

D. To remove the waste for harmless deposition at another location. For example, use of domestic waste as a soilconditioner.

All 4 methods are not equally applicable to all environmental problems, but they will be mentioned because it is significant in environmental technology to be receptive to new and non-traditional solutions. The relationship between environmental issues and the other serious problems mankind is facing - shortage of resources, the energy crisis and ever-increasing population growth - make it absolutely essential to seek new ways.

CHAPTER 6

WATER AND WASTE WATER PROBLEMS

6.1. INTRODUCTION TO THE PROBLEMS OF WATER AND WASTE WATER.

Water pollution problems were surveyed in section 5.5, and this chapter is devoted to linking these problems with the various water treatment methods available.

Selection of the optimum waste water and water treatments is a very complex problem. Quantitative management requires the application of the principles presented in Part A of this book, often in the form of an ecological model. The search for an optimum solution also requires a comprehensive knowledge of the treatment methods available. Furthermore, the problem is complicated by the interdependence of water supply and waste water disposal.

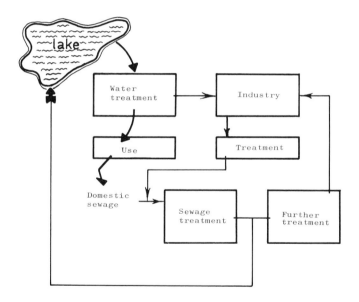

Fig. 6.1. Example of good water management practice, where 1) a part of the treated waste water is reclaimed for industries after an additional treatment, 2) The industrial waste water is treated before being discharged into the public sewage system. The sewage treatment chosen pays regards to as well the lake as the water treatment.

Good water management practice should therefore consider not only the ecological effects in the receiving body of water, but also the effect on the quality and economy of the water supply. Fig. 6.1 illustrates these considerations.

Effluents fall into seven groups, but many have, to a greater or lesser degree the polluting properties of at least two of these categories:

1. **Organic residues,** including domestic sewage, effluent from food-processing-industries, ensilage, manure heaps and cattle yards, laundries, paper mills, etc. These effluents vary a great deal, but they have much in common. They all contain complex organic compounds in solution and/or suspension, sometimes with toxic substances and various salts. Their basic property, however, is that they contain unstable compounds, which are readily oxidized and so use up the dissolved oxygen in the water. Some of these compounds are more readily decomposed than other: for example, slaughterhouse wastes oxidize rapidly while wood pulp is comparatively stable. Section 6.2 focuses on the technical solutions to this problem, while section 2.6 in Part A deals with the effects of organic residues on ecosystems.

2. **Nutrients,** including ammonia, nitrates, other nitrogenous compounds, orthophosphates, other phosphorous compounds, silica and sulphates. The main sources are domestic sewage and effluents from fertilizer manufacture. Discharge of nutrients may cause undesirable eutrophication as described in section 2.8. Section 6.3 discusses the methods available for nutrient removal.

3. **Poison** in solution occur in the waste waters from many industries. They include acids, alkalis, oil, heavy metals and toxic organic compounds, mainly from chemical industries, gas works and use of insecticides. Their effects on ecosystems are described in section 2.13. 6.5 reviews the methods available for the removal of heavy metals and 6.4 deals with removal of toxic organic compounds.

4. **Inert suspensions** of finely divided matter result from many types of mining and quarrying and from washing processes, such as those of coal and root crops. The effect of these pollutants on the ecosystem can be evaluated from the principles mentioned in chapter 2. Removal of inert finely divided matter can be carried out by mechanical treatment methods mentioned in 6.2.2.

5. **Other inorganic agents,** such as salts or reducing agents (e.g. sulphides, sulphites and ferrous salts) occur as constituents of the effluent of several types of industry. Minor discharges of salts are generally harmless to the environment, but reducing compounds use up the oxygen in the receiving body of water, and have the same effect as organic residues (see 2.6). This effect can, however, easily be eliminated by aeration, a process which will only be discussed in relation to

biological treatment methods (see 6.2.3). For a more comprehensive account, see Jørgensen (1979).

6. **Hot water** is produced by many industries that use water for cooling purposes. They often use river water, which is pumped through the cooling system and sometimes raised to very high temperatures during part of its journey. The effects of this process are described in section 2.4. The methods available to meet this problem are a) use of cooling towers, b) use of heat exchangers, c) use of alternative technology, d) use of alternative receiving water body which is less susceptible to damage. These solutions will not be discussed in this context, as they either must be considered as purely industrial engineering problems, or have already been mentioned in chapters 2 and 4.

7. **Bacteriological contamination** of waters originates mainly from domestic sewage, but food-processing industries, manure heaps and cattle yards are also sources of this type of pollution. Methods to meet this problem are used in the production of potable water, for process water in industry and in the treatment of waste water. The available methods are mentioned in 6.6.5.

Waste waters emanate from four primary sources: 1) municipal sewage, 2) industrial waste waters, 3) agricultural runoff and 4) storm water and urban runoff. The problem of the first group is related mainly to organic residues and nutrients, while the second group encompasses the entire spectrum of pollution problems, although discharge of heavy metals and organic compounds are the most serious ones. It is not possible to give comprehensive review on industrial waste water problems here; for detailed discussion see Jørgensen (1979).

As municipal and industrial waste waters receive treatment, increasing emphasis is being placed on the pollutional effects of urban and agricultural runoff. The range of pertinent characteristics of these waste waters is given in Table 6.1.

In many places sewage continues to be discharged into systems of drains intended also for the removal of surface runoff from rainstorms and melting snow or ice. This is called combined sewerage. However, in most modern developments, sewage and runoff are each colleted into a separate system of sanitary sewers and storm drains in order to avoid pollution of water course by the occaaional spillage of sewage and stormwater mixtures. This is called separate sewerage.

Often the receiving body of water also serves as an important source of supply for many purposes. It is this multiple use of natural waters that creates the most impelling reasons for sound water-quality management, as already mentioned above. This part of the problem is discussed in section 6.6.

Agricultural pollution problems are related to:

1. the extensive use of natural and industrially produced fertilizers to increase yield,
2. use of pesticides to eliminate damage by pests
3. waste from domestic animals.

It seems only possible to solve problems 1) and 2) by use of sound ecological engineering, which has been treated in section 4.10. Waste-water and solid-waste problems related to 3) are, in principle, not different from municipal and industrial waste problems and are therefore touched on in this part of the book.

TABLE 6.1
Pollution from urban and agricultural runoff

Constituent	Urban runoff (Storm water)	Agricultural runoff
Suspended solids (mg/l)	5 - 1200	-
Chemical oxygen demand COD (mg/l)	20 - 610	-
Biological oxygen demand BOD (mg/l)	1 - 173	
Total phosphorus (mg/l)	0.02- 7.3	1.1- 0.65
Nitrate nitrogen (mg/l)	-	0.03- 5.0
Total nitrogen (mg/l)	0.3 - 7.5	0.5- 6.5
Chlorides (mg/l)	3 - 35	-

TABLE 6.2
Operations

Operation	Detailed description, see
Screening	6.2.2
Settling	6.2.2
Filtration	6.2.2
Flotation	6.2.2
Biological decomposition	6.2.3
Irrigation	6.2.5
Chemical precipitation	6.3.2 + 6.5.2
Flocculation	6.3.2 + 6.5.2
Nitrification	6.3.3
Denitrification	6.3.3
Stripping	6.3.4
Ion exchange	6.3.6 + 6.5.3
Algae ponds	6.3.7
Adsorption	6.4.3
Chemical oxidation	6.4.4
Chemical reduction	6.4.4 + 6.5.2
Extraction	6.5.4
Membrane processes	6.5.5

Almost all unit operations applied for water and waste water treatment are mentioned in this chapter. Some of the individual operations can, however, be used to solve more than one pollution problem. The general description of these operations are given in context with their area of major application, and can be found in accordance with Table 6.2.

6.2. REDUCTION OF THE BIOLOGICAL OXYGEN DEMAND.

6.2.1 The BOD_5-problem and its sources.

When organic matter is added to an aquatic ecosystem it is immediately attacked by bacteria, which break it down to simpler substances, using up oxygen in the process, see section 2.6. The rate of which a particular type of effluent is able, in the presence of ample oxygen, to satisfy its oxygen demand depends on what it contains. Industrial effluents which contains only chamical reducing agents, such as ferrous salts and sulphides, take up oxygen purely by chemical reactions. They do this very rapidly, exerting what is called immediate oxygen demand. Organic substances, such as carbohydrates, proteins, etc., become oxidized by the activities of bacteria. The rate at which they are broken down therefore depends first on the presence of suitable bacteria and second on how satisfactory and balanced a food they are for microorganisms. Compounds which are more or less refractory are decomposed at a low or very low rate. They will therefore not make a significant contribution to the BOD_5.

These compounds are more or less toxic to the aquatic flora and fauna, and section 6.5 deals with methods of removing toxic organics, which obviously cannot be removed by the same methods as biodegradable organic matter.

Readily biodegradable matter has a biological oxygen demand, which can be calculated theoretically by using the reaction scheme. Waste water containing 100 mg l^{-1} of glucose will have a BOD_5 of 110 mg l^{-1}, which is in accordance with the following reaction:

$$C_6H_{12}O_6 + 6O_2 \; - \; 6CO_2 + 6H_2O \qquad\qquad (6.1)$$

Glucose is easily broken down to CO_2 and H_2O, oxidation being complete in less than 5 days. However, other components readily oxidized but at a lower rate will not give a BOD_5 in accordance with the reaction of decomposition because a part of their mass is synthesized into new bacterial substances, which will not be broken down during the 5-days period.

For a more complex mixture of organic compounds the decomposition reaction might be written as follows:

$$COHN + O_2 \xrightarrow{\text{(cells)}} CO_2 + H_2O + NH_3 + cells \qquad (6.2)$$

As mentioned in 6.1 the major sources of organic residues are domestic sewage and the food industry. Table 6.3 gives the range of concentration of pertinent characteristics of domestic waste water in the case of separate sewerage.

TABLE 6.3
Concentration of pertinent characteristic of domestic waste water (mg l^{-1})

BOD_5	150 - 300
N_{total}	25 - 45
P_{total}	6 - 12

The consumption of water in the food industry is shown in Table 6.4, which demonstrates that the food industry makes a significant contribution to the overall pollution problem due to its high water consumption and relatively high BOD_5.

The composition of domestic sewage varies surprisingly little from place to place, although, to a certain extent, it reflects the economic status of the society. A typical organic composition of domestic waste water is given in Table 6.5. A more detailed analysis reveals that domestic sewage is a very well balanced food for microorganisms. It contains sufficient amounts of essential amino acids, nutrients and vitamins. Consequently its biological decomposition causes few problems; most difficulties with biological treatment plants are related to the discharge of more or less uncontrolled amounts of industrial waste water (see also 6.4.1 and 6.5.1).

TABLE 6.4
Pollution from the food-processing industry in Denmark (Denmark is a country with a highly developed food industry)

Branch	Production tons/year (round figures)	m^3 waste water/ton of product	Million m^3 waste water/year	BOD_5	10^5 *) person equivalents
Abattoirs	600,000	20	12	1500	8.0
Dairies	5,000,000	1.5	7.5	1800	6.0
Fish filletting plants	100,000	2	0.2	5500	0.5
Potato starch production	25,000	30	0.75	5500	1.9
Breweries	600,000	10	6	1500	4.0

*) 60 g BOD_5/24h per inhabitant

TABLE 6.5
Organic composition of domestic waste water

Organic constituent	Concentration		
	Soluble (mg/l)	Particulate (mg/l)	Total (mg/l)
Total Carbohydrate	30.5	13.5	44.0
Free Amino Acids	3.5	0	3.5
Bound Amino Acids	7	21.25	28.25
Higher Fatty Acids	0	72.5	72.5
Soluble Acids	22.75	5	27.75
Esters	0	32.7	32.7
Anionic Surfactants	11.5	4	15.5
Amino Sugars	0	0.7	0.7
Amide	0	1.35	1.35
Creatinine	3.1	0	3.1
Fraction sum	78.35	151	229.35
Present in Waste Water	94	211.5	305.5

Reduction of BOD_5 in domestic waste water and other types of waste with a similar composition is carried out by a combination of mechanical and biological treatment methods.

P.6.1. Mechanical methods remove suspended matter in one or more steps, with the aim of not only removing the coarser inorganic pariticles, but also reducing the BOD_5 in accordance with the amount of suspended organic matter.

In principle, the entire BOD_5-reduction could be carried out using biological treatment methods alone, but this will often prove more expensive, than a combined mechanical and biological treatment. The results of the combined treatment are outlined in 6.2.4.

6.2.2   Mechanical treatment methods.

P.6.2. Mechanical treatment methods comprise screening, sand trap, sedimentation, filtration and flotation.

Screening.   Screens are used for removing larger particles, such as branches, rage, etc. They are made from iron bars or gratings and can be classified according to the distance between the bars as coarse (40-100 mm) and fine (10-40 mm). Screens need to be cleaned frequently, either manually or mechanically by rakes. The purpose of the screens is to protect pumps and other mechanical equipment. The removed material is either composting or by incineration.

Sand traps. The sand trap or grit chamber removes mainly inorganic particles 0.1-3 mm in size. It operates by sedimentation, but due to a short retention time (10-20 minutes) and air stirring, the finer organic particles are prevented from settling. Fig. 6.2 shows a sand trap at a municipal sewage plant.

Fig. 6.2. A sand trap at a municipal sewage plant.

Sedimentation. Sedimentation is used to remove suspended solids from waste water. In principle, it is the same process as is used in nature, see section 2.6.

A settling tank has three main functions:
1. It must provide for effective removal of suspended solids so that its effluent is clear.
2. It must collect and discharge the subnatant stream of sludge.
3. It must thicken the sludge to a certain concentration of solid.

Three distinct types of sedimentation may be considered:
1. Discrete settling. This is the settling of a dilute suspension of particles which have little or no tendency to flocculate.

2. Flocculent settling, which occurs when the settling velocity of the particles increases as they fall to the bottom of the tank, due to coalescence with other particles.
3. Zone settling, which happens when interparticle forces are able to hold the particles in a fixed position relative to each other. In this case the particles sink as a large mass rather than as discrete particles.

Plug flow is never achieved in practice. Some of the particles will be short circuited and will therefore be held in the tank for a time less than V/Q, where V = the tank volume (m^3), and Q = the flow rate (m^3/h).

Wind effects, hydraulic disturbances and density and temperature effects will all result in a deviation from the ideal plug flow. Short circuiting in a tank can be characterized by tracer techniques. Dye, salt or radioactive materials are introduced into the inlet and the concentration distribution in the effluent stream indicates the flow patterns. Some typical curves for effluent concentration versus time are shown in Fig. 6.3.

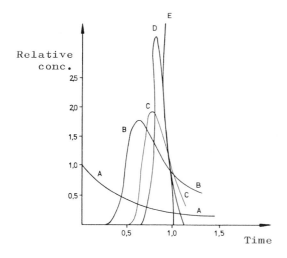

Fig. 6.3. Effluent concentration versus time.

The relative concentration 1.0 corresponds to the concentration achieved by complete mixing, and curve A shows the results of such an experiment, where the tank content was completely mixed. Curve B is typical of a wide shallow regular tank, while curve C represents the situation in a long narrow tank. Curve D represents a baffled tank and is, as shown, close to the ideal case for plug flow E.

The results of studies by Dague and Baumann (1961) are shown in Fig. 6.4. Centre and peripheral feed circular clarifiers were examined my means of dye dispersion. As can be seen, the difference between the two feeding methods is significant.

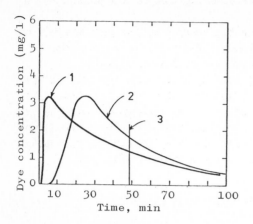

Fig. 6.4. 1) Feed in centre, 2) Feed in periphery, 3) Theoretical retention time.

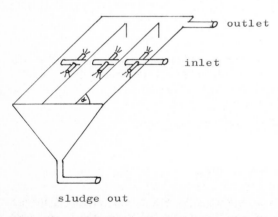

Fig. 6.5. Steeply inclined tube settler.

It is frequently possible to improve the performance in an existing settling tank by making modifications based on the results of a dispersion test. The addition of stream-deflecting baffles, inflow dividing mechanism and velocity dispersion feed wells may decrease short circuiting and increase efficiency.

Fig. 6.5 illustrates the principle of tube settlers. The design incorporates the use of very small diameter tubes in an attempt to apply the shallow depth principle as suggested by Camp (1946).

Flow through tubes with a diameter of 5-10 cm offers optimum hydraulic conditions and maximum hydraulic stability. Culp et al. (1968) have reported excellent results using tube settlers with a retention time of less than 10 minutes. The retention time can be calculated in accordance with the following equation:

$$Y_A = v_s \left(\frac{L}{S} \cos \beta + 1\right) \tag{6.3}$$

where

$$Y_A = \frac{Q}{A} = \frac{\text{flow rate}}{\text{area of tube settler}}$$

L = length of tube
S = distance between the tubes (the diameter of the tubes)
β = the angle of the tube to the horizontal (see Fig. 6.5)
v_s = direct settling rate

As can be seen from this equation, Q/A will increase as β decreases. It should therefore be an advantage to place the tubes as near as possible to horizontal. However, the horizontal settler is not self-cleaning and must be back-washed. Therefore, the steeply inclined 60° tube settler is more commonly used. Continuous gravity draining of settled solid might be achieved from tubes inclined of angles between 45 and 60°.

The clarifier may be designed as a rectangular or circular tank, and may utilize either centre or peripheral feed. The tank can be designed for centre sludge withdrawal or for withdrawal over the entire tank bottom. The different types of tank are shown in Figs. 6.6 and 6.7. The first one is designed for small flows, where the height of the tank is only moderate in spite of the angle of the cone.

The second clarifier is made of concrete and is able to deal with a considerably larger flow rate. An inlet device is designed to distribute the flow across the width and the depth of the settling tank, and correspondingly an outlet device is designed to collect the effluent uniformly at the outlet end of the tank.

It is very difficult to design a full-scale sedimentation tank based on settling experiments, as several important factors influencing particle behaviour in a full-scale operation are neglected in such experiments. Tanks are subject to eddies, currents, wind action, resuspension of sludge, etc. A full-scale clarifer will therefore show a slightly reduced efficiency compared to settling experiments, but this can be taken into consideration by incorporating a safety factor. The choice of an acceptable safety factor requires experience. The practical factor might vary form 1.5 when the tank is very small, baffled and protected form wind, to 3.0 in the case of a large tank, unbaffled and unprotected from wind. Even with the use of the safety factor, however, perfect performance should not be expected.

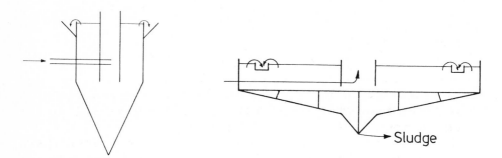

Fig. 6.6. Settling tank for small flow rattes.   **Fig. 6.7.** Settling tank for large flow rates.

Filtration. Water treatment by filtration uses principally either deep granular filters or precoat filters.

Deep granular filters are either silica sand or a dual medium or multi-media filters. A dual medium filter of coal over sand is widely used, and multi-media filters consisting, for example, of coal over silica sand over garnet sand, are finding increasing application. Precoat filters use diatomaceous earth, perlite or powdered activated carbon.

Sand filters were developed in England in the middle of the 19th century. These filters operated at a relatively low rate, between 0.1 and 0.3 m/h. Today the same filters are used at rates of up to 0.6 m/h, and are known as slow sand filters in contrast with the rapid sand filters developed later in the 19th century in the U.S.A., which operate with a filtration velocity of 3 to 10 m/h.

The present filters, which consist of a number of porous septa in a filter housing, have found wide application since the second World War. The septa support is a thin-layer filter medium, which is deposited on the outside of the septa at the beginning of the filtration cycle.

As mentioned above, sand filters can be divided into two classes - slow filters and rapid filters. There are two main differences between them:

1. As shown in Table 6.6, the properties of the filter media are different. The effective grain size is the diameter of the largest grain of sand in that 10% of the sample by weight which contains the smallest grains. The uniformity coefficient is the ratio of the largest grain in th 60% of the sample by weight which contains the smallest grain, to the effective size.

As can be seen rapid filters operate with a higher effective size and a smaller uniformity coefficient. The finer the sand used, the smaller will be the turbidity of the treated water and the flow rate.

TABLE 6.6
Typical properties of filter media

	Slow sand filter	Rapid sand filter
Effective size (mm)	0.45-0.60	0.6-1.0
Uniformity coefficient	1.50-1.80	1.2-1.8
Material	sand and/or cru-shed anthracite	multi media

2. Slow filters operate for 10 to 30 days. By then the head loss will be 1 m of water or more. The filtration is interrupted and 1.5 to 4 cm of the filter sand is removed. When the sand layer reaches a height of about 40 cm, new or washed sand is added to replace up to 30 cm of the sand layer removed. In rapid filtration, impurities are removed by back-washing, usually by reversing the flow of water through the filter at a rate adequate to lift the grains of the filter medium into suspension. The deposited material thus flushed up through the expanded bed is washed out of the filter.

The rapid filter can be either an open filter or a pressure filter. Open filters are mainly built of concrete, whereas pressure filters are water-tight steel tanks which are usually cylindrical and may stand either horizontally or vertically. The most common use of pressure filters is in small cities treating ground water supplies for iron and manganese removal, in swimming pool filtration or for polishing industrial water.

The filtration cycle by use of precoat filtration consists of three steps:
1. Precoating
2. Filtration
3. Removal of the spent filter cake

A precoat thickness of 1.3-3 mm is generally used. During filtration the suspended solids are removed on the precoat surface resulting in an increasing pressure drop across the filter. Due to the hydraulic compression of the solid, the filtration cycle may be very short unless additional filter aid is used during filtration. The amount required varies with the type and concentration of suspended solids in the treated water. A typical pressure filter flow is shown in Fig. 6.8.

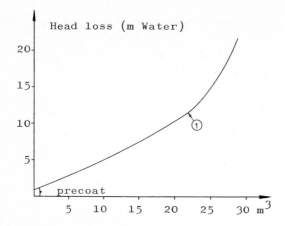

Fig. 6.8. Head loss plotted against volume for a precoat filter. Filtration should be interrupted at (1).

Flotation. Flotation is used to remove suspended solid from waste water and to concentrate sludge. Thus flotation offers an alternative to sedimentation, especially when the waste water contains fat and oils.

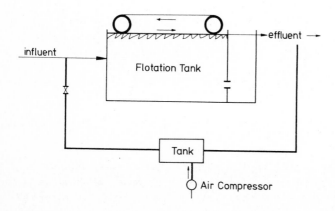

Fig. 6.9. Flotation unit.
Either a portion of the waste water or the clarified effluent is

pressurized at 3-6 atm. When the pressurized water is returned to normal atmospheric pressure in a flotation unit air bubbles are created. The air bubbles attach themselves to particles and the air-solute mixture rises to the surface, where it can be skimmed off, while the clarified liquid is removed form the bottom of the flotation tank.

Fig. 6.9 shows a flotation system with partial recirculation of the effluent. Generally it is necessary to estimate the flotation characteristics of the waste water by use of a laboratory flotation cell:

1. The rise of the sludge interface must be measured as a function of time. An example is shown in Fig. 6.10.
2. The retention time must be varied and the corresponding saturation of pressurized water determined, see Fig. 6.11.
3. The effluent quality must be determined as a function of the air/solids ratio, see Fig. 6.12.

Based on such results it is possible to scale up.

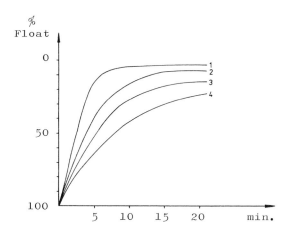

Fig. 6.10. Rise characteristics of paper fibres at four different air/solids ratios. 1) 0.03, 2) 0.06, 3) 0.16 and 4) 0.30.

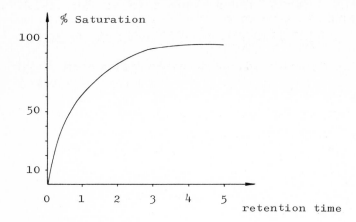

Fig. 6.11. Saturation plotted to retention time. (Paper fibres, pressure = 9 atm.). Retention time is indicated in minutes.

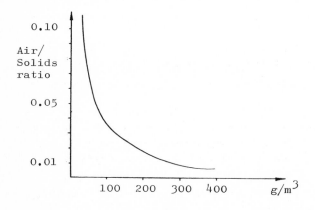

Fig. 6.12. Effluent quality versus air/solids ratio. Results of laboratory experiments on 550 mg/l waste water from a paper machine.

6.2.3   Biological treatment processes.

P.6.3.   Many types of biological processes are active in the breakdown of organic matter.

A nutritional classification of organisms is given in Table 6.7.

TABLE 6.7
Nutritional classification of organisms

Class	Nutritional requirements
Autotrophic	The organisms depend entirely on inorganic compounds
Heterotrophic	Organic compounds are required as nutrient
Phototrophic	Use radiant energy for growth
Chemotrophic	Use dark redox reaction as energy source
Lithotrophic	Use inorganic electron donors (e.g. hydrogen gas, ammonium ions, hydrogen sulphate and sulphur)
Organotrophic	Require organic compounds as electron donors
Strictly aerobic	Cannot grow without molecular oxygen, which is used as oxidant
Strictly anaerobic	Use compounds other than oxygen for chemical oxidation. Sensitive to the presence of minor traces of molecular oxygen
Facultative anaerobic	Can grow either in the presence or absence of air.

P.6.4. **Most biological systems used to treat organic waste depend upon heterotrophic organisms, which use organic carbon as their energy source.**

As seen in the table, the organisms can be either strictly aerobic, strictly anaerobic or facultative anaerobic. Anaerobic breakdown is used in the treatment of sludge, or denitrification where nitrate is the oxygen source. Chemolithotrophic organisms are also used in biological treatment processes. These comprise specialized groups of bacteria which are able to oxidize inorganic compounds such as those of hydrogen, sulphur or ammonia.

P.6.5. **Of the various types of metabolism in which the redox reaction provides the ultimate source of energy, there are three major classes of energy-yielding processes:**

Fermentation, in which organic compounds serve as the final electron acceptors;
Respiration (aerobic), in which molecular oxygen is the ultimate electron acceptor;
Respiration (anaerobic), in which inorganic compounds - not oxygen - are ultimate electron acceptors.

These reactions can be described by the following overall process:

$$\text{Organic matter} + O_2 + NH_3 + \text{cells} \rightarrow CO_2 + H_2O + \text{new cells} \qquad (6.4)$$

Nitrification results from a two-step oxidation process. First, ammonia is oxidized to nitrite by Nitrosomonas. Second nitrite is oxidized to nitrate by Nitrobacter:

$$2NH_4^+ + 3O_2 \rightarrow 2NO_2^- + 2H_2O + 4H^+ \qquad (6.5)$$

$$2NO_2^- + O_2 \rightarrow 2NO_3^- \qquad (6.6)$$

Respiration and nitrification are the same processes as those used in nature to oxidize organic matter and ammonia. The processes are of importance for the oxygen balance of streams, as presented in section 2.6.

Nitrate can be used as an oxygen source for the biological decomposition of organic matter. The reaction - called denitrification - is:

$$2NO_2^- + H_2O \rightarrow N_2 + 2OH^- + 5O \qquad (6.7)$$

In comparison with molecular oxygen supplied by the aeration method, the use of nitrate as an oxygen source is undoubtedly easier because of its extremely high solubility. Further, it can be expected that satisfactory biodegradation of organic matter may be carried out with microorganisms and waste water containing nitrate. Industrial waste water, especially from petrochemical plants, sometimes contains a large amount of nitrate as well as highly concentrated organic matter. The application of a biological treatment method for treating such waste water using nitrate as the oxygen source is therefore attractive.

Studies by Miyaja et al. (1975) have shown that the amount of BOD_5 removed by biological treatment with nitrate as an oxygen source is linearly related to the amount of nitrate removed in the reaction tank. Studies have shown that one of the microorganism involved is Pseudomonas denitrificans.

P.6.6. **Cellular growth can often be described as a first-order reaction:**

$$\frac{dX}{dt} = \mu_m \, {}^* \, X \qquad (6.8)$$

where
X = concentration of volatile biological solid matter

μ_m = the maximum growth rate

t = time

Integration of this equation where $X = X_0$ and $t = 0$, gives:

$$\ln \frac{X}{X_0} = \mu_m * t \tag{6.9}$$

This equation is only valid during the so-called logarithmic growth phase in which the substrate (the organic matter) is unlimited. When the substrate becomes the limiting factor, the growth rate can be described by means of the following equation:

$$\frac{dX}{dt} = \mu_m X * S * \frac{1}{K} \tag{6.10}$$

where S = substrate concentration and K_S = a constant.

These two expressions can be combined by means of the Michaelis-Menten equation:

$$\mu = \mu_m \frac{S}{S + K_S} \tag{6.11}$$

where

μ = growth rate $(= \dfrac{dX * 1}{dt * X})$

μ_m = maximum growth rate

S = substrate concentration

K_S = Michaelis-Menten constant

As can be seen, when $S \longrightarrow \infty$ the equation becomes (6.8) and when $K_S \gg S$ the equation is transformed into (6.10).

Fig. 6.13 shows the Michaelis-Menten relationship; the growth rate is plotted against the substrate concentration.

It is often convenient to illustrate the Michaelis-Menten equation by means of a Lineweaver-Burk plot. The reciprocals of the growth rate and of the substrate concentration are plotted against each other (see Fig. 6.47). The relationship is linear as can be seen from equation (6.11) which can be transformed to:

$$\frac{1}{\mu} = \frac{K_S}{\mu_m} * \frac{1}{S} + \frac{1}{\mu_m} \tag{6.12}$$

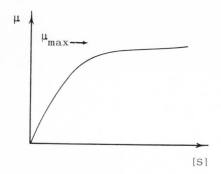

Fig. 6.13. Michaelis-Menten relationship.

Equation 6.8 is incomplete without an expression to account for depletion of biomass through endogenous decay (respiration). A first-order expression could be used:

$$\frac{dX}{dt}\text{(end)} = -k_d * X \qquad (6.13)$$

Incorporation of endogenous decay and (6.11) into (6.8) results in:

$$\frac{dX}{dt} = \mu_m \frac{S}{K_S + S} * X - k_d * X \qquad (6.14)$$

By use of the yield constant, a (a = mg biomass produced per mg of substrate used), dX can be expressed in terms of substrate removal:

$$dX = -a * dS \qquad (6.15)$$

Combining equations (6.15) and (6.11) gives:

$$\frac{-dS}{dt} = \frac{\mu_m}{a} * \frac{S}{S + K_S} * X \qquad (6.16)$$

The set of equations (6.8) to (6.16) is not valid for complex substrate mixtures, but in many cases the equations can be used as good approximations.

Equation (6.16) is related to the first-order kinetic equation (2.35) in section 2.6. If X can be considered constant and $K_S >> S$, which is a good approximation in many water courses, (6.16) will correspont to a first-order decomposition equation: $L \sim S$ and K_1 to $\dfrac{\mu_m * X}{a}$

Temperature influences these processes significantly. The effect of temperature on the reaction rate can be expressed by the following relationship:

$$\mu_T = \mu_{20°C} * \pi^{T-20°C} \qquad (6.17)$$

where π = a constant.

π is listed in Table 6.8 for various types of processes.

TABLE 6.8
Temperature effects on biological processes

Process	π
Activated sludge (low loading)	1.00 - 1.01
Activated sludge (high loading)	1.02 - 1.03
Trickling filter	1.035
Lagoons	1.05 - 1.07
Nitrification	1.143

The progress of aerobic biological purification is shown in Fig. 6.14. Note the slow BOD_5 reduction after 10 hours, and the decline in the amount of synthesized material after 5 hours.

Example 6.1
The BOD_5 of a waste water is determined to be 150 mg I^{-1} at 20°C. Find BOD_7 (which is used in Sweden to indicate biological oxygen demand) at 15°C. As natural conditions are simulated, the constants from section 2.6 are used not those valid for biological treatment methods.

Solution.
Since it is waste water, $K_1 = 0.35$ day^{-1} and $K_T = 1.05$ are used.
The following equation is used (see section 2.6 equation 2.40):
$L_t = L_0 * e^{-k_1 * t}$

150 mg I^{-1} corresponds to $L_0 - L_5 = L_0 (1-e^{-0.35*5})$, which gives:

$$L_0 = \frac{150}{1 - e^{-0.35 \cdot 5}} = 181 \ \text{mg} \ l^{-1}$$

$$K_{15} = (1.05^{15-20}) \cdot 0.35 = 0.27$$

At 15°C:

$$BOD_7 = L_0 - L_7 = L_0(1 - e^{-0.27 \cdot 7}) = 154 \ \text{mg} \ l^{-1}$$

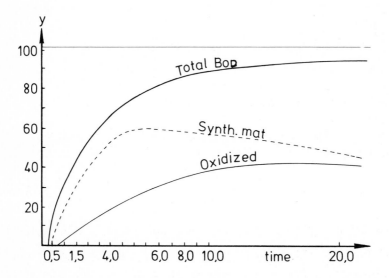

Fig. 6.14. Removal of organic inbalances by biomass in a batch operation. Reduction of total carbonanceous oxygen demand (%) = y versus time (hours).

The various biological treatment processes can be summarized as follows:

1. The conventional **activated-sludge process** (see Fig. 6.15) is defined as a system in which flocculated biological growth is continuously circulated and contacted with organic waste water in the presence of oxygen, which is usually supplied in the form of air bubbles injected into the liquid sludge mixture. The process involves an aeration step followed by sedimentation. The separated sludge is partly recycled back to be removed with the waste water. The following processes occur:
a. Rapid adsorption and flocculation of suspended organics,
b. Oxidation and decomposition of adsorbed organics, and
c. Oxidation and dispersion of sludge particles.

Sometimes, depending on the retention time and amount of oxygen introduced, ammonium ions are oxidized to nitrate by nitrifying organisms. This is seen particularly during the summer, and is due to the influence of the temperature on the rate constant for the nitrification process (see section 2.6 and Table 6.8).

Activated sludge usually provides an effluent with a soluble BOD_5 of 10-20 mg/l. The process necessitates the treatment of excess sludge before disposal.

An activated sludge plant can be designed by use of mass-balance principles.

If we assume that
1. the activated sludge reactor can be considered a mixed-flow reactor (see section 2.3),
2. that the influent and effluent biomass concentrations are negligible compared to biomass in the reactor or in the waste sludge,
3. all reactions occur in the reactor, i.e. neither biomass production or food utilization occurs in the clarifier,

then at steady-state conditions, we have: Biomass growth = biomass out or by use of (6.14):

$$V * X \left(\mu_m \frac{S}{K_S + S} - k_d \right) = Q_W * X_W \qquad (6.18)$$

where V is the volume, Q_W flow of waste sludge and X_W the biomass concentration in this flow (kg m^{-3}).

A mass balance at steady-state for the substrate gives:

Food in - food out = food consumed

or by use of (6.16):

$$Q_0(S_0-S) = \frac{\mu_m * S}{a(S + K_S)} * X * V \qquad (6.19)$$

where Q_0 is the flow rate (m^3 d^{-1}) and S_0 the concentration of food (substrate, BOD_5) in the influent (kg m^{-3}).

By rearrangement of (6.18) and (6.19) we obtain

$$\mu_m * \frac{S}{K_S + S} = \frac{Q_W * X_W}{V * X} + k_d = \frac{Q_0 * a}{V * X} (S_0 - S) \qquad (6.20)$$

V/Q_0 is the hydraulic retention time π and $V*X/Q_W*X_W$ is called the mean all-residence time, π_c.

By substituting π and π_c into (6.20), we get:

$$\frac{1}{\pi_c} = \frac{a(S_0 - S)}{X * \pi} - k_d \qquad (6.21)$$

This equation can be used to give design data for an acitvated sludge reactor. a is usually 0.3-0.7 and k_d usually 0.01-0.1 d^{-1}. a = 0.5 and k_d = 0.05 d^{-1} are often used for municipal waste water. X is often chosen to be 2-8000 mg l^{-1}.

Fig. 6.15. Conventional activated-sludge plant.

Example 6.2

Design a activated sludge system for secondary treatment of 10,000 m^3 d^{-1} of municipal waste water with BOD_5 = 150 mg l^{-1}. A completely mixed reactor is to be used to obtain an effluent of 10 mg BOD_5 l^{-1}. $X = 3000$ mg l^{-1} is chosen. Find also the mass of waste solid per day. High nitrification efficiency is desirable.

Solution.
A $\pi_c = 10$ days is chosen to give a high nitrification efficiency (see Fig. 6.35 in 6.3.3, where the relationship between sludge age and nitrification is shown). From (6.21) we have:

$$\frac{1}{10} = \frac{0.5\ (0.15 - 0.01)}{3 * \pi} - 0.05 \qquad \text{or } \pi = 0.155$$

Reactor volume: $10,000 * \pi = 1550\ m^3$

$$\text{As } \pi_c = 10\ d = \frac{V * X}{Q_w * X_w} = \frac{1550 * 3}{Q_w * X_w}$$

$$QX_w = 465\ kg\ d^{-1}$$

2. **The extended aeration** process works on the basis of providing sufficient aeration time for oxidizing the biodegradable portion of the sludge produced form the organics removed from the process. Fig. 6.16 shows the process schematically. The excess sludge in the process contains only non-bio- degradable residue remaining after total oxidation. The total BOD_5 provided by this process is 20 mg/l or less.

Oxidation ditches have been developed as self-sufficient structures for the extended aeration of waste waters from small communities. Fig. 6.17 demonstrates the principles of an oxidation from 1 to 3 days. No effluent is withdrawn until the water level in the channel has built up to the highest operation level. The influent is then cut off and the rotor stopped. Solids are allowed to settle for an hour or two, then the clarified supernatant is withdrawn through an effluent launder, and, if desired, excess sludge is lifted from a section of the ditch to drying beds. The effluent is then cut off and the operation routine is repeated. Because, the solids are well stabilized during the long aeration time, they are no longer putrescent and water is readily removed from them. The fact that the sludge is well stabilized may allow it to be discharged into the receiving water with the effluent, provided the overall loading resulting from effluent and sludge can be accepted.

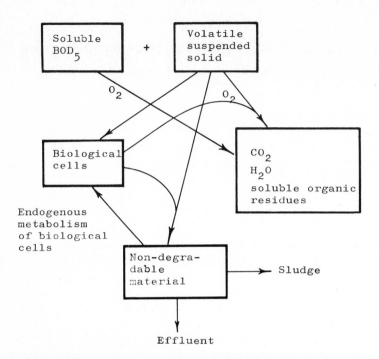

Fig. 6.16. Extended aeration process.

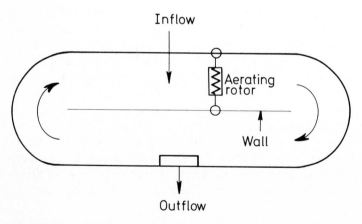

Fig. 6.17. Oxidation ditch.

3. In **the contact stabilization process** the waste water is aerated with stabilized sludge for a short period of 1/2 - 1 hour. The mixed liquid is then separated by sedimentation and when settled the sludge is transferred to a sludge stabilization tank where aeration is continued to complete the oxidation. This process is used to advantage when a high percentage of BOD is removed rapidly by bioadsorption after contact with the stabilized sludge. The extent of removal depends on the characteristics of the sludge and of the waste water. As general rule the process should give an efficiency of 85% BOD_5 removal.

4. A **trickling filter** is a bed packed with rocks, although, more recently, plastic media have been used. The medium is covered with a slimy micro-biological film. The waste water is passed through the bed, and oxygen and organic matter diffuse into the film where oxidation occurs. In many cases recirculation of the effluent improves the BOD removal, especially when the BOD of the effluent is relatively high. A high-rate trickling filter provides an 85% reduction of BOD for domestic sewage, but 50-60% is the general figure for BOD_5 reduction in the treatment of organic industrial waste water.

A plastic-packed trickling filter will require substantially less space than a stone-packed one due to its bigger depth and specific surface area. According to Wing et al. (1970), plastic media packed to a depth of 6.5 m in a trickling filter will require less than one-fifth of the land required by those packed with stones to the usual depth of 2-4 m.

The specific surface area of rock-trickling filters is 40-70 m^2/m^3 and void space 40-60%, while plastic filters have a specific surface of 80-120 m^2/m^3 or even more, with a void space of 94-97%.

The design of biotowers is based on the following equation:

$$\frac{S_e}{S_a} = \frac{e^{-k \cdot D/Q^n}}{(1 + R) - R * e^{-k \cdot D/Q^n}} \tag{6.22}$$

where S_e is the effluent substrate concentration BOD_5 (mg l^{-1}), S_a is the BOD_5 of the mixture of raw and recycled mixture applied, D is the depth, Q the hydraulic loading ($m^2/(m^3$ min)), k treatability constant related to the waste water (min^{-1}), n is a coefficient related to medium characteristics. R is ratio of the recycled flow to the influent flow. S_a is found from a simple mass balance as:

$$S_a = \frac{S_0 + R * S_e}{1 + R} \tag{6.23}$$

where S_0 is the BOD_5 of the raw sewage (influent).

k is from 0.01 to 0.1 at 20°C. For municipal waste water, 0.055 min^{-1} (20°C) is often used. n can be taken as 0.5 if not known from pilot-plant analysis. k is temperature dependent, see equation (6.17) and Table 6.8.

Example 6.3.

Design a biotower (trickling filter) composed of a modular plastic medium. 10,000 m^3 d^{-1} municipal waste water with a BOD_5 of 150 mg l^{-1} must be treated. The temperature is 25°C, the depth of the tower is 6.5 m and the unit should be designed to produce an effluent of 10 mg l^{-1}.

Solution.

R = 2 is chosen. (R = 1.5 or 3 could also be tested)

$$S_a = \frac{150 + 2 *10}{1 + 2} = 56.7 \text{ mg l}^{-1}$$

$$k_{25} = k_{20} * (1.035)^{25-20} = 0.055 * (1.035)^5 = 0.065 \text{ min}^{-1}$$

Equation (6.22) is used:

$$\frac{10}{56.7} = \frac{e^{-0.065*6.5/Q^{0.5}}}{(1+2) - 2*e^{-0.065*6.5/Q^{0.5}}}$$

Q is found to be 0.20 m^2/m^3 min. 10,000 m^3 d^{-1} = 6.95 m^3 min^{-1}

Surface area must therefore be $\frac{6.95}{0.2}$ m^2 = 34.8 m^2.

The unit has a depth of 6.5 m and if it is square, the dimensions are $\sqrt{34.8}$ = 5.9 m. The tower is 5.9 * 5.9 * 6.5 m^3.

5. **Lagoons** are the most common methods of organic waste treatment, when suficient area is available. They can be divided into four classes:
a. *Aerobic algal ponds,* which depends upon algae to provide sufficient oxygen.
b. *Facultative ponds,* which have an aerobic surface and an anaerobic bottom.
c. *Anaerobic ponds,* which are loaded to such an extent that anaerobic conditions exist throughout the liquid volume.
d. *Aerated lagoons,* which are basins where oxidation is accomplished by mechanical or diffused aeration units and induced surface aeration. The turbulence is usually insufficient to maintain solids in suspension, thus

most inert solids settle to the bottom where they undergo anaerobic decomposition. The basin (2-4 m deep) may include a sedimentation compartment to yield a more clarified effluent. If the turbulence level in the basin is increased to maintain solids in suspension, the system becomes analogous to an activated-sludge system.

An aerated lagoon can provide an effluent with less than 50 mg/l BOD_5, depending on the temperature and the characteristics of the waste water. Post-treatment is necessary when a highly clarified effluent is desirecd, and large areas are required compared with the activated sludge process.

A pond can be designed in accordance with the mass-balance equations given in section 2.3. Equation (2.12) can be applied, or if several ponds are arranged in series, equation (2.13) is valid. If 50% efficiency of aeration is foreseen, what would be the needed oxygen supply?

Example 6.4.
A waste water flow of 500 m^3/d from a small community is treated by use of 4 ponds in series. How large must the ponds be to remove 90% BOD_5 when the average BOD_5 is 200 mg l^{-1} and the temperature is 18°C?

Solution.
K_1 (20°C) = 0.35 day^{-1} (see Table 2.18)

O = 1.06   (see areated lagoons, Table 6.8)

$$K_T = K_{18°} = 0.35 * 1.06^{-2} = 0.31$$

$$\frac{C_{im}}{C_{io}} = \frac{1}{10} = (\frac{1}{1 + 0.31 * tr})^4$$

$$(1 + 0.31 * tr)^4 = 10$$

$$1 + 0.31 * tr = 1.78$$

$$tr = 2.52$$

Each of the 4 ponds should have a volume of 2.52 * 500 = 1258 m^3.

If a depth of 1.5 m is used, the area will be 839 m^2 * 4 = 3356 m^2.

The oxygen must most probably be supplied by aeration, i.e. a minimum of

180 * 5000 = 900 kg/day or, if an efficiency of only 50% is assumed, 1800 kg/day.

6. **Anaerobic digestion.** The anaerobic breakdown of organic matter to harmless end-products is very complicated. Fig. 6.18 summarizes some of the more general processes.

```
Carbohydrate              Acetic acid                 CH₄
Fats            ─────     and other organic   ─────   CO₂
Proteins                  acids                       H₂S
```

Fig. 6.18.   Anaerobic degradation of organics.

Methane-producing organisms convert long-chain volatile acids to methane, carbon dioxide and other volatile acids with a short carbon chain, which are then fermented in a similar fashion. Acetic acid is directly converted into carbon dioxide and methane. The rate of methane fermentation controls the overall reaction rate. Sufficient time must be available in the reactor to permit growth of the organisms or they will be washed out of the system. This means that the retention time must be greater than that corresponding to the growth rate of the methane-producing organisms. It is possible, by the use of extracellular enzymes, to cut down the resistance time considerably, but the use of such enzymatic processes is still only in its infancy.

Many factors, such as the composition of the sludge and the waste water. pH and temperature, influence the reaction rate, but it is generally shown that the overall rate is controlled by the rate of conversion of volatile acids to methane and carbon dioxide. Digestion fails to occur when there is an inbalance in the rate of the successive processes, which might result in a build-up of volatile acids. The optimum conditions can be summarized as follows: pH 6.8-7.4; redox potential -510 to -540 mV; concentration of volatile acids 50-500 mg/l; alkalinity (as calcium carbonate) 1500-5000 mg/l; temperature 35-40°C.

It should be possible to obtain effective digestion with a retention period as low as 5 days, but increasing the retention time to 10 days should assure 90% degradation of organic matter. Anaerobic digestion is used for the treatment of sludge form biological processes as well as for the treatment of industrial waste water with an extremely high BOD_5, e.g. industrial waste water from the manufacture of yeast.

The major part of the gas produced by anaerobic treatment processes comes from the breakdown of volatile acids. The gas is composed of methane, carbon dioxide, hydrogen sulphide and hydrogen. The higher the resistance time, the lower the percentage of carbon dioxide and the higher the percentage of methane in the gas produced.

Lawrence and McCarty (1967) have shown that methane gas production, at a good approximation, is 0.4 m^3 gas per kg BOD removed. This value must be considered as the maximum obtained by complete conversion of the solid into methane.

7. **Effective nitrification**   occurs when the age of the sludge is greater than the growth rate of the nitrofying micro-organisms. Further details on this process, see 6.3.3.

8. **Nitrate can be reduced to nitrogen**   and dinitrogen oxide by many of the heterotrophic bacteria present in activated sludge. For further details, see 6.3.3.

6.2.4   Mechanico-biological treatment systems.

Many variations in flow pattern for mechanico-biological systems, are used in the treatment of municipal waste water.

Fig. 6.19 shows a flow diagram of a conventional activated-sludge plant.

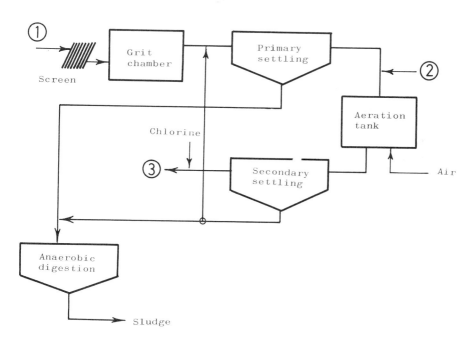

Fig. 6.19.   Flow diagram of a conventional activated-sludge plant. (see Table 6.9).

In the process sewage is mixed with a portion of returned activated-sludge to facilitate the primary settling (improved flocculation). For typical municipal sewage an aeration time of 4-6 hours must be applied, if a BOD_5-removal of 90% or more is required. Table 6.9 gives some typical results.

Trickling filters are generally used at smaller sewage treatment plants (1000 - 10,000 inhabitants). They are classified according to the hydraulic and organic loading used, see Table 6.10.

TABLE 6.9
Typical results obtained from activated-sludge plant (mg/l)

	1	2	3
BOD_5	150 - 300	100 - 180	15 - 20
Suspended matter	50	5 - 10	2 - 8
N_{total}	25 - 45	22 - 40	18 - 32
P_{total}	6 - 12	6 - 12	5 - 10

Numbers refer to Fig. 6.19

TABLE 6.10
Trickling filters

	Low-rate operation	High-rate operation
Hydraulic loading ($m^3/m^2/24h$)	1 - 4	8 - 40
Process loading ($kg/m^3/24h$)	0.2 - 1.5	2 - 20
Depth (m)		
Single stage	1.5 - 3	1 - 3
Multi stage	0.7 - 1.5	0.5 - 1.5
Relative recirculation in term of inflow		0.5 - 10
BOD_5-removal	85 - 90%	75 - 85%
Suspended solid removal	90 - 95%	80 - 90%

A combination of an activated-sludge plant and a trickling filter is often applied where high quality effluent is required. In addition to a 95% or more BOD_5-removal, 80-90% nitrification of ammonia will take place, except perhaps during the winter (the nitrification process is highly dependent on temperature, see Table 6.8). Fig. 6.20 illustrates an example of a sewage plant using such a two-step biological treatment.

Lagoons are only used where considerable space is available, and as, in most cases, they do not give an effluent of sufficient quality by today's standards, this method is now used less and less. Table 6.11 summarizes design factors and results for lagoons.

The final selection of biological treatment method is highly dependent on the receiving water, which determines the acceptable BOD_5 of the effluent, see section 2.6.

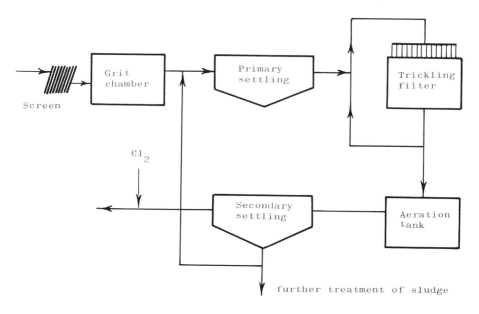

Fig. 6.20. Sewage plant combining trickling filter and activated sludge treatment.

TABLE 6.11
Design factors for lagoons

	Aerobic	Facultative	Anaerobic	Aerated
Depth (m)	0.2 - 0.4	0.75 - 2	2.5 - 4	2 - 4
Retention time (days)	2 - 6	7 - 30	30 - 50	2 - 10
BOD loading (kg/ha/day)	100 - 200	20 - 50	300 - 500	depending upon waste characteristics and aeration
BOD removal (%)	80 - 90	75 - 85	50 - 70	50 - 90

6.2.5   Other methods used for BOD-removal.

Mechanico-biological systems are widely used for BOD removal, but other methods are also available for the treatment of municipal and industrial waste waters - combinations of physico and chemical methods and irrigation. They are relatively more attractive for the treatment of waste water with high BOD_5, such as waste water from the food industry, but can also be used on municipal waste water.

It is difficult to provide guidelines on when physico-chemical methods are more advantageous than mechanico-biological ones, but the following issues should be considered:

1. Physico-chemical methods are not *susceptible to shock loadings* and the presence of toxic compounds.
2. *The space* required for physico-chemical methods is, in most cases, less than for mechanico-biological treatment plants.
3. *Recovery of fat, grease and proteins can be achieved* using physico-chemical methods in the treatment of waste water from the food-processing industry.
4. Although *operation costs are slightly higher* for physico-chemical methods, *investment cost will generally be lower.*

P.6.7.   The following physico-chemical methods are used for BOD-reduction: chemical precipitation, ion exchange, adsorption and reverse osmosis.

Apart form obtaining a substantial reduction of the phosphate concentration organic matter is also precipitated, by using lime, aluminium sulphate or iron(III) chloride, due to a reduction of the zetapotential of organic flocs (Balmer et al., 1968). Normally a direct precipitation on municipal waste wate4r will reduce the potassium permanganate number and the BOD_5 by 50-65%, which has to be compared with the effect obtained by plain settling without the addition of chemicals (Davidson and Ullman, 1971). Furthermore, precipitants (lignosulphonic acid, activated bentonite and glucose sulphate) have been developed for precipitation of proteins, permitting their recovery from food-processing industries (see Tønseth and Berridge, 1968 and Jørgensen, 1971).

Ion exchangers are able to take up ionic organic compounds, such as polypeptides and amino acids. Macroporous ion exchangers are even able to remove protein molecules from waste water. Cellulose ion exchangers designed for protein removal have been developed (Jørgensen, 1969, 1970 and 1973). They are inexpensive compared with other ion exchangers and are highly specific in uptake of high molecular weight ions.

Adsorption on activated carbon can remove not only refractory organic compounds (see 6.4.3), but also biodegradable material. This has been used for the treatment of municipal waste water (see below).

Reverse osmosis is an expensive waste water treatment process, which explains why its application has been limited to cases in which other processes are inappropriate or where recovery is possible. The process has been used to recover proteins from whey, which could not be treated by conventional methods, because of its high concentration of proteins and carbohydrates. Table 6.12 summarizes the results of this application.

TABLE 6.12
Reverse osmosis of whey

	Dry matter	Protein	Lactose	Ashes
Whey (%)	6.4	0.8	4.4	0.7
Concentrate (%)	29.0	3.6	23.0	2.2
Permeate (%)	0.15	0.06	-	0.1

The applications of purely physico-chemical methods for BOD-removal are best illustrated by some examples. *The so-called Guggenheim process uses chemical precipitation with lime and iron sulphate followed by a treatment on zeolite for ammonia removal* (Gleason and Loonam, 1933). A 90% removal of BOD_5, phosphate and ammonia is achieved (Culp, 1967 and 1967a).

A combination of chemical precipitation by aluminium sulphate and ion exchange on cellulose ion exchangers and clinoptilolite also seems promising for the treatment of municipal waste water. 90% removal of BOD_5 and ammonia is obtained in addition to 98% removal of phosphate and suspended matter (see Jørgensen, 1973 and 1976).

The so-called AWT system is *a combination of precipitation with lime and treatment on activated carbon* . Zuckerman and Molof (1970) claim that by using lime the larger organic material is hydrolysed, giving better adsorption because activated carbon prefers molecules with a molecular weight of less than 400. A flowchart of this process is shown in Fig. 6.21.

Fig. 6.22 is a flow diagram of *the combination of chemical precipitation and ion exchange* used in the treatment of waste water from the food industry (Jørgensen, 1968, 1969, 1973 and 1978). This process allows recovery of fat, grease and proteins. Table 6.13 gives the analytical data obtained when this process was used on waste water from herring filetting after centrifugation of the raw waste water to recover fish oil. Table 6.14 gives a before and after analysis of this process for waste water from an abattoir. For comparison the table also includes the results obtained from using a biological plastic filter.

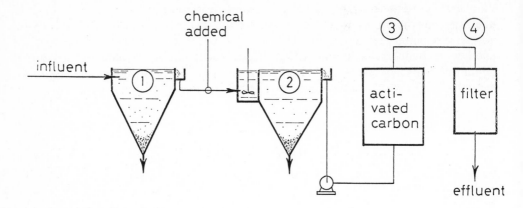

Fig. 6.21. The AWT system consists of 1) mechanical treatment, 2) settling after chemical precipitation, 3) treatment on activated carbon, 4) filtration.

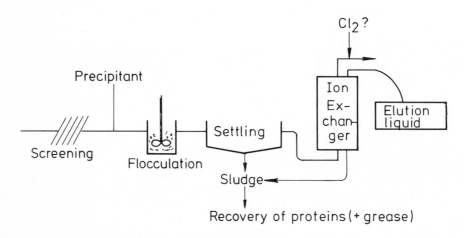

Fig. 6.22. Recovery of proteins (+ grease).

In addition to mechanico-biological treatment or a combination of physico-chemical methods

P.6.8. the use of spray irrigation can also be considered as a solution to the problem of the biological oxygen demand of waste water.

Spraying and irrigation of waste water onto fields and meadows is widely considered to be the best and cheapest way of treating waste water from the food-processing industry. The industry favours this solution as it ensures maximum utilization of the fertilizing properties of these waste waters (Sanborn, 1953).

TABLE 6.13
Analytical data of waste water from herring filetting

	Raw waste water	After centrifugation 1. step	After chem. precipitation 2. step	After Cellulose ion-exchanger 3. step
BOD$_5$ (mg/l)	11000	5800	2000	1100
N (mg/l)	180	162	60	23
Suspended matter (mg/l)	400	170	40	2
KMnO$_4$ (mg/l)	8000	4000	1200	600

TABLE 6.14
Analysis of waste water from an abattoir (mg/l)

	Raw water	After biological plastic filter	After chem. precipitation (glucose sulphate is used)	After chem. precipitation and ion exchange
BOD$_5$	1500	400	600	50
KMnO$_4$	950	350	460	60
Total N	140	42	85	15
HN$_3$-N	20	15	18	2
NO$_3$-N	4	5	4	1
P	45	38	39	1.5

As early as 1942 an outline of the agricultural use of waste water from dairies was published.

Raw waste water was stored for 5-6 hours in a tank situated on an irrigated area. Irrigation was recommended at a rate of 150-400 mm/year or 75-200 m^3/day of effluent per ha of irrigated fields.

The storage tank should be cleaned continuously, by means of compressed air for example. After irrigation the tanks and pipes should be rinsed with fresh water in an amount equal to 1/3 of the daily discharge of the waste water.

The method is not usually recommended in areas subject to persistent frosts, or where the soil has a natural high moisture content or a high ground water level (see further below).

Spraying the waste water on pastures carries the risk of affecting cattle with tuberculosis. Therefore a period of 14 days is recommended between irrigation and putting cows out to pasture. The limitations of this process can be summarized as follows (Eckenfelder et al., 1958).

1. **Depth of the ground water.**
   The quantity of waste water that can be sprayed onto a given area will be proportional to the depth of the soil through which the waste water must travel to the ground water. A certain soil depth must be available if the contamination of the ground water is to be avoided.
2. **Initial moisture content.**
   The capacity of the soil to absorb is dependent on its initial moisture content.
   Sloping sides will increase the runoff and decrease the quantity of water which can be absorbed by a given area.
3. **Nature of the soil.**
   Sandy soil will give a high filtration rate, while clay will pass very little water. A high filtration rate will give unsufficient biological degradation of the organic material and too low a filtration rate will reduce the amount of water which can be absorbed by a given area.

P.6.9. **The capacity is proportional to the coefficient of permeability of the soil and can be calculated by the following equation:**

$$Q = K * N \tag{6.24}$$

where Q = the quantity of water (m^3) which can be absorbed per m^2 per 24h; K = the permeability coefficient, expressed as m/24h; N = the saturation of the soil for which a value of 0.8-1.0 can be used in most cases.

If the soil has different characteristics at different depths an overall coefficient can be calculated:

$$K = \frac{H}{H_1/K_1 + H_2/K_2 + \ldots\ldots\ldots H_n/K_n} \tag{6.25}$$

where H = the total depth and H_1, H_2 .... H_n = the depth of the different types of soil with permeability coefficients K_1, K_2, etc.

Table 6.15 gives the permeability coefficients for different soil types (Eckenfelder et al., 1958).

In addition to soil properties several characteristics require consideration in a spray irrigation system. Suspended matter must be removed by screening or by sedimentation before the water is sprayed. Otherwise the solids will clog the spray nozzles and may mat the soil surface rendering it impermeable to further percolation (Canham, 1955). The pH must be adjusted as excess acid or alkali may be harmful to crops.

High salinity will also impair crop growth. A maximum salinity of 0.15% has been suggested to eliminate this problem (U.S. Department Agricultural Handbook, 1954).

TABLE 6.15
Permeability coefficients for different soil types

Description	Permeability (m per 24h)
Fine sand	100 - 500
Trace silt	15 - 300
Light agricultural soil	1 - 5
50% clay and 50% organic soils	0.15 - 0.40
Predominating clay soil	< 0.1

Spray irrigation has been succesfully used for waste water from dairies, pulp and paper industries (Gellman et al., 1959; Wisneiwski, 1956), cannery waste (Luley, 1963; Williamson, 1959) and fruit and vegetable processing plants.

The data from various spray irrigation plants are given in Table 6.16.

TABLE 6.16
Data from spray irrigation plants

Waste water from the manufacture of:	Loading kg BOD_5 per year	Application rate (m/year)
Asparagus and beans	30	5
Tomatoes	600	3
Starch	750	3
Cherries	700	1.6
Paperboard and hardboard	500	1.6

Disposal of industrial waste water by irrigation can be carried out in several ways:
1. Distribution of the water over sloping land with a runoff to a natural water source.
2. Distribution of the water through spray nozzles over relatively flat terrain.
3. Disposal to ridge and furrow irrigation channels.

6.3.   NUTRIENT REMOVAL.

6.3.1.   The eutrophication problem and its sources.

The word eutrophication means nutrient rich, which is generally considered to be undesirable due to:

1. the green colour of eutrophic lakes makes swimming and boating more unsafe,
2. an aesthetic point-clear water is preferred to turbid water,
3. the oxygen content of the hypolimnion is reduced due to decomposition of algae, especially in the autumn, but also in the summer, when the stratification of deeper lakes is most pronounced.

The eutrophication is, however, not controlled merely by determining the limiting element or elements and then reducing their concentration in the discharged waste water - the problem is far more complex (see section 2.7).

It is important to consider the sources of the nutrients and for a balanced overview of the situation a mass balance must be set up for each case (see principle 2.23).

However, in some situations it is not possible to obtain the necessary data for setting up a mass balance. In these cases the information given in Table 6.17 can be used. This table also illustrates the relative importance of the three sources of nutrient input to receiving waters, showing that a significant input of nitrogen originates from land and precipitation. This is the general picture, but it is not always valid and should only be used in a data-poor situation.

The role of forest fertilization as a source of nutrients has been examined by Rouger (1981). His results we summarized in appendix 9.

Schindler (1971) has shown that midsummer maximum productivity and midsummer maximum biomass increase linearly with $(A_d + A_1)/V$, where A_d is the watershed area, A_1 the lake area and V the lake volume. This is in accordance with Table 6.17, as the P- and N-loading is proportional to the watershed area and lake area and by division by the volume, the nutrient concentration results, which again determines the maximum productivity and biomass.

Phosphorus is mainly derived from sewage, although the input from agriculture in some cases may also be significant. It is often difficult to control the input of nitrogen to an ecosystem. Such processes as nitrogen fixation by algae may also play a major role.

Control of the eutrophication of lakes requires removal of one or more nutrients from waste water, and the final decision as to the right management strategy - what nutrient should be removed and to what extent

- implies the use of ecological models, see Jørgensen (1976, 1980 and 1981) and Jørgensen et al. (1978).

A simple model might, however, be the right model in a data-poor situation.

TABLE 6.17
Sources of nutrients

A: Export scheme of phosphorus E_P and nitrogen E_N (mg m^{-2}y^{-1}) [1]

Land use	E_P Geological classification		E_N Geological classification	
	Igneous	Sedimentary	Igneous	Sedimentary
Forest runoff				
Range	0.7 - 9	7 - 18	130 - 300	150 - 500
Mean	4.7	11.7	200	340
Forest + pasture				
Range	6 - 16	11 - 37	200 - 600	300 - 800
Mean	10.2	23.3	400	600
Agricultural areas				
Citrus	18		2240	
Pasture	15 - 75		100 - 850	
Cropland	22 - 100		500 - 1200	

B: Nutrient concentration in rain water (mg l^{-1}) [2]

	C_{PP}	C_{NP}
Range	0.015 - 0.1	0.3 - 1.6
Mean	0.07	1.0

C: Artificial P- and N-loads

The discharge per capita per year is 800-1800 g P and 3000-3800 g N in industrialized countries. If possible, local values should be applied.

The figures are based on an interpretation of the following references: [1] Dillon and Kirchner (1975), Lønholt (1973) and (1976), Vollenweider (1968) and Loehr (1974). [2] Schindler and Nighswander (1970), Dillon and Rigler (1974), Lee and Kluesener (1971) and Jørgensen et al. (1973).

Example 6.5
A lake has a surface area of 5 km^2 and a catchment area of 60 km^2, mainly agricultural. The average depth of the lake is 5 m. It receives an annual precipitation of 0.8 m and mechanical-biologically treated waste

water from 10,000 inhabitants.

Set up an approximate mass balance for the lake and suggest, based on Fig. 2.39, what should be done to reduce its eutrophication.

Solution.

E_P and E_N are chosen to be 80 mg $m^{-2}y^{-1}$ and 900 mg $m^{-2}y^{-1}$, respectively (see Table 6.17).

This gives an annual input of:

A. Runoff   : P-input 80 * 60 * 10^6 mg =   4800 kg
                                    N-input 900 * 60 * 10^6 mg =   54000 kg

B. Precipitation   : P-input 0.07 * 0.8 * 5 * 10^6 g =   280 kg
                                      N-input 1 * 0.8 * 5 * 10^6 =   4000 kg

C. Waste water   : 100 m^3/inh. 10 mg l^{-1} P 30 mg l^{-1} N
                                   corresp. to 1000 g P/inh. and 3000 g N
                                   (compare with Table 6.17 C)
                                   P-input: 10000 kg
                                   N-input: 30000 kg

About 2/3 of P-input comes from waste water, while only about 1/3 of N-input comes from this source.

The total P-loading at present corresponds to:

$$\frac{(4800 + 280 + 10000) * 10^3}{5 * 10^6} \text{ g P/m}^2 \text{ y} \quad \text{(see Fig. 2.39)}$$

$\approx$ 3 g P/m^2 y: A very high value.

A considerable improvement could be expected if the P-input from waste water were to be reduced by 90-99% (see Fig. 2.39).

The removal of nitrogen would not reduce the eutrophication correspondingly. A **simultaneous** removal of nitrogen would not improve the conditions, as:

1. 58,000 kg would still be the input and this is more than 10 times as much as the P-input after P-removal. The use of P and N by algae is estimated to be in the ratio of 1:9 (see Table 2.22).
2. Nitrogen-fixing algae might be present in the lake.

In addition to the eutrophication problem discharge of nitrogen

compounds into receiving waters can cause two environmental problems:
1. *Ammonium compounds may be nitrified, influencing the oxygen balance in accordance with the following equation:*

$$NH_4^+ + 2O_2 \rightarrow NO_3^- + H_2O + 2H^+ \qquad (6.26)$$

Municipal waste water, which is treated by a mechanico-biological method, but not nitrified, will normally contain approximately 28 mg ammonium-nitrogen per litre. In accordance with equation 6.18, this concentration might cause an oxygen consumption of 128 mg oxygen (see also section 2.7 and principle 2.16).
2. *Ammonia is toxic to fish* as mentioned in sections 2.4 and 4.7.

P.6.10. **Under all circumstances methods for the removal of both nitrogen and phosphorus must be available,**

and these methods are discussed in the following paragraphs.

P.6.11. **Phosphorus can be removed by chemical precipitation or ion exchange, while nitrogen can be removed by nitrification + denitrification, ammonia-stripping, ion exchange or oxidation by chlorine followed by treatment on activated carbon. Both nutrients can be removed using algal ponds.**

6.3.2.   Chemical precipitation of phosphorous compounds.

Chemical precipitation is widely used as a unit process for the treatment of waste water.

Municipal waste water is treated by precipitation with:

aluminium sulphate $(Al_2(SO_4)_3, 18H_2O)$

calcium hydroxide $(Ca(OH)_2)$ or

iron chloride $(FeCl_3)$

for removal of phosphorus.

When chemical precipitation is combined with a mechanico-biological treatment, the precipitation can be applied after the sand-trap using the primary settling for sedimentation, simultaneously with the biological treatment or after a complete mechanico-biological treatment. In these cases the precipitation is described respectively as, *direct precipitation, simultaneous precipitation and a post-treatment,* see Figs. 6.23, 6.24 and 6.25.

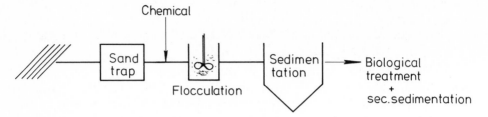

Fig. 6.23. Direct precipitation.

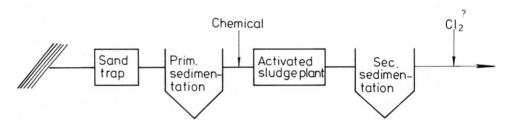

Fig. 6.24. Simultaneous precipitation. Recirculation of sludge is not shown.

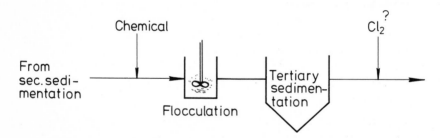

Fig. 6.25. Post-treatment.

Lately chemical precipitation has become popular in the treatment of municipal waste water in Sweden, U.S.A., Switzerland and other countries. However, the process was already in use for this purpose in Paris in 1740 and in England more than 100 years ago. Just before the Second World War, Guggenheim introduced the process as a treatment preceeding an activated-sludge plant (Rosendahl, 1970). However, the method did not find favour until the late 1960s due to the cost of chemicals and to the low reduction in BOD_5, which was supposed to be the major concern.

The main objective of applying chemical precipitation is the removal of phosphorus for the control of eutrophication. The efficiency of phosphorus removal is generally between 75 and 95% (Statens Naturvårdsverk, 1969). However, *in addition to this effect, chemical precipitation can also achieve the following:*

1. Substantial *reduction in the number of microorganisms,* particularly when calcium hydroxide is used (Buzell, 1967).
2. *Reduction in BOD_5* of 50-65% (Buzell, 1967).
3. *Reduction of non-biodegradable material* (so-called refractory material) in the same ratio as bio-degradable material (Jørgensen, 1976).
4. *Reduction in the concentration of nitrogenous compounds* due mainly to the removal of organic compounds.
5. *A reduction in heavy metal concentration,* which is essential (see also 6.5.2 and Nilsson, 1971).

P.6.12. **The removal of phosphorus entails the use of three chemicals: aluminium sulphate, iron(III) chloride and calcium hydroxide.**
As the composition of waste water and the cost of the chemicals vary from place to place, it is difficult to give a general recommendation on the choice of chemicals. It seems necessary to examine each case separately, although it is possible to draw some conclusions and set up some general guidelines.

Municipal waste water contains orthophosphate, and other phosphorus species. Table 6.18 shows the approximate concentration of the major phosphorous compounds. pH also influences the phosphorous species as demonstrated in Fig. 6.26.

TABLE 6.18
Approximate concentrations in typical raw domestic sewage

Form	mg P l^{-1}
Total	10
Orthophosphate	5
Tripolyphosphate	3
Pyrophosphate	1
Organic phosphate	≤ 1

Source: Jenkins et al. (1971)

The chemical precipitations form insoluble compounds with phosphorus species, but the solubility is naturally dependent on pH, as shown for the most important precipitated compounds in Fig. 6.27.

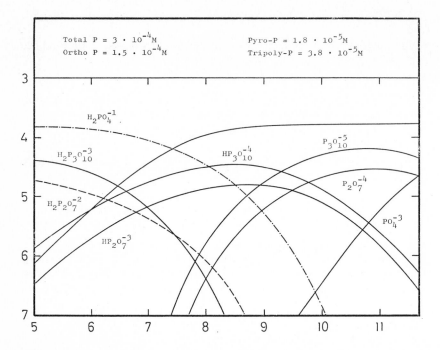

Fig. 6.26. -log conc. mol/l, phosphates. Plotted to pH. Reproduced from Jenkins et al., 1971.

For aluminium and iron phosphates the pH dependence is probably explained to a certain extent by the following process:

$$(MePO_4)_n + 3OH^- \longrightarrow Me_n(OH)_3(PO_4)_{n-1} + PO_4^{3-} \qquad (6.27)$$

Increasing OH^- concentration will effect this process to the right causing dissolution of the precipitated phosphate.

The precipitation of phosphates using calcium hydroxide is based on the formation of $Ca_{10}(PO_4)_6(OH)_2$, which will involve a high pH value: 10.5 - 12.0 (Nilsson, 1969).

However, many of the processes interact, and the phosphate species are able to form complexes with some metal ions, thereby significantly increasing phosphate stability. Table 6.19 shows some of the more important processes which may influence precipitation.

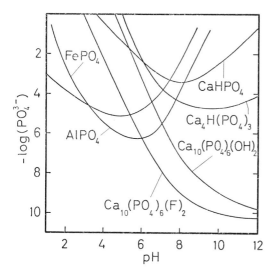

Fig. 6.27. The solubility of phosphate salts as a function of pH. Reproduced from Stumm and Leckie, 1970.

Aluminium sulphate is the precipitant most widely used in Scandinavia, where a technical grade named AVR is used. In Table 6.20 the composition of AVR is compared with an iron-free quality (Boliden, 1967 and 1969).

TABLE 6.19
Phosphate equilibria (Source: Sillen and Martell (1964))

Equilibrium	Log equilibrium constant 298°K
$H_3PO_4 = H^+ + H_2PO_4^-$	-2.1
$H_2PO_4^- = H^+ + HPO_4^{2-}$	-7.2
$HPO_4^{2-} = H^+ + PO_4^{3-}$	-12.3
$H_3P_2O_7^- = H^+ + H_2P_2O_7^{2-}$	-2.5
$H_2P_2O_7^{2-} = H^+ + HP_2O_7^{3-}$	-6.7
$HP_2O_7^{3-} = H^+ + P_2O_7^{4-}$	-9.4
$H_3P_3O_{10}^{2-} = H^+ + H_2P_3O_{10}^{3-}$	-2.3
$H_2P_3O_{10}^{3-} = H^+ + HP_3O_{10}^{4-}$	-6.5
$HP_3O_{10}^{4-} = H^+ + P_3O_{10}^{5-}$	-9.2
$Ca^{2+} + PO_4^{3-} = CaPO_4^-$	6.5
$Ca^{2+} + HPO_4^{2-} = CaHPO_4$	2.7
$Ca^{2+} + H_2PO_4^- = CaH_2PO_4^+$	1.4
$Ca^{2+} + P_2O_7^{4-} = CaP_2O_7^{2-}$	5.6
$Ca^{2+} + HP_2O_7^{3-} = CaHP_2O_7^-$	3.6
$Ca^{2+} + P_3O_{10}^{5-} = CaP_3O_{10}^{3-}$	8.1
$Ca^{2+} + HP_3O_{10}^{4-} = CaHP_3O_{10}^{2-}$	3.9
$Ca^{2+} + H_2P_3O_{10}^{3-} = CaH_2P_3O_{10}^-$	3.9

TABLE 6.20
Composition of AVR compared with an iron-free quality

| | Composition percentage | |
	Iron-free	AVR
Al_2O_3 total	17.1	15.1
Fe_2O_3	0.01	2.1
Fe	0	0.2
Water insoluble	0.05	2.5
H_2O, chemical	43.7	43.9
pH 1% water soluble	3.6	3.6
Size (mm)	0.25 - 2.0 granular	0.25 - 2.0 granular
Colour	white	brownish
Weight kg/l	0.90	0.95

The relationship between the addition of AVR-quality aluminium sulphate and the phosphorus concentration in municipal waste water is shown in Fig. 6.28 for different concentrations of phosphorus in the treated water. The dose of the chemical depends on the phosphorus concentration in the waste water and on the required concentration in the treated water (Nilsson and Isgard, 1971).

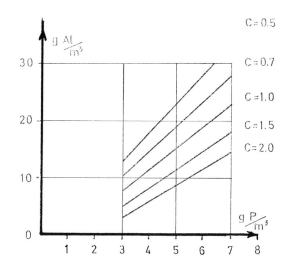

Fig. 6.28. The dependence of AVR addition on the P concentration of the waste water and the desired P concentration, indicated as C.
Source: Gustafson and Westberg (1968).

P.6.13. **The relationship, that describes the effect of chemical precipitation, can be interpreted by the Freundlich Adsorption isotherm:** (see Fig. 6.28)

$$\frac{C_o - C}{n} = a * C^b \qquad (6.28)$$

where

C_o = initial concentration of P (mg l^{-1})

C = final concentration of P (mg l^{-1})

n = dose of chemical mg l^{-1} Fe or Al

a and b = characteristic constants

The constants a and b can be found in Table 6.21.

TABLE 6.21
a and b in Freundlich's adsorption isotherm for aluminium sulphate and iron(III) chloride

Precipitation with	a	b
Aluminium sulphate	0.63	0.2
Iron(III) chloride	0.26	0.4

Example 6.6.

Find the dose of Al required to remove 90% P from waste water with a P-concentration of 10 mg l^{-1} and hypolimnic water with a P-concentration of 1 mg l^{-1}.

Solution.

$$\frac{C_0 - C}{n} = a * C_b$$

or

$$n = \frac{C_0 - C}{a * C^b} = \frac{C_0 - C}{0.63 * C^{0.2}}$$

<u>A. Waste water</u> $C_0 = 10$ $C = 1$

$$n = \frac{10 - 1}{0.63 * 1^{0.2}} = \frac{9}{0.63} = 14.3$$

<u>B. Hypolimnic water</u> $C_0 = 1$ $C = 0.1$

$$n = \frac{1 - 0.1}{0.63 * 0.1^{0.2}} = \frac{0.9}{0.63 * 0.63} = 2.3$$

As seen, about 6 times as much precipitant is needed in case A, but 10 times more phosphorous is removed. In practice, even relatively more precipitant will be used in case B than the theoretical calculations show here.

The values given in Fig. 6.28 and Table 6.19 are based on the assumption that the pH is close to the optimum (for aluminium sulphate 5.5 - 6.5, for iron(III) chloride 6.5 - 7.5). The relationship between the efficiency and the pH is plotted in Fig. 6.29 under what we could call normal conditions, that is when the phosphorus concentration is about 10 mg l^{-1}, the concentration of different phosphorus species is as shown in Table 6.18 and the calcium concentration is 20 - 50 mg l^{-1}.

Some references mention pH optima other than that shown in Fig. 6.29. This discrepancy can probably be explained by a difference in calcium concentration (Gustafson and Westberg, 1968).

Iron(III) chloride is more expensive to use than aluminium sulphate, but iron(II) sulphate is a waste product of the chemical industry and can easily be oxidized to iron(III) by the use of chlorine or by aeration, for instance in the activated-sludge process (Särkka, 1970). Therefore iron(II) sulphate has been widely used as a precipitant, especially for simultaneous precipitation.

The relationship between dose, phosphorus concentration and the final concentration in the treated water is similar to that for aluminium sulphate (see Fig. 6.30). The precipitation in this case also follows the Freundlich adsorption isotherm, but the constants a and b are different from those for aluminium sulphate.

As mentioned above the efficiency is dependent on pH and this relationship is also included in Fig. 6.29.

The efficiencies reported in the literature differ significantly more than those for aluminium sulphate, probably due to the difference in optimum pH (Thomas, 1965), but these observations might also be explained by the formation of complexes or a different ability to absorb or react chemically with the more comples phosphorus species.

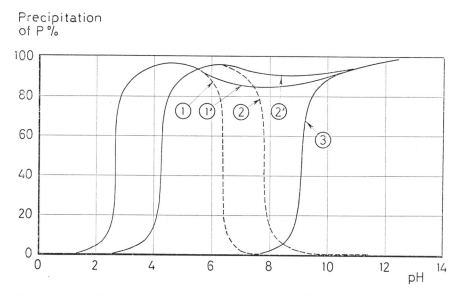

Fig. 6.29. The relationship between the efficiency of precipitation with coagulants and the pH. (1) for aluminium sulphate, (2) for iron(III) chloride, (3) for calcium hydroxide. (1') and (2') is (1) respectively (2) by presence of calcium ions. Reproduced from Gustafson and Westberg (1969).

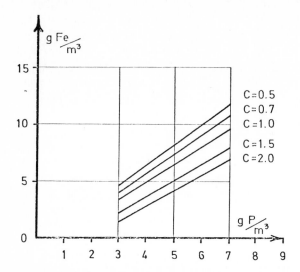

Fig. 6.30. The dependence of iron(III) salt addition on the P concentration of the waste water and the desired P concentration, indicated as C. Source: Gustafson and Westberg (1968).

Hydrated lime or calcium hydroxide is the cheapest of the three precipitants used for removal of phosphorus. However, greater amounts of calcium hydroxide are required to obtain a given efficiency. While 120 - 180 mg per litre of aluminium sulphate or iron(III) chloride are sufficient in most cases to achieve 85% removal or more, 2 to 5 times more calcium hydroxide will be needed.

For the precipitation of phosphorus compounds in waste water with aluminium sulphate and iron(III) chloride more than the stoichiometric amount is used, as pure orthophosphate is not precipitated, but adsorption is also involved. However, for calcium hydroxide precipitation a pH value of at least 10 is required to achieve a sufficiently high OH^- concentration.

P.6.14. Thus the amount of calcium hydroxide is determined by the alkalinity of the water.

(Alkalinity is a measure of the concentration of alkaline ions, mainly hydrogen carbonate and carbonate; it can be expressed as mg l^{-1} of calcium carbonate).

The practical relationship between the alkalinity and the mg of calcium hydroxide required to reach pH 11, which assure effective precipitation, is illustrated in Fig. 6.31. Most waste water has an alkalinity in the range of

150 - 300 mg calcium carbonate per litre, which, as can be seen from the figure, corresponds to about 300 to 480 mg calcium hydroxide per litre.

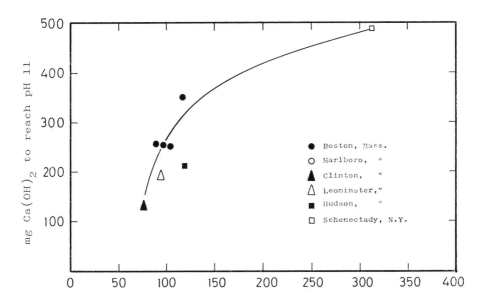

Fig. 6.31. Alkalinity mg $CaCO_3$ l^{-1}. From Jenkins et al. (1971)

Generally higher efficiencies are obtained by precipitation with hydrated lime than by the other two precipitants - 90 - 95% efficiency being common in practice. The orthophosphate is precipitated with an efficiency of almost 100%, while polyphosphates are precipitated with more modest efficiency giving an overall efficiency close to 95%.

The disadvantage of precipitating with lime is the high pH, which makes an adjustment after the precipitation necessary. Carbon dioxide produced by incineration of the sludge can be used for this purpose (see Fig. 6.32), where the calcium oxide produced by the incineration is also reused to certain extent at the same time as the carbon dioxide is applied to adjust the pH. In this process the calcium oxide can be recycled 3 - 5 times corresponding to a more stoichiometric ratio between calcium and phosphate.

The calcium hydroxide produced in this manner can be used as fertilizers (Dryden and Stern, 1968).

The amount of sludge produced by chemical precipitation is dependent on several factors, of which the major factor is the dose of chemicals (see Table 6.22).

P.6.15. The sludge formed from the precipitation of phosphorus with aluminium sulphate can be treated together with anaerobic biological sludge, without difficulty, while that formed from the precipitation with iron compounds might be dissolved due to reduction of iron(III) to iron(II) phosphate, which can react in accordance with the following process:

$$Fe_3(PO_4)_2 + 3H_2S \ — \ 3FeS + 2H_2PO_4^- + 2H^+ \qquad (6.29)$$

The same process occurs, when ironrich sediment releases phosphorus due to anaerobic conditions, see section 4.9.

TABLE 6.22
Additional sludge by precipitation
D = dose expressed as g/m^3 Al, Fe or hydrated lime

Chemical	Additional sludge in g per m^3 waste water
Aluminium sulphate	4 D
Iron(III) salt	2.5 D
Hydrated lime	1 - 1.5 D

P.6.16. Without a pH-adjustment the sludge produced by the calcium hydroxide precipitation can hardly be treated by anaerobic digestion. If the chemical sludge is mixed with a large volume of biological sludge the pH might be sufficiently low to allow an anaerobic treatment.

The same considerations are valid for an aerobic sludge treatment with the exception that the above-mentioned reduction of iron, of course, cannot take place.

Filtration, centrifugation or drying of chemical sludge do not cause any additional problems. Recirculation of sludge is practised in many plants and often means a reduced consumption of chemicals. Recovery of aluminium sulphate has even been suggested. By treatment of filtered sludge with sulphuric acid after aluminium sulphate precipitation the aluninium ions go into solution as aluminium sulphate, which can be reused. However, the economical advantage is in most cases too modest to pay for this extra process.

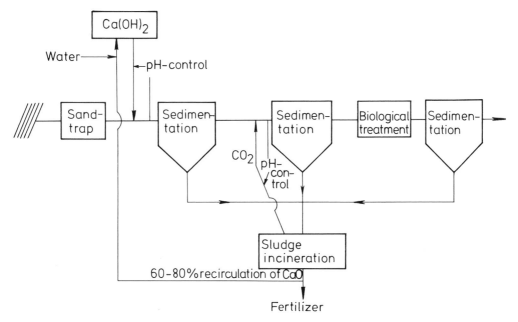

Fig. 6.32. Chemical precipitation with partial recirculation of calcium oxide and use of CO_2 from incineration for adjustment of pH.

Several heavy metals are removed quite effectively by precipitation. Aluminium sulphate is able to precipitate lead, copper and chromium with high efficiency, while cadmium and zinc are only precipitated partially. Iron-based chemicals produce the same effects at the same pH, but as the pH is often lower with this precipitation, in practice iron salts often give a lower efficiency than aluminium sulphate. Calcium hydroxide removes almost all heavy metals very efficiently, as most metal hydroxides have a very low solubility at pH 10-11. Calcium hydroxide is used (see 6.4.2) to remove heavy metals from industrial waste water.

The relationship between pH and the solubility for metal ions + metal hydroxide complexes is illustrated in Fig. 6.33 for 6 important metal ions. However, municipal waste water will often contain compounds that are able, through formation of complexes, to increase the solubility for most metal ions. This is demonstrated in Fig. 6.34, where the solubility of the same 6 metal ions is plotted against pH for a water containing NTA (Nitrilo Three Acetate), chloride and calcium ions.

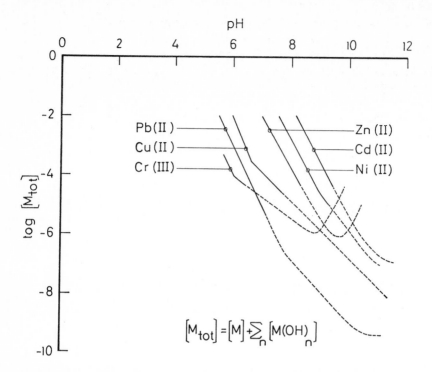

Fig. 6.33. The solubility of pure metal hydroxides as a function of pH. (From Nilsson, 1971).

P.6.17. **These additional effects of the precipitation process must not be overlooked. For treatment of municipal waste, it might also be important to precipitate toxic organics and heavy metals in addition to achieving a reduction of BOD_5 of approximately 60% and a 80-95% removal of phosphorus compounds.**

If the municipal waste water contains more or less treated industrial waste water, in some cases it might be a better solution to use chemical precipitation rather than mechanical-biological treatment, depending on the receiving water. It seems worthwhile, at least as a primary step, to consider the chemical precipitation treatment of waste water for discharges to lakes or the sea.

The mentioned effects can, however, only be obtained by a current process control. An unacceptable efficiency results from:

1. **Too low a dose.** Recirculation of the chemical sludge might be an

attractive remedy.
2. **Wrong pH**. In some cases it might be necessary to adjust the pH by the addition of calcium hydroxide or sulphuric acid.
3. **The flocculation is insufficient**. Too small flocs that are unsettleable are formed. It might be necessary to change the design of the flocculator (Jørgensen, 1979).

P.6.18. The flocculation can be further improved by the application of synthetic polyflocculants.

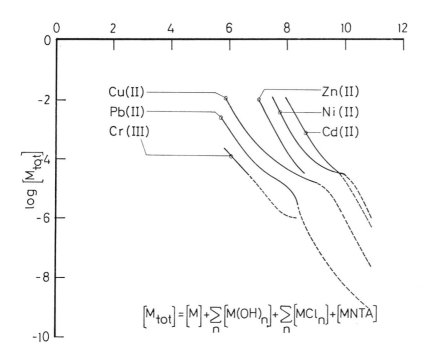

Fig. 6.34. The solubility of pure metal hydroxides, as a function of pH, in water containing NTA ($10^{-4.7}$mol/l), Cl$^-$ ($10^{-2.4}$mol/l) and Ca^{2+} ($10^{-2.7}$mol/l) at pH 7.0 and $10^{-2.3}$mil/l at pH 11.0). (From Nilsson, 1971).

Clay, starch and gelatine can be used to accelerate and stimulate the flocculation, but the synthetic polyelectrolytes that have appeared on the market during the last decade or two seem to give a better efficiency (Black, 1960). They work mainly on flocs in the range of 1-50 μm.

TABLE 6.23
Synthetic organic polymeric flocculants

A. <u>Cationic polyelectrolytes</u>
  Polydiallyldimethylammonium

$$
\cdots \!-\! \underset{\underset{H_2C}{|}}{CH} \overset{CH_2}{\diagup} \underset{\underset{CH_2}{|}}{CH} \!-\! CH_2 \!-\! \cdots
$$

(structure: diallyldimethylammonium ring with H_2C and CH_2 connected to N^+, bearing H_3C and CH_3)

B. <u>Anionic polyelectrolytes</u>
1. Polyacrylic acid

$$
\cdots \!-\! CH_2 \!-\! \underset{\underset{\underset{O^-}{\diagdown}}{C \,=\, O}}{CH} \!-\! \cdots
$$

2. Hydrolyzed polyacrylamide: A mixture of subunits B 1 and C 1
3. Polystyrene sulphonate

$$
\cdots \!-\! CH_2 \!-\! CH \!-\! \cdots
$$

(benzene ring bearing SO_3^-)

C. <u>Nonionic polymers</u>
1. Polyacrylamide

$$
\cdots \!-\! CH_2 \!-\! \underset{\underset{\underset{NH_2}{\diagdown}}{C \,=\, O}}{CH} \!-\! \cdots
$$

2. Polyethylene oxide

$$
\cdots \!-\! CH_2 \!-\! CH_2 \!-\! O \!-\! \cdots
$$

The long-chain polymers are able to collect many small flocs and agglomerate them into larger groups. The polymers are able to form a bridge between the small particles. For flocculation of municipal waste water anionic polyelectrolytes are chiefly used, while the cationic polyelectrolytes are used for flocculation of different types of industrial waste water or even for improving the dewatering of municipal sludge.

Several synthetic polymers are on the market, and non-ionic, anionic or cationic ones are available, see Table 6.23. For waste water treatment polyacrylamide (see below) is generally used (Robert, 1970).

$$---- \quad CH_2 \quad - \quad CH_2 \quad ---- \\ \qquad\qquad\qquad | \\ \qquad\qquad\qquad C=O \\ \qquad\qquad\qquad | \\ \qquad\qquad\qquad N \\ \qquad\qquad / \quad \backslash \\ \qquad\quad H \qquad H$$

Polyacrylamide

It is possible to produce polyacrylamides with a molecular weight of 4-10 million. By a copolymerization of acrylamide and acrylic acid it is possible to prepare an anionic polymer of this type.

The main effects of using polymers for flocculation are (Boeghlin, 1972 and Ericsson and Westberg, 1968):
1. Gaining a higher settling rate,
2. Reduction in turbidity, caused by small flocs.

6.3.3.   Nitrification and denitrification.

P.6.19. The nitrification process eliminates oxygen consumption related to the oxidation of ammonium

$$NH_4^+ + 2O_2 \quad - \quad NO_3^- + H_2O + 2H^+ \qquad\qquad (6.30)$$

As seen it is the same process as otherwise would occur in the ecosystem (see section 2.6).

Effective nitrification occurs when the age of the sludge produced by biological treatment is greater than the reciprocal rate constant of the nitrifying micro-organisms.

Bernhart (1975) has demonstrated that it is possible to oxidize ammonia in a complex organic effluent by biological nitrification.

The sludge age, *f*, is defined as (see also 6.2.3):

$$\frac{X}{\Delta X} = f \qquad\qquad\qquad (6.31)$$

where X = the mass of biological solid in the system and ΔX = the sludge yield.

The relationship between nitrification and sludge age is shown in Fig. 6.35.

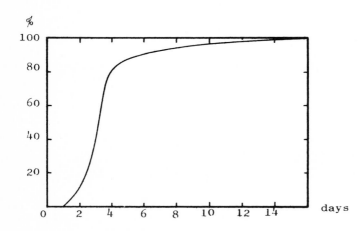

Fig. 6.35.   Relationship between nitrification and sludge age.

Nitrification results from a two-step oxidation process, see 6.2.3, equations (6.5) and (6.6).

The optimum pH range for Nitrosomonas is 7.5-8.5 and for Nitrobacter 7.7-7.9. The rate seems to be dependent on the ammonium ion concentration at concentrations in excess of 0.5 mg/l, which is considerably lower than those generally found in waste water containing ammonium ions. Heavy metals are toxic at rather low concentrations. Toxic levels of about 0.2 mg/l are reported for chromium, nickel and zinc.

Temperature exerts a profound effect on nitrification (see also sections 2.4, 2.6 and Table 6.8). Downing (1966) has reported that the influence of temperature on the rate coefficient can be expressed as follows:

$$K_N = 0.18 * 1.128^{T-15} \quad (24h^{-1}) \tag{6.32}$$

As can be seen, $K_N = 0.18$ at 15°C. Compare with Table 6.8.

However, the amount of nitrogen which can be remoced per m^3 is limited. This is illustrated in Fig. 6.36, where it is seen that it is not possible to remove more than 21 mg/l per 24h. Even the ammonium concentration in the inflow is increased to about 1000 mg/l.

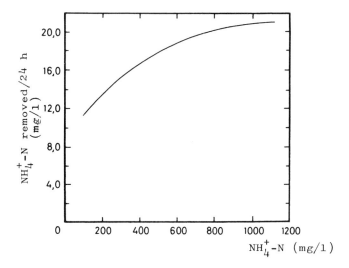

Fig. 6.36. Ammonium removal per 24h plotted against concentration.

P.6.20. Nitrate can be reduced to nitrogen and dinitrogen oxide by many of the heterotrophic bacteria present in activated sludge, but the process requires anaerobic conditions.

The pH effects the process rate, the reported optimum being above 7.0. The same process is observed in anaerobic sediment.

As the denitrifying organisms are heterotrophic, they require an organic carbon source. It is possible either to add the carbon source, e.g. by using methanol or molasses, or to use the endogenous by-product as the food supply.

If acetate is used as a carbon source the process is:

$$5CH_3COO^- + 8NO_3^- \; - \; 4N_2 + 7HCO_3^- + 3CO_3^{2-} + 4H_2O \tag{6.33}$$

Fig. 6.37 illustrates the influence of the COD/NO$_3^-$-N ratio on the denitrification efficiency. The plots shown in this figure are based on experiments for the denitrification of waste water from the manufacture of nitrogen fertilizers.

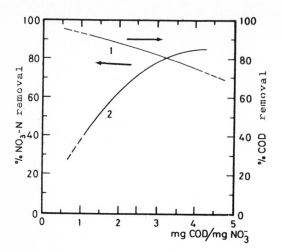

Fig. 6.37.   Efficiency plotted against the ratio COD/NO$_3^-$-N.

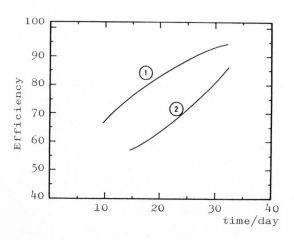

Fig. 6.38.   Denitrification efficiency plotted against retention time. (1) COD removal and (2) denitrification.

The denitrification efficiency as a function of the retention time is shown in Fig. 6.38. As seen, a minimum of 10 days retention time is required for the anaerobic denitrification process.

However, the ratio between COD and the nitrate that is removed decreases with increasing retention time (see Fig. 6.39).

The rate of denitrification increases with increasing concentrations of a available carbon and of nitrate. Francis et al. (1975) report a successful denitrification of waste water that contained more than 1000 ppm nitrate-nitrogen, which should be compared with the concentration of nitrate-nitrogen in municipal waste water of 20-40 ppm.

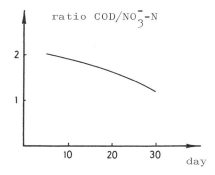

Fig. 6.39. Ratio COD removal per unit of time to NO_3^--N. Removal per unit of time plotted against retention time.

6.3.4. Stripping.

P.6.21. The stripping process is used to remove volatile gases, such as hydrogen sulphide, hydrogen cyanide and ammonia by blowing air through the waste water.

The removal of ammonia by stripping is used in the treatment of municipal waste water, but it has also been suggested for the treatment of industrial waste water or for the regeneration of the liquid used for elution of ion exchangers (Jørgensen, 1976).

P.6.22. The rate at which carbon dioxide, hydrogen sulphide, ammonia and hydrogen cyanide can be removed by air stripping is highly dependent on pH, since all four of these volatile gases are acids or bases.
Ammonia stripping is based on the following process:

$$NH_4^+ \ \text{—} \ NH_3 + H^+$$

The equilibrium constant for this process is $10^{-9.25}$ at 18°C, which means that:

$$\frac{[NH_3] \ [H^+]}{[NH_4^+]} = 10^{-9.25}$$

By separating H^+ in this equation and converting to a logarithmic form, we get:

$$pH = 9.25 + \log \frac{[NH_3]}{[NH_4^+]} \hspace{3cm} (6.34)$$

The same considerations are used, when the ammonium concentration is found in an aquatic ecosystem to estimate the toxicity level of the water, see section 2.4 and Table 2.13.

From equation (6.34) we can see that at pH 9.25, 50% of the total ammonia-nitrogen is in the form of ammonia and 50% in the form of ammonium. Correspondingly the ratio between ammonia and ammonium is 10 at pH 10.25 and 100 at pH 11.25. Consequently it is necessary to adjust the pH to 10 or more before the stripping process is used. Due to the very high solubility of ammonia in water a large quantity of air is required to transfer ammonia effectively from the water to the air.

The efficiency of the process depends on:
1. **pH,** in accordance with the considerations mentioned above.
2. **The temperature**. The solubility of ammonia decreases with increasing temperature. The efficiency at three temperatures - 0°C, 20°C and 40°C - is plotted against the pH in Fig 6.40.
3. **The quantity of air per m^3 of water treated**. At least 3000 m^3 of air per m^3 of water are required (see Fig. 6.41).
4. **The depth of the stripping tower**. The relationship between the efficiency and the quantity of air is plotted for three depths - Fig. 6.41.
5. **The specific surface of the packing** (m^2/m^3). The greater the specific surface the greater the efficiency.

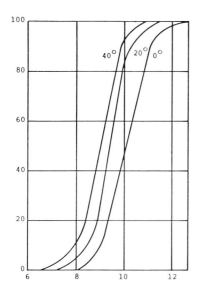

Fig. 6.40. Stripping efficiency as function of pH at three different temperatures.

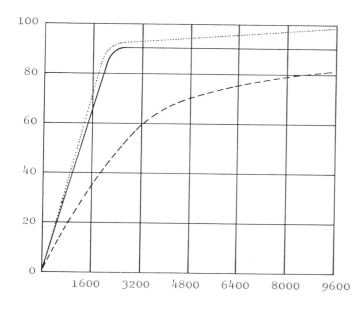

Fig. 6.41. Efficiency as function of m³ of air per m³ of water for three different depths. ..... line = 8 m, ———— line = 6.7 m, --- line = 4 m.

    Fig. 6.42 demonstrates the principle of a stripping tower. The waste water treatment plant at Lake Tahoe, California, includes a stripping process. 10,000 m³ of waste water is treated per 24h at a cost of approximately 3 US cents per m³. The investment cost is in the order of 5 US cents per m³ (based on 16% depreciation and interest per year of the investment).

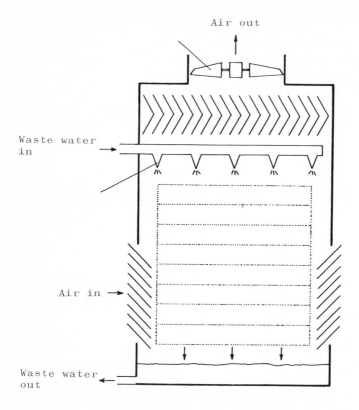

Fig. 6.42. The principle of a stripping tower.

The cost of stripping is relatively small, but the process has two crucial limitations:
1. It is practically impossible to work at temperatures below 5-7°C. The large quantity of air will cause considerable evaporation, which means that *the water in the tower is freezing.*
2. *Deposition of calcium carbonate* can reduce the efficiency or even block the tower.

Due to limitation 1) it will be necessary to use warm air for the stripping during winter in temperate climates, or to instal the tower indoors. This makes the process too costly for plants in areas with above 10,000 inhabitants and limits the application for treatment of bigger volumes to tropical or maybe subtropical latitudes.

However, it seems advantageous to combine the process with ion exchange. By ion exchange it it possible to transfer ammonium from 100 m^3 of waste water to 1 m^3 of elution liquid. As elution liquid is used as a base (see 6.3.6), a pH adjustment is not required. Furthermore, the ratio of air to water is not increased although the elution liquid has a 100 times greater ammonia concentration. Consequently a stripping tower 100 times smaller with a preconcentration by ion exchange can be used.

Recovery of ammonia, for use as a fertilizer, also seems easier after the preconcentration. The stripped ammonia could, for example, easily be absorbed into a sulphuric acid solution, for the production of ammonia sulphate.

As mentioned in the introduction, the shortcoming of some of the technological solutions is, that *they do not consider a total environmental solution, as they solve one problem but create a new one.* The stripping process is a characteristic example, since *the ammonia is removed from* the waste water *but transferred to the atmosphere, unless recovery of* ammonia is carried out. In each specific case it is necessary to assess whether the air pollution problem created is greater that the water pollution problem solved.

6.3.5. Chlorination and adsorption on activated carbon.

P.6.23. **Chlorine can oxidize ammonia in accordance with the following reaction sheme:**

$$Cl_2 + H_2O \rightarrow HOCl + HCl$$

$$NH_3 + HOCl \rightarrow NH_2Cl + H_2O \qquad (6.35)$$

$$NH_2Cl + HOCl \rightarrow NHCl_2 + H_2O$$

$$NHCl_2 + HOCl \rightarrow NCl_3 + H_2O$$

Activated carbon is able to adsorb chloramines, and so a combination of chlorination and adsorption on activated carbon can be applied for removal of ammonia.

The most likely reaction for chloramine on activated carbon is:

$$C + 2NHCl_2 + H_2O - N_2 + 4H^+ + 4Cl^- + CO \qquad (6.36)$$

Further study is needed, however, to show conclusively that surface oxidation results from this reaction. Furthermore, it is important to know that the Cl_2/NH_3-N oxidized mole ratio is 2:1, which is required for ammonium oxidation by this pathway.

The monochloramine reaction with carbon appears more complex. On fresh carbon the reactio is most probably:

$$NH_2Cl + H_2O + C \rightarrow NH_3 + H^+ + Cl^- + CO \qquad (6.37)$$

After this reaction has proceeded to a certain extent, partial oxidation of monochloramine is observed, possibly according to the reaction:

$$2NH_2Cl + CO \rightarrow N_2 + H_2O \quad 2H^+ + 2Cl^- + C \qquad (6.38)$$

It has been observed that acclimation of fresh carbon is necessary before monochloramine can be oxidized.

In the removal of ammonia with a dose of chlorine followed by contact with activated carbon, pH control can be used to determine the major chlorine species. The studies reported herein indicate that a pH value near 4.5 should be avoided, because $NHCl_2$ predominates and thus 10 parts by weight of chlorine are required for each part of NH_3-N oxidized to N_2. At a slightly higher pH and acclimated carbon, the portion of monochloramine increases and the chlorine required per unit weight of NH_3-N oxidized should approach 7.6 parts, ignoring the chlorine demand resulting from other substances. However, further testing should be used to verify this conclusion in each individual case.

When accidental overdosing of chlorine has occurred or after an intentional addition of large quantities of chlorine to accelerate disinfection, it will be desirable to remove the excess chlorine. This is possible using a reducing agent, such as sulphur dioxide, sodium hydrogen sulphite or sodium thiosulphate.

Complete removal of the 25-40 mg per litre ammonium-N is far too costly by this method. Chlorine costs about 25-30 US cents per kg, which means that the chlorine consumption alone will cost about 6 US cents per m^3 waste. When the capital cost and the other operational costs are added the total treatment cost will be as high as 13-17 US cents per m^3, which is considerably more expensive than other removal methods.

It is possible to use chlorine to oxidize ammonium compounds to free nitrogen, but this process involves even higher chlorine consumption and, therefore, is even more expensive.

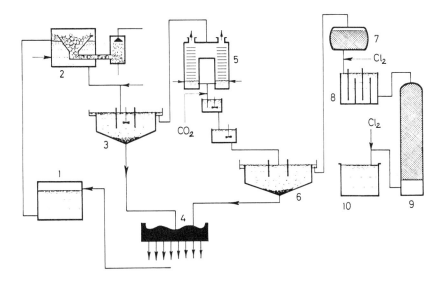

Fig. 6.43. Waste water treatment plant Pretoria. After mechanical-biological treatment: 2) aeration, 3) lime precipitation, 4) sludge drying, 5) ammonia stripping, 6) recarbonization, 7) sand filtration, 8) chlorination, 9) adsorption on activated carbon, 10) chlorination.

The method has, however, one advantage: by using sufficient chlorine it is possible to obtain a very high efficiency. This has meant that the method has found application *after other ammonium removal methods,* where high efficiencies are required. This is the case when the waste water is reclaimed, for example in the two plants shown in Figs. 6.43 and 6.44. As seen it is necessary to use several treatment processes to achieve a sufficient water quality after the treatment. Chlorination and treatment on activated carbon are used as the last treatment to assure good ammonium removal and sufficient disinfection of the water. An additional chlorination is even used after the treatment on activated carbon to ensure a chlorine residue in the water supply system.

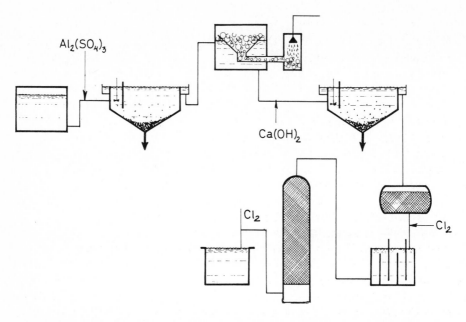

Fig. 6.44. Production of potable water from waste water (Windhoek).

6.3.6. Application of ion exchange for removal of nutrients.

P.6.24. Ion exchange is a process in which ions on the surface of the solid are exchanged for ions of a similar charge in a solution with which the solid is in contact.
Ion exchange can be used to remove undesirable ions from waste water. Cations (positive ions) are exchanged for hydrogen or sodium and anions (negative ions) for hydroxide or chloride ions.

The cation exchange on a hydrogen cycle can be illustrated by the following reaction, using, in this example, the removal of calcium ions, which are one of the ions (Ca^{2+} and Mg^{2+}) that cause hardness of water:

$$H_2R + Ca^{2+} \rightarrow CaR + 2H^+ \tag{6.39}$$

where R represents the resin.

The anion exchange can be similar illustrated by the following reactions:

$$SO_4^{2-} + R(OH)_2 \rightarrow SO_4R + 2OH^- \qquad (6.40)$$

When all the exchange sites have been replaced with calcium or sulphate ions, the resin must be regenerated. The cation exchanger *can* be regenerated by passing a concentrated solution of sodium chloride through the bed, while the anion exchanger, which in this case is of hydroxide form, must be treated by a solution of hydroxide ions, e.g. sodium hydroxide.

Ion exchange is known to occur with a number of natural solids, such as soil, humus, metallic minerals and clay, see also section 4.9. *Clay*, and in some instances other natural materials, can be used for demineralization of drinking water. In the context of adsorption, the ability of aluminium oxide to make a surface ion exchange should be mentioned, but the natural clay mineral, *clinoptilolite*, can also be used for waste water treatment as it has a high selectivity for removal of ammonium ions.

Synthetic ion exchange resins consist of a network of compounds of high molecular weight to which ionic functional groups are attached. The molecules are cross-linked in a three-dimensional matrix and the degree of the cross-linking determines the internal pore structure of the resin. Since ions must diffuse into and out of the resin, ions larger than a given size may be excluded from the interaction through a selection dependent upon the degree of cross-linking. However, the nature of the groups attached to the matrix also determines the ion selectivity and thereby the equilibrium constant for the ion exchange process.

The cation exchangers contain functional groups such as *sulphonic* $R-SO_3-H$ - *carboxylic*, $R-COOH$ - *phenolic*, $R-OH$ and *phosphonic*, $R-PO_3H_2$ (R represents the matrix). It is possible to distinguish between strongly acidic cation exchangers derived from a strong acid, such as H_2SO_4, and weakly acidic ones derived from a weak acid, such as H_2CO_3. It is also possible to determine a pK-value for the cation exchangers in the same way as it is for acide generally.

This means:

$$R-SO_3H \rightarrow R-SO_3^- + H^+$$

$$\frac{[H^+] \, * \, [R-SO_3^-]}{[R-SO_3H]} = K \qquad pK = -logK \qquad (6.41)$$

Anion exchange resins contain such functional groups as *primary amine,* R-NH$_2$, *secondary amine,* R-R$_1$NH, and *tertiary amine* R-R$_1$-R$_2$N groups and the *quaternary ammonium group* R-R$_1$R$_2$R$_3$N$^+$OH$^-$.

It can be seen that the anion exchanger can be divided into weakly basic and strongly basic ion exchangers derived from quaternary ammonium compounds.

It is also possible to introduce ionic groups onto natural material. This is done using cellulose as a matrix, and due to the high porosity of this material it is possible to remove even high molecular weight ions.

Preparation of cation exchange resin, using hydrocarbon molecules as a matrix, is carried out by polymerization of such organic molecules as styrene and metacrylic acid.

The degree of cross-linking is determined by the amount of divinylbenzene added to the polymerization. This can be illustrated by the example shown below.

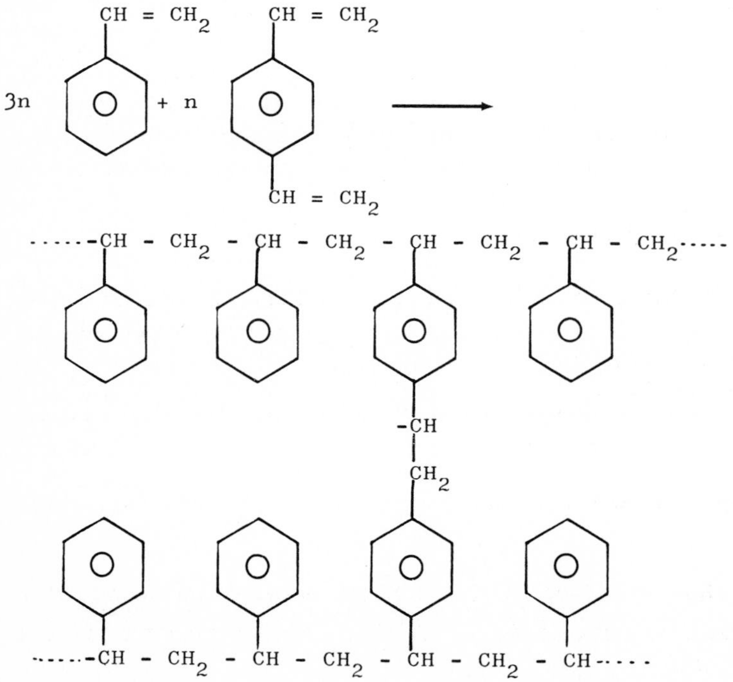

P.6.25. It is characteristic that the exchange occurs on an

equivalent basis. The capacity of the ion exchanger is usually expressed as equivalent per litre of bed volume.

When the ion exchange process is used for reduction of hardness, the capacity can also be expressed as kg of calcium carbonat per m³ of bed volume. Since the exchange occurs on an equivalent basis, the capacity can be found based either on the number of ions removed or the number of ions released. Also, the quantity of regenerant required can be calculated from the capacity. However, neither the resin nor the regeneration process can be utilized with 100% efficiency.

The Fig. 6.45 plot is often used as an illustration of the preference of an ion exchange resin for a particular ion. As seen, the percentage in the resin is plotted against the percentage in solution.

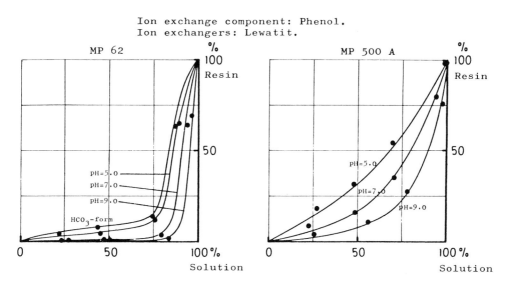

Fig. 6.45. Illustration of the preference of an ion exchange resin for a particular ion.

The selectivity coefficient, K_{AB}, is not actually constant, but is dependent upon experimental conditions. A selectivity coefficient of 50% in solution is often used $= a_{50\%}$.

If we use concentration and not activity, it will involve, for monocharged ions:

$$c_B = c_A$$

$$a_{50\%} = K_{AB,50\%} = \frac{c_{RA}}{c_{RB}} \qquad (6.42)$$

The plot in Fig. 6.45 can be used to read $a_{50\%}$.

The selectivity of the resin for the exchange of ions is dependent upon the ionic charge and the ionic size. An ion exchange resin generally prefers counter ions of high valence. Thus, for a series of typical anions of interest in waste water treatment one would expect the following order of selectivity:

$$PO_4^{3-} > SO_4^{2-} > Cl^-.$$

Similar for a series of cations:

$$AL^{3+} > Ca^{2+} > Na^+.$$

But this is under circumstances where the internal pore structure of the resin does not exclude the ions mentioned from reaction. Organic ions are often too large to penetrate the matrix of an ion exchange, which is, of course, more pronounced when the resins considered have a high degree of cross-linking. As most kinds of water and waste water contain several types of ions besides those which must be removed it is naturally a great advantage to have a resin with a high selectivity for the ions to be removed during the ion exchange process.

The resin utilization is defined as the ratio of the quantity of ions removed during the actual treatment to the total quantity of ions that could be removed at 100% efficiency; this is the theoretical capacity. The regeneration effeciency is the quantity of ions removed from the resins compared to the quantity of ions present in the volume of the regererant used.

Weak base resin has a significant potential for removing certain organic compounds from water, but the efficiency is highly dependent upon the pH.

It seems reasonable to hypothesize that an adsorption is taking place by the formation of a hydrogen bond between the free amino groups of the resin and hydroxyl groups of the organic substance taken up. As pH decreases, so that the amino groups are converted to their acidic form, the adsorption capacity significantly decreases.

The exchange reaction between ions in solution and ions attached to the resin matrix is generally reversible. The exchange can be treated as a simple stoichiometric reaction. For cation exchange the equation is:

$$A^{n+} + n(R^-)B^+ \rightharpoonup nB^+ + (R^-)_nA^{n+} \qquad (6.43)$$

The ion exchange reaction is selective, so that the ions attached to the fixed resin matrix will have preference for one counter ion over another. Therefore the concentration of different counter ions in the resin will be different from the corresponding concentration ration in the solution.

According to the law of mass action, the equilibrium relationship for reaction (6.43) will give:

$$K_{AB} = \frac{a_B^n * a_{RA}}{a_A * a_{RB}^n} \qquad (6.44)$$

where a_B and a_A are the activity of the ions B^+ and A^{n+} in the solution and correspondingly a_{RB} and a_{RA} are the activities of the resin in B- and A-form, respectively.

As mentioned above the clay mineral, clinoptilolite, can take up ammonium ions with a high selectivity. This process is used for the removal of ammonium from municipal waste water in the U.S.A., where good quality clinoptilolite occurs.

Clinoptilolite has less capacity than the synthetic ion exchanger, but its high selectivity for ammonium justifies its use for ammonium removal. The best quality clinoptilolite has a capacity of 1 eqv. or slightly more per litre. This means that 1 litre of ion exchange material can remove 14 g ammonium -N from waste water, provided all the capacity is occupied by ammonium ions. Municipal waste water contains approximately 28 g (2 eqv.) per m^3, which means that 1 m^3 of ion exchange material can treat 500 m^3 waste water (which represents a capacity of 500 bed volumes). The practical capacity is, however, considerably less - 150-250 bed volumes - due to the presence of other ions that are taken up by the ion exchange material, although the selectivity is higher for ammonium that for the other ions present in the waste water. The concentration of sodium, potassium and calcium ions might be several eqv. per litre, compared with only 2 eqv. per litre of ammonium ions.

Clinoptilolite is less resistant to acids or bases than synthetic ion exchangers. A good elution is obtained by the use of sodium hydroxide, but as the material is dissolved by sodium hydroxide a very diluted solution should be used for elution to minimize the loss of material. A mixture of sodium chloride and lime is also suggested as alternative elution solution.

The flow rate through the ion exchange column is generally smaller for clinoptilolite than for synthetic material resin - 10 m as against 20-25 m.

The elution liquid can be recovered, as mentioned in 6.3.4, by air stripping. The preconcentration on the ion exchanger makes this process attractive - the sludge problem is diminished and the cost of chemicals reduced considerably. For further details about this method of recovery, see Jørgensen (1975).

Phosphate can also be removed by an ion exchange process. Synthetic ion exchange material can be used, but due to the presence of several eqv. of anions, it seems a better solution to use activated aluminium, which is a selective adsorbant (ion exchanger) for phosphate.

In accordance with Neufeld and Thodos (1969) the phosphorus adsorption on aluminia is a dual process of ion exchange and chemical reaction, although many authors refer to the process as an adsorption.

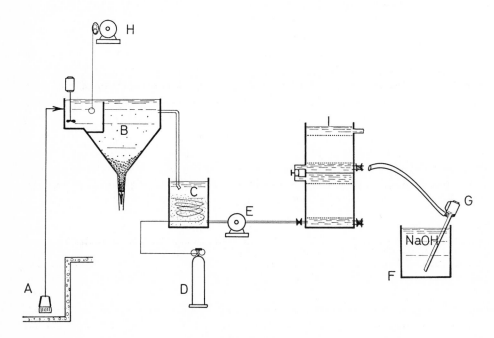

Fig. 6.46. Flowchart of a combination of chemical precipitation and ion exchange. (A) a submersible pump, (B) the settling basin, (C) an intermediate vessel, where carbon dioxide is added, (D) a carbon dioxide bomb (50 atm., 25 lit.), (E) a pump feeding the ion exchangers, (F) elution liquid, (G) a handpump, (H) a dosing pump.

The selectivity for phosphorus uptake is very high (Yee, 1960), and only a minor disturbance from other ions is observed. Municipal waste water contains about 10 mg phosphorus or 1/3 eqv. P per 1 m^3. If the aluminia is activated by nitric acid before use, the theoretical capacity will be as high as 0.561 eqv. per litre or 1500-3000 bed volumes. The practical capacity is lower - about 1000 bed volumes (Ames, 1969).

Also activated aluminia can be regenerated by sodium hydroxide, but again the use of a low concentration is recommended to reduce loss of material (Ames, 1970).Aluminia has a high removal efficiency for all inorganic phosphorus species in waste water, but removal efficiency for organic phosphorus compounds is somewhat lower. Jørgensen (1978) has suggested the use of an anionic cellulose exchanger together with activated aluminia as a mixed bed ion exchange column to improve the overall efficiency for removal of all phosphorus species present in waste water, including the organic ones.

A combination of chemical precipitation and ion exchange has developed as an alternative to the mechanical-biological-chemical treatment method. A flowchart of such a plant is shown in Fig. 6.46.

After the chemical precipitation the waste water is treated on two ion exchangers (which, however, could be in one mixed bed column). The first ion exchanger is cellulose based for removing proteins and reducing BOD_5. The second column could be either clinoptilolite or activated aluminia + the above mentioned cellulose anion exchanger. A plant using this process has been in operation since 1973 in Sweden, giving results comparable with or even better than the generally applied 3 steps treatment (see Table 6.24). The capital cost and operational costs are approximately the same as for a 3 steps plant. However, the plant produces 2-4 times less sludge than the normal 3 step plant, giving a correspondingly lower sludge treatment cost.

TABLE 6.24
Analysis of municipal waste water after chemical precipitation + ion exchange (mg l^{-1}) (flowchart see Fig. 6.46)

BOD_5	10 - 18
COD	30 - 45
P	< 0.1
N	10 - 20

6.3.7.   Algal ponds.

P.6.26.   The uptake of nutrients by algae can be utilized as a waste water treatment process. Per 100 g of dry matter phytoplankton will, on average, contain 4-10 g nitrogen and 0.5-2.5 g phosphorus.

The process is known from the entrophication problem, see section 2.7. It can be considered a biological nutrient removal.

Algal ponds will, under favourable conditions be able to produce 100 g phytoplankton per m^3 during 24 hours, which means that as much as 25% of the nitrogen and phosphorus can be removed from solution into phytoplankton (i.e. into suspension) during 24 hours. It a retention time of 3-5 days is applied, as much as 40-60% of the nutrient in municipal waste water will be removed from solution.

If a chemical precipitation follows the treatment in algal ponds the suspended phytoplankton can easily be removed.

The cost of the method is moderate, and the removal of nitrogen especially is often a great advantage. The method is used in the plant in Windhoek (see Fig. 6.44) prior to the chemical precipitation with aluminium sulphate.

The method has, however, some pronounced disadvantages that limit its applications:
1. *Only in tropical or subtropical regions* is solar radiation sufficient throughout the year to ensure an acceptable efficiency.
2. *The areas needed are large* - 3-5 days retention time.
3. The growth of phytoplankton is *sensitive to many organic compounds,* e.g. mineral oil.

Another biological nutrient-removal method should be mentioned in this context: **a root-zone plant**. This has been discussed only recently in Europe, but quite a wide range of results from various studies has been published. More than 90% removal of P and N with 2 m^2 root zone area per pers. equiv. has been quoted, but other studies refer to removal efficiencies under same conditions of less than 10%. Further studies are needed before general conclusions can be drawn.

The dimension of a root-zone plant can be found from the following equation:

$$\log (C/C_o) = -A_i \, V/Q \tag{6.45}$$

where
C is the concentration in the treated water (e.g. as BOD_5 mg/l),
C_o is the concentration in the inflow.
A_i is a constant (found in several experiments to be about 0.13 l/24h)
V is the plant volume (m^3) and
Q is the flowrate (m^3/24h)

The dimensions should also account for the soil permeability. The flow, Q_m, which can pass through soil is

$$Q_m = b * d * K_o * i$$

(6.46)

where
b   is the width of the plant
d   the depth of the plant
K_o the soil permeability (m/24h) and
i   is the slope (m/m)

d is usually in the order of 0.5-1.0 m and since $V = l * b * d$, where l is the length of the plant, b and l are selected in such a way that $Q_m > Q$. Often Q_m is chosen to be 3-4 times Q to account for precipitation. For further details on this promising method, see Boyt et al. (1977); Reddy et al. (1982); Bucksteeg et al. and Finlayson and Chic (1983).

6.4.   REMOVAL OF TOXIC ORGANIC COMPOUNDS.

6.4.1.   The problem and source of toxic organic compounds.

The presence of toxic organic compounds in waste water is causing severe problems:
1. Since toxic organics are scarcely decomposed in the biological plant, life in the receiving water will suffer from their toxic effect. The fate of toxic substances in aquatic ecosystems is already covered in sections 2.13, 5.2 and 5.4.
2. Inhibition of the biological treatment might reduce the efficiency of this process considerably.
   Here the inhibition of biological treatment will be considered in more detail, in relation to heavy metals as well as toxic organic compounds.

P.6.27.   The biological reactions are influenced by the presence of inhibitors. In the case of competitive inhibition, the Michaelis-Menten equation (see 6.2.3) becomes:

$$\mu = \mu_m \frac{S}{S + K_s * (K_{s,I} + I)/K_{s,I}}$$

(6.47)

where $K_{S,I}$ = inhibition constant, and I = concentration of the inhibitor(s).

Competitive inhibition occurs when the inhibitor molecule has almost the same structure as the substrate molecule, which means that the micro-organism is able to break down the inhibitor and the substrate by the same, or almost the same, biochemical pathway. The resulting influence on the Lineweaver-Burk plot (equation 6.12 in 6.2.3) is also shown in Fig. 6.47.

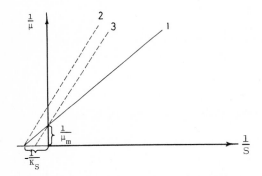

Fig. 6.47. Lineweaver-Burk plot (1). (2) the plot when a toxic compound is present. (3) the plot when a competitive inhibition takes place.

P.6.28. **If a toxic compound (non-competitive inhibitor) is present, only the maximum growth rate will be influenced, and it is reduced according to the relationship:**

$$\mu_m^+ = \mu_m * \frac{K_{S,I}}{K_{S,I} + I} \qquad (6.48)$$

In this case the Lineweaver-Burk plot is also changed (see Fig. 6.47). Heavy metals and cyanide are examples of toxic materials that inhibit non-competitively. The approximate values of $K_{S,I}$ for some inhibitors are given in Table 6.25.

Uptake of toxic substances (organic compounds as well as heavy metals) by the biomass depends on several factors, including pH and the concentration of organic matter and metals present in the system. A higher initial concentration of the toxic substance or sludge increases the overall uptake. In general, the uptake capacity increases with increasing pH. Although the affinity of the biomass for the toxic substances is relatively small, it is generally higher than that of competing organics in a supernatant. The large-scale accumulation of some toxic substances by activated sludge with its subsequent removal in a secondary clarifier explains the significant reduction observed in many treatment plants.

TABLE 6.25
Effect of inhibitors

Non-competitive inhibitor	$K_{S,I}$ (mg/l)
Hg	2
Ag	5
Co	10
Cu	20
Ni	40
Cr^{6+}	200
CN^-	200-2000

In general industrial waste water contains toxic organics, which consequently **must be removed** before this waste water is discharged into the municipal sewage system. Only in this way can the two problems mentioned above be solved simultaneously.

Compounds of minor toxicity and with a biodegradation rate slightly slower than average for municipal waste water can be tolerated in municipal waste water at low concentrations. However, industrial waste water containing medium or high levels of toxic compounds should not be discharged into the municipal sewage system without effective treatment.

As a first estimation of the relationship between the composition of an organic compound and its biodegradability, the following rough rules can be used:
1. **Polymer compounds** are generally *less biodegradable* than monomer compounds.
2. Aliphatic compounds are *more biodegradable than* **aromatic compounds.**
3. **Substitutions**, especially with *halogens and nitro groups,* will *decrease* the biodegradability.
4. Introduction of **double bonds** will generally mean *an increase in the biodegradability.*
5. Introduction of an **oxygen bridge** -O- or a **nitrogen bridge** -N- in an molecule will *decrease the biodegradability.*
6. **Tertiary or secondary compounds** (defined as compounds that contain tertiary or secondary carbon atoms) *are less biodegradable* than the corresponding primary compounds.

Many solvents, for instance alcohol (C_2H_5OH) and acetone (CH_3COCH_3), have a biodegradability similar to that of municipal waste water, but still they should not be discharged into the municipal sewer except in very small amounts as they have a high oxygen consumption. As seen from:

$$C_2H_5OH + 3O_2 - 2CO_2 + 3H_2O \qquad (6.49)$$

100 kg 24 hr^{-1} of ethanol has an oxygen consumption of 209 kg 24 hr^{-1} corresponding to approximately 4175 inhabitants (1 person eqv. to 50 g BOD_5 24 hr^{-1}).

TABLE 6.26
Survey of biodegradable and non-biodegradable organic compounds

Biodegradable organic compounds	Non-biodegradable organic compounds
Aliphatic acids	Ethers
Aliphatic alcohols	Ethylene chlorine hydrine
Aliphatic primary and secondary alcohols	Isoprene
Aliphatic aldehydes	Butadiene
Aliphatic esters	Methylvinyl keton
Alkylbenzene sulphonates	Naphtalene
Amines	Various polymeric compounds
Mono- and dichlorophenols	Polypropylene benzene sulphonates
Glycols	Certain carbon hydrides, especially of aromatic structures, including alkyl-aryl compounds
Nitriles	Tertiary benzene sulphonate
Phenols	Tri-, tetra- and pentachlorophenols
Styrene	
Phenyl acetate	

A survey of biodegradable and non-biodegradable organic compounds is given in Table 6.26 (Ludsack and Ettinger, 1960).

Apart from biological treatment with acclimated sludge, which is applicable in some cases, the following methods might come into consideration for treatment of toxic organics:
1. **Separators** for removal of oil.
2. **Flocculation.** With flocculation it is possible to remove a wide range of organic colloids.
3. **Extraction.** This method is generally rather costly, but recovery of chemicals will often justify the high cost. The method is mostly used when high concentrations are present in the waste water.
4. **Flotation** is of special interest when impurities with a specific gravity of less than 1 are present in the waste water.
5. **Adsorption.** A wide range of organic compounds, such as insecticides and dyestuffs can be adsorbed by activated carbon.
6. **Sedimentation** is only used in conjunction with removal of suspended matter or in combination with a chemical precipitation or flocculation

and biological treatment.

7. **Oxidation and reduction** are mainly used for the treatment of cyanides, some toxic dyestuffs and chromate. The process might be rather costly.

8. **Distillation** of waste water is used when the recovery of solvents is possible or when other methods are not available; for the treatment of radioactive waste water, for example. The method is very costly.

9. Organic acids and bases can be removed by **ion exchange.**

10. **Filtration** can be used for the removal of suspended matter from small quantities of waste water.

11. **Neutralization**. In all circumstances it is necessary to discharge the waste water with a pH between 6 and 8. Calcium hydroxide, sulphuric acid and carbon dioxide are used for the neutralization process.

In the following paragraphs, process numbers 2, 5 and 7 will be discussed in more detail, as they are the most important processes for treatment of industrial waste water containing toxic organics.

6.4.2. Application of chemical precipitation for treatment of industrial waste water.

Chemical precipitation for the removal of phosphorus compounds has been mentioned in 6.3.2. At the same time as it precipitates phosphorus compounds, the process will reduce the BOD_5 by approximately 50-70%. This effect is mainly a result of adsorption on the flocs formed by the chemical precipitation. As this is a chemico-physical effect it is understandable that the COD is also reduced by 50-70%, while the biological treatment will always show a smaller effect on COD removal than on BOD_5 reduction.

The chemical precipitation used for municipal waste water - precipitation with aluminium sulphate, iron(III)chloride and calcium hydroxide - is therefore also able to remove many organic compounds, toxic as well as non-toxic, which means that the process is applicable for the treatment of some types of industrial waste water.

In addition to these three precipitants

P.6.29. **a wide range of chemicals is used for the precipitation of industrial waste water: hydrogen sulphide, xanthates, sodium hydroxide, bentonite, kaoline, starch, polyacrylamide, lignin sulphonic acid, dodecylbenzensulphonic acid, glucose trisulphate.**

Table 6.27 gives a survey, with references, of the use of chemical precipitation for different types of industrial waste water.

TABLE 6.27
Use of chemical precipitation for treatment of industrial waste water

Type of waste water	Chemical used	Reference
Metal plating and finishing industry	Lime	Schjødtz-Hansen, 1968
Iron industry and mining	Lime, Aluminium sulphate	S.E. Jørgensen, 1973
Electrolytic industry	Hydrogen sulphide	S.E. Jørgensen, 1973
Coke and tar industry	Lime or sodium hydroxide	S.E. Jørgensen, 1973
Cadmium mining	Xanthates	Hasebe & Yamamoto, 1970
Manufacturing of glass- and stone wool	Sodium hydroxide	Schjødtz-Hansen & Krogh, 1968
Oil refineries	Aluminium sulphate, iron(III) chloride	S.E. Jørgensen, 1973
Manufacture of organic chemicals	Aluminium sulphate, iron(III) chloride	S.E. Jørgensen, 1973
Photochemicals	Aluminium sulphate	S.E. Jørgensen, 1973
Dye industry	Iron(II) salts, aluminium sulphate, lime	S.E. Jørgensen, 1973
Fertilizer industry	PO_4^{3-}: Iron(II) salts, aluminium sulphate, lime, NH_4^+: magnesium sulphate + phosphate	S.E. Jørgensen, 1973
Plastics industry	Lime	S.E. Jørgensen, 1973
Food industry	Lignin sulphonic acid, dodecylbenzen-sulphonic acid, glucose trisulphate, iron(III) chloride, aluminium sulphate	S.E. Jørgensen, 1973
Paper industry	Bentonite, kaoline, starch, polyacrylamide	S.E. Jørgensen, 1973
Textile industry	Bentonite, aluminium sulphate	S.E. Jørgensen, 1973

6.4.3. Application of adsorption for treatment of industrial waste water.

P.6.30. Adsorption involves accumulation of substances at an interface, which can either be liquid-liquid, gas-liquid, gas-solid or liquid-solid. The material being adsorbed is termed the adsorbate and the adsorbing phase the adsorbent.

The word sorption, which includes both adsorption and absorption, is generally used for a process where the components move from one phase to another, but particularly in this context when the second phase is solid.

A solid surface in contact with a solution has the tendency to accumulate a surface layer of solute molecules, because of the imbalance of surface forces, and so an adsorption takes place. The adsorption results in the formation of a molecular layer of the adsorbate on the surface. Often *an equilibrium concentration is rapidly formed at the surface* and is generally followed by a slow diffusion onto the particles of the adsorbent.

P.6.31. **The rate of adsorption is generally controlled by the rate of diffusion of the solute molecules. The rate varies reciprocally with the square of the diameter of the particles and increases with increasing temperature (Weber and Morris, 1963). For practical application either Freundlich's isotherm or Langmuir's isotherm provide a satisfactory relationship between the concentration of the solute and the amount of adsorbed material.**

The Freundlich isotherm is expressed in the following equation (compare with equation (6.28)):

$$a = k * C^n \tag{6.50}$$

where
k and n = constants
a = the amount of solute adsorbed per unit weight
C = the equilibrium concentration of the solute in the liquid phase

The values of k and n are given for several organic compounds in Table 6.28 (Rizzo and Shepherd, 1977).

TABLE 6.28
Freundlich's constant for adsorption of some organic compounds on activated carbon

Compound	k	n
Aniline	25	0.322
Benzene sulphonic acid	7	0.169
Benzoic acid	7	0.237
Butanol	4.4	0.445
Butyraldehyde	3.3	0.570
Butyric acid	3.1	0.533
Chlorobenzene	40	0.406
Ethylacetate	0.6	0.833
Methyl ethyl ketone	24	0.183
Nitrobenzene	82	0.237
Phenol	24	0.271
Phenol	24	0.271
TNT	270	0.111
Toluene	30	0.729
Vinyl chloride	0.37	1.088

*) a in mg/g when C in mg/l

Langmuir's adsorption isotherm is based on the following expression:

$$a = \frac{A_o * C}{1 + b * C}$$ (6.51)

where
a and C = as defined above
b and A_o = constants

As can be seen, $a = A_o/b$ when $C = \infty$.

Adsorption of nutrients, heavy metals and organics in sediment and soil is in principle the same process, see also section 4.10.

TABLE 6.29
Langmuir's constant for adsorption of some organic compounds on activated carbon

Compound	A_o	b (l/mg)
Phenol	0.118	1.15
p-Nitrochloro-benzene	0.286	0.714
Dodecylbenzene sulphonate	1.83	13.2

*) a in mg/l when C in mg/l

The Langmuir constant for several organic compounds that can be adsorbed on activated carbon has been found by Weber and Morris (1964). Most types of waste water contain several substances that will be adsorbed, and in this case a direct application of Langmuir's adsorption isotherm is not possible.

Weber and Morris (1965) have developed an equation (6.52) and (6.53) for **competitive adsorption of two substances** (A and B). In other words, **competitive adsorption can be described in the same way as a competitive enzymatic reaction:**

$$a_A = \frac{A_{Ao} * C_A}{1 + b_A * C_A + b_B * C_B}$$ (6.52)

$$a_B = \frac{A_{Bo} * C_B}{1 + b_A * C_A + b_B * C_B}$$ (6.53)

Langmuir's constant for some organic compounds generally found in waste water are shown in Table 6.29.

When it is necessary to find whether Freundlich's adsorption isotherm or

Langmuir's adsorption isotherm gives the best fit to a set of data, the two plots in Figs. 6.48 and 6.49 should be used.

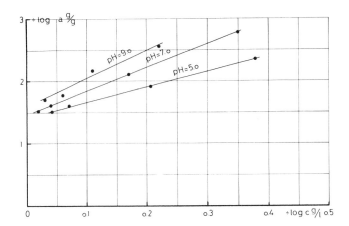

Fig. 6.48. Adsorption of lysine on a cellulose cation exchanger at three different pH values. (Jørgensen, 1976).

When a linear relationship between log a and log C is obtained, Freundlich's adsorption isotherm is a good description of a set of dats, as:

$$\log a = \log k + n \log C \qquad (6.54)$$

When the reciprocal values of a and C give a linear equation (see Fig. 6.49) Langmuir's adsorption isotherm, equation (6.51), gives a good description of the set of data, since

$$\frac{1}{a} = \frac{1 + b * C}{A_o * C}$$

or

$$\frac{1}{a} = \frac{1}{A_o} * \frac{1}{C} + \frac{b}{A_o}$$

Notice that Fig. 6.49, is a parallel to Lineweaver-Burk's plot used for enzymatic processes, see also Fig. 6.47.

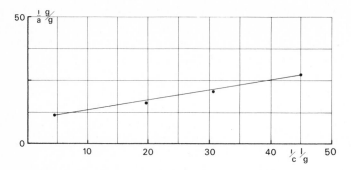

Fig. 6.49. Adsorption of uric acid on activated carbon (pH = 9.0. Jørgensen, 1976).

It is possible, to a certain extent, to predict the adsorption ability of a given component. The solubility of the dissolved substance is by far the most significant factor in determining the intensity of the driving forces.

P.6.32. **The greater the affinity of a substance for the solvent, the less likely it is to move towards an interface to be adsorbed. For an aqueous solution this means that the more hydrophilic the substance is, the less likely it is to be adsorbed. Conversely, hydrophobic substances will be readily adsorbed from aqueous solutions.**

Many organic components, e.g. sulphonated alicylic benzenes, have a molecular structure consisting of both hydrophilic and hydrophobic groups. The hydrophobic parts will be adsorbed at the surface and the hydrophilic parts will tend to stay in the water phase.

The sequential operation is frequently called "contact filtration", because the typical application includes treatment in a mixing tank followed by filtration, but more frequently settling is used for removal of the used adsorbents in industrial waste water engineering. The sequential adsorption operation is limited to treatment of solutions where the solute to be removed is adsorbed relatively strongly when compared with the remainder of the solution. This is often the case when colloidal substances are removed from aqueous solutions using carbon, as in the production of process water.

The method for dealing with the spent adsorbent depends upon the system under consideration. If the adsorbate is valuable material, it might be desorbed by contact with a solvent other than water. If the adsorbate is volatile, it may be desorbed by reduction of the partial pressure of the adsorbate over the solid by passing stram or air over the solid. In the case of most sequential adsorption operations in the context of waste water treatment, the adsorbate is of no value and it is not easily desorbed. The adsorbent may then be regenerated by burning off the adsorbate, followed by reactivation.

In the continuous operation, the water and the adsorbent are in contact throughout the entire process without a periodic separation of the two phases. The operation can either be carried out in strictly continuous steady-state fashion by movement of the solid as well as the fluid, or in a semicontinuous fashion characterized by moving fluid but stationary solid, the so-called fixed adsorption.

Due to the inconvenience and relatively high cost of continuously transporting solid particles, it is generally found more economical to use a stationary bed of adsorbent for waste water treatments.

The design of a fixed bed adsorber and the prediction of the length of adsorption cycle requires knowledge of the percentage approach to saturation at the break point. Fig. 6.50 shows an idealized breakthrough curve.

P.6.33. **The extent of adsorption is proportional to the surface area.**

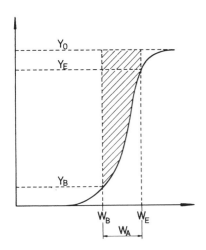

Fig. 6.50. Idealized breakthrough curve. Concentration (Y) versus volume (W)

In order to compare different adsorbents a specific surface area, defined as that portion of the total surface area that is available for adsorption per unit of adsorbent, is used. This means that *the adsorption capacity of a non-porous adsorbent should vary inversely with the particle diameter, while in the case of highly porous adsorbent, the capacity should be almost independent of the particle diameter.* However, for some porous material, such as activated carbon, the breaking up of large particles to form smaller ones opens some tiny sealed channels in the column, which might the become available for adsorption (Weber and Morris, 1964).

The nature of the adsorbate also influences the adsorption. *In general an inverse relationship can be anticipated between the extent of adsorption of a solute and its solubility in the solvent (water) from which adsorption occurs. This is the so-called Lundilius' rule,* which may be used for semiquantitative prediction of the effect of the chemical character of the solute on its uptake from solution (water) (Lundilius, 1920).

P.6.34. **Ordinarily, the solubility of any organic compound in water decreases with increasing chain length because the compound becomes more hydrophobic as the number of carbon atoms increases. This is Traube's rule.**

Adsorption from aqueous solution *increases as homologous series are ascended,* largely because the expulsion of increasingly large hydrophobic molecules from water permits an increasing number of water-water bonds to form. Fig. 6.51 shows the effect of molecular weight on the capacity for adsorption for several sulphonated alkylbenzenes. As seen, the figure illustrates very well the above-mentioned Traube's rule. The molecular weight is also related to the rate of uptake of solutes by activated carbon, if the rate is controlled by intraparticle transport. Data are plotted in Fig. 6.52 for the rates of adsorption of a series of sulphonated alkylbenzenes of different molecular size. It can be seen that the molar rate of uptake decreases with increasing molecular weight.

pH strongly influences the adsorption as hydrogen and hydroxide ions are adsorbed, and the charges of the other ions are influenced by the pH of the water. For typical organic pollutants from industrial waste water, the adsorption increases with increasing pH.

Normally, the adsorption reactions are exothermic, which means that the adsorption will increase with decreasing temperature, although small variations in temperature do not tend to alter the adsorption process to a significant extent.

P.6.35. **Adsorption can be used to remove several organic compounds, such as phenol, alkylbenzene-sulphonic-acid,**

dyestuffs and aromatic compounds from waste water by
the use of activated carbon.

μ mol/g

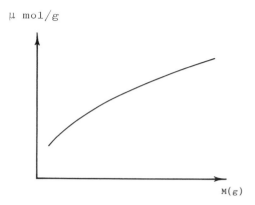

M(ɡ)

Fig. 6.51. Effect of molecular weight on capacity for adsorption for several
sulphonated alkylbenzenes.

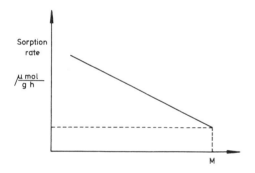

Fig. 6.52. Sorption rate plotted against the molecular weight.

Scaramelli et al. (1973) have examined the effect (on effluent quality) of
adding powdered activated carbon to an activated sludge system.

They found that 100 to 200 mg/l were able to reduce TOC from about 20
mg/l to 7 mg/l.

Activated carbon has also been suggested as an adsorbent for the
removal of refractory dyestuffs (Eberle et al., 1976).

In very small plants is may be feasible to use granular carbon on a **use
and throw away basis**, althugh economics probably favour the use of

powdered carbon in a sequential operation. The use of granular activated carbon involves the regeneration and reuse of carbon, with some exceptions. This **regeneration can be carried out with sodium hydroxide** provided that high molecular weight colloids are removed before the treatment of the activated carbon. It is possible with this **chemical regeneration** to *recover phenols,* for example (Jørgensen, 1976), and *to remove coloured bodies* (Chamberlin et al., 1975; Mulligan et al., 1976). Also **solvent** can be used for regeneration of activated carbon, as indicated by Rovel (1972).

When sodium hydroxide or solvent is used, the adsorbate is passed through the carbon bed in the opposite direction to that of the service cycle, until all is removed. The bed is then drained and the regenerated carbon is ready to go back into the stream.

Juhols and Tupper (1969) have studied the **thermal regeneration** of granular activated carbon, which consists of three basic steps, (1) **drying**, (2) **baking of adsorbate** and (3) **activation by oxidation** of the carbon residues from decomposed adsorbates. Drying requires between 100 and 700°C and activation a temperature above 750°C. All three steps can be carried out in a direct fired, multiple hearth furnace. This is the best commercial equipment available for regeneration of carbon for use in combination with waste water treatment. The capacity of the activated carbon *will generally decrease by approximately 10% during the first thermal regeneration and another 10% during the next 5-10 regenerations.*

6.4.4. Application of chemical Oxidation and Reduction for Treatment of industrial waste water.

By chemical oxidation and reduction the oxidation stage of the substance is changed. Oxidation is a process in which the oxidation stage is increased, while chemical reduction is a process in which the oxidation stage is decreased.

P.6.36. An oxidation-reduction (redox) reaction is defined as a process which electrons are transferred from one substance to another.

An oxidation means loss of electrons and a reduction involves gain of electrons. The concept of electron exchange is very useful because it affords a simple means of balancing redox reactions. With regard to this balancing, the concept of oxidation stage is introduced. The definition of this concept is based on the following rules:
1. All elements have an oxidation stage of zero.
2. All ions have the same oxidation stage as their charges.

3. Hydrogen has the oxidation stage +1 in all its compounds.
4. Oxygen has the oxidation stage -2 in all its compounds, except hydrogen peroxide and its derivates.
5. The sum of the oxidation stages of all atoms in an uncharged molecule is zero.

To illustrate this point a chlorination can be considered,

e.g. $Cl_2 + C_2H_6 - C_2H_5Cl + HCl$.

In accordance with Pauling (1960), the oxidation number of each atom in a covalent compound is the charge remaining on the atom when each electron pair is assigned completely to the more electro-negative of the atoms sharing the electrons. The three elements involved in the reaction are H, C and Cl. The electronegativity of these three elements is, respectively, 2.1, 2.5 and 2.8. This means that C in C_2H_6 has the oxidation stage -3, while in C_2H_5Cl it has the oxidation stage -2. In both C_2H_5Cl and HCl, Cl has the stage -1.

The application of oxidation is largely limited, mainly for the reasons of economics.

Several aspects are considered in the selection of a suitable oxidizing agent for industrial waste water treatment. They are:

1. Ideally, *no residue of oxygen* should remain after the treatment and there should be *no residual toxic or other effects.*
2. *The effectiveness of the treatment must be high.*
3. *the cost must be as low as possible.*
4. *The handling should be easy.*

P.6.37. **Only a few oxidants are capable of meeting these requirements. The following oxidants are in use today for treatment of water and waste water:**
   1. **Oxygen or air.**
   2. **Ozone.**
   3. **Potassium permanganate.**
   4. **Hydrogen peroxide.**
   5. **Chlorine.**
   6. **Chlorine dioxide.**

Oxygen has its significance in **biological oxidation**, but also plays an important role in **chemical oxidation**.

The primary attraction of oxygen is that it can be applied in the form of air. Organic material, including phenol, can be catalytically oxidized **by the use of suitable catalysts**, such as oxides of copper, nickel, cobalt, zinc, chromium, iron, magnesium, platinum and palladium, but, as yet, this process has not been developed for use on a technical scale.

Ozone is a more powerful oxidant than oxygen and is able to react rapidly with a wide spectrum of organic compounds and microorganisms present in waste water.

It is produced from oxygen by means of electrical energy, which is a highly attractive process since air is used. One of the advantages of using ozone is that it does not impart taste and odour to the treated water. Ozone is used in the following areas of water treatment:

1. **Removal of colour, taste and odour** (Holluta, 1963).
2. **Disinfection.**
3. **For the oxidation of organic substances, e.g. phenol, surfactants** (Eisenhauer, 1968; Wynn et al., 1972) **and cyanides** (Anon, 1958; Khandelwal et al., 1959).

The solubility of ozone is dependent upon the temperature. Henry's coefficient as a function of the temperature for ozone and oxygen is shown in Table 6.30.

Henry's coefficient, H, is used in Henry's law (see section 2.5, principle 2.14):

$$p = H * x \qquad (6.55)$$

where

p = the partial pressure in atmospheres

x = the mole fraction in solution

Ozone has low thermodynamic stability at normal temperature and pressure (Kirk-Othmer, 1967). It decomposes both in the gas phase and in solution. Decomposition is more likely in aqueous solutions, where it is strongly catalyzed by hydroxide ions, see Table 6.31 (Stumm, 1958).

The oxidation of phenol by gaseous ozone has been studied by Gould et al. (1976) under a number of conditions.

TABLE 6.30
Henry's coefficient for oxygen and ozone in water

Temperature (°C)	H * 10^{-4}	
	Oxygen	Ozone
0	2.5	0.25
5	2.9	0.29
10	3.3	0.33
15	3.6	0.38
20	4.0	0.45
25	4.4	0.52
30	4.8	0.60
35	5.1	0.73
40	5.4	0.89
45	5.6	1.0
50	5.9	1.2

TABLE 6.31
Influence of pH on half-life of ozone in water (Stumm, 1958)

pH	Half-life (min)
7.6	41
8.5	11
8.9	7
9.2	4
9.7	2
10.4	0.5

Virtually complete removal of phenol and its aromatic degradation products is realized when 4-6 moles of ozone have been consumed for each mole of phenol originally present. At this point approximately 1/3 of the initial organic carbon will remain and 70-80% reduction of the COD-number will have been achieved. Concentrations of the non-aromatic degradation products will be less than 0.5 mg/l. Subsequent dilution of the discharge of this effluent should reduce the concentration of the various components in the receiving body of water to tolerable levels.

Ozone is extremely toxic, having a maximum tolerable concentration of continuous exposure of 0.1 ppm. However, the half-life of ozone is reduced by high pH as demonstrated in Table 6.31.

The ozone concentration can be analyzed either by ultraviolet absorption spectroscopy or by iodide titration:

$$O_3 + 3I^- + H_2O - I_3^- + O_2 + 2OH^- \tag{6.56}$$

The liberated iodine is titrated with thiosulphate (Kirk-Othmer, 1967). As indicated above, ozone is generated from dry air by a high voltage electric discharge. A potential of 5000 to 40,000 volts between the electrodes is used. Cooling is usually employed to minimize ozone decomposition in the reactor. Theoretically 1058 g of ozone can be produced per kWh of electrical energy, but in practice a production of 150 g/kWh is more usual.

Permanganate is a powerful oxidizing agent and is widely used by many municipal water plants for taste and odour control and for the removal of iron and manganese. Furthermore, it can be used as an oxidant for the removal of impurities such as Fe^{2+}, Mn^{2+}, S^{2-}, CN^- and phenols present in industrial waste water.

In strongly acidic solutions permanganate is able to take up five electrons:

$$MnO_4^- + 8H^+ + 5e^- - 4H_2O + Mn^{2+} \tag{6.57}$$

While in the pH range from approximately 3 to 12 only three electrons are transferred and the insoluble manganese dioxide is formed:

$$MnO_4^- + 4H^+ + 3e^- - 2H_2O + MnO_2 \qquad (6.58)$$

or

$$MnO_4^- + 2H_2O + 3e^- - 4OH^- + MnO_2 \qquad (6.59)$$

The stoichiometry for the oxidation of cyanide in a hydroxide solution of pH 12-14 is:

$$2MnO_4^- + CN^- + 2OH^- - 2MnO_4^{2-} + CNO^- + H_2O \qquad (6.60)$$

In a saturated solution of calcium hydroxide (Posselt, 1966) the reaction takes the form:

$$2MnO_4^- + 3CN^- + H_2O - 3CNO^- + 2MnO_2 + 2OH^- \qquad (6.61)$$

The presence of calcium ions affects the rate of manganate(IV) production disproportionately:

$$3MnO_4^{2-} + 2H_2O - 2MnO_4^- + 4OH^- + MnO_2 \qquad (6.62)$$

The permanganate concentration can be determined by spectrophotometry (absorption maximum at 526 µm or by a titrimetric method.

Table 6.32 gives a survey of the oxidation of organic compounds by permanganate.

Chlorine is known to be a successful **disinfectant** in waste water treatment, but it is also able to oxidize effectively such compounds **as hydrogen sulphide, nitrite, divalent manganese and iron and cyanide**. The oxidation effectiveness usually increases with increasing pH.

Cyanide, which is present in a number of different industrial waste waters, is typically oxidized with chlorine at a high pH.

The oxidation to the much less toxic cyanate (CNO^-) is generally satisfactory, but in other cases complete degradation of cyanide to carbon dioxide and nitrogen is required.

The disadvantage of chlorine is that it forms aromatic chlorocompounds, which are highly toxic, e.g. chlorophenols, when phenol-bearing water is treated with chlorine (Aston, 1947).

This fact has resulted in the more widespread use of chlorine dioxide. Chlorine dioxide is as unstable as ozone and must therefore be generated in situ.

TABLE 6.32
Permanganate oxidation of organic compounds

-CH - CH-	--------	$\overset{\displaystyle OH}{\underset{\displaystyle \vert}{-CH}}$ - $\overset{\displaystyle OH}{\underset{\displaystyle \vert}{CH}}$-
$R - CH_2OH$	--------	$R - COOH$
$R - CHO$	--------	$R - COOH$
$R_2 - CHOH$	--------	$R_2 - \underset{\displaystyle H}{C} = O$
$R - SH$	--------	$R - SO_3H$
Alkyl amines	--------	$R - COOH + NH_3$
$R_1 - S - R_2$	--------	$R_1 - SO_3 - R_2$
$R_1 - S - S - R_2$	--------	$R_1 - SO_3H + R_2 - SO_3H$
$R_1 - SO - R_2$	--------	$R_1 - SO_2 - R_2$

The industrial generation of chlorine dioxide is carried out by means of a reaction between chlorine and sodium chlorite in acid solution (Granstrom and Lee, 1958):

$$Cl_2 + 2NaClO_2 - 2ClO_2 + 2NaCl \qquad (6.63)$$

As chlorine dioxide is a mixed anhydride of chlorous and chloric acid, disproportionation to the corresponding anions occurs in basic solutions:

$$2ClO_2 + 2OH^- - ClO_2^- + ClO_3^- + H_2O \qquad (6.64)$$

This process becomes negligible under acidic conditions. The equilibrium:

$$2ClO_2 + H_2O - HClO_2 + HClO_3 \qquad (6.65)$$

shifts to the left at lower pH.

According to Myhrstad and Samdal (1969) chlorine dioxide can be analysed with acid chrome violet K, and determined spectrophotometrically without interference by Cl_2, ClO^-, ClO_2^- and ClO_3^-.

As iodometric titration gives Cl_2:

$$Cl_2 + 2I^- \; - \; 2Cl^- + I_2 \quad (pH = 7.0) \tag{6.66}$$

and iodometric titration at pH = 2.5-3.0 gives the total amount of $Cl_2 + ClO_2 + ClO_2^-$:

$$Cl_2 + 2I^- \; - \; 2Cl^- + I_2 \tag{6.67}$$

$$2ClO_2 + 8H^+ + 10I^- \; - \; 2Cl^- + 4H_2O + 5I_2 \tag{6.68}$$

$$ClO_2^- + 4H^+ + 4I^- \; - \; Cl^- + 2H_2O + 2I_2 \tag{6.69}$$

it is possible to make a simultaneous determination of ClO_2, ClO_2^- and Cl_2 independently.

Chlorine dioxide is used for **taste and odour control**. It has been reported to be a selective oxidant for industrial waste water containing **cyanide, phenol, sulphides and mercaptans** (Wheeler, 1976).

Hydrogen peroxide can be used as an oxidant for **sulphide** in water (Cole et al., 1976). Recently this oxidizing ability has been applied to control odour and corrosion in domestic and industrial waste water.

6.5.   REMOVAL OF (HEAVY) METALS.

6.5.1.   The problem of heavy metals.

Heavy metals are the most harmful metals, but in the title the word heavy is shown in brackets to indicate that toxic metals other than heavy metals are dealt with in this chapter.

1. **Toxic metals are harmful to aquatic ecosystems**. (For further details see section 2.1 and 5.4). Treatment by a mechanico-biological process is, to a certain extent, able to remove metals, but the efficiency is only in the order of 30-70%, depending on the metal. The heavy metals removed are concentrated in the sludge, which, even for plants treating solely municipal waste water, shows measurable concentrations (see section 5.4).

As mentioned in 6.3.2, chemical precipitation offers a high efficiency in the removal of heavy metals.

2. **Heavy metals are harmful to biological treatment and are example of non-competitive inhibitors**. (For further details see 6.4.1).

It is possible, however, according to Neufeld et al. (1975), to grow cultures of activated biota in the presence of mercury, cadmium and zinc levels that are higher than those that would previously have been thought possible. Mercury, cadmium and zinc are rapidly removed from aqueous solutions by biological flocculation . Although the eventual equilibrium was only achieved after about 2-3 weeks, three hours of contact were sufficient almost to reach that equilibrium. The ratio of the weight of metal in the surrounding aqueous phase for the metals mercury, cadmium and zinc at equilibrium, ranges from 4000 to 10,000.

In the main industrial waste water contains harmful concentrations of heavy metals, and consequently these **must be removed** before such waste water is discharged into the municipal sewage system. Only in this way can the two problems mentioned above be solved simultaneously.

P.6.38. **Minor amounts of the least toxic metals may be discharged provided their contribution to the concentration in the municipal waste water is of little importance, but as demonstrated in 6.4.1 only a small concentration of toxic metals can be tolerated.**

Chiefly, four processes are applied for removal of heavy metals from industrial waste water:
1. **Chemical precipitation**
2. **Ion exchange**
3. **Extraction**
4. **Reverse osmosis**

Adsorption has also been suggested for the removal of mercury by Logsdon et al. (1973). 1 mg per litre of powdered carbon is needed for each 0.1 µg per litre of mercury to be removed.

It is often economically as well as feasibly, preferable from the resource and ecological point of view, to recover heavy metals. The increasing cost of metals, foreseen in the coming decades, will probably provoke more industries to select such solutions to their waste water problems. *Partial or complete recirculation of the waste water* should also be considered, since a decrease in water consumption and in loss of material is achieved simultaneously; in many cases there is even a considerable saving in water obtained by reorganization of processes.

6.5.2. The application of chemical precipitation for removal of heavy metals.

One of the processes most used for the removal of metal ions from water is precipitation, as metal hydroxide.

Table 6.33 lists the solubility products for a number of metal hydroxides.

TABLE 6.33

pK_s values at room temperature for metal hydroxides
$pK_s = -\log K_s$, where $K_s = [Me^{z+}] [OH^-]^z$

Hydroxide	z = charge of metal ions	pK_s
AgOH(1/2 Ag$_2$O)	1	7.7
Cu(OH)$_2$	2	20
Zn(OH)$_2$	2	17
Ni(OH)$_2$	2	15
Co(OH)$_2$	2	15
Fe(OH)$_2$	2	15
Mn(OH)$_2$	2	13
Cd(OH)$_2$	2	14
Mg(OH)$_2$	2	11
Ca(OH)$_2$	2	5.4
Al(OH)$_3$	3	32
Cr(OH)$_3$	3	32

P.6.39. From the solubility product it is possible to find the pH value at which precipitation will start for a given concentration of the metal ions.

Fig. 6.53 shows the solubilities of metal ions at various pH values. By means of this diagram it is possible to find the concentration in solution at any given pH. For example, at pH 6.0 the concentration of Cr^{3+} is 10^{-6} M. The same concentration for Zn^{2+} is obtained at pH 8.0.

The slopes of the lines in Fig. 6.53 correspond to the valency of the metal ions. The slope of the Cr^{3+} and Al^{3+} lines, for example, is +3, while the other ions in the figure have lines with a slope of +2. This is obtained from the solubility product:

$$[Me^{n+}] [OH^-]^n = K_s$$

$$\log [Me^{n+}] = -pK_s + n * pOH = -pK_s + 14 - n * pH \qquad (p = \text{"-log"})$$

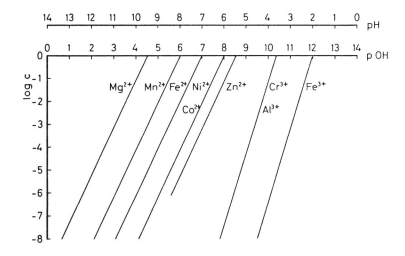

Fig. 6.53. log(solubility) against pH and pOH.

From the basis of the solubility product it is possibel to find one point though which the line pass. For example, for the solubility product of iron(II) hydroxide, which is 10^{-15}, we have

$$[Fe^{2+}] \; [OH^-]^2 \; = \; 10^{-15}$$

The line will therefore go through the point -log c = 5 and pOH = 5 (c = concentration of Fe^{2+}), since $(10^{-5})^3 \; = \; 10^{-15}$.

Table 6.34 gives the pH value at which the solubility is 10 mg or less of the metal ion per litre and 1 mg or less per litre, respectively.

TABLE 6.34
The pH value indicates where the solubility is $\leq$ 10 mg/l and $\leq$ 1 mg/l

| Metal ion | Solubility | |
	$\leq$ 10 mg/l	$\leq$ 1 mg/l
Mg^{2+}	11.5	12.0
Mn^{2+}	10.1	10.6
Fe^{2+}	8.9	9.4
Ni^{2+}	7.8	8.3
Co^{2+}	7.8	8.3
Zn^{2+}	7.2	7.7
Cr^{3+}	5.1	5.4
Al^{3+}	5.0	5.3
Fe^{3+}	3.2	3.5

From these considerations it is presumed that other ions present do not influence the precipitation, but in many cases it is necessary to consider the ionic strength:

$$I = \Sigma \ 1/2 \ C \ Z^2 \qquad\qquad (6.70)$$

where C = the molar concentration of the considered ions and Z = the charge.

On the basis of the ionic strength, it is possible to find the activity coefficient, f, from

$$-\log f = \frac{0.5 * Z^2 * \sqrt{I}}{\sqrt{I} + 1} \qquad\qquad (6.71)$$

where
I = ionic strength, Z = charge and f = activity coefficient

TABLE 6.35
Activity coefficient f at different ionic strengths

I	$\dfrac{\sqrt{I}}{1 + \sqrt{I}}$	f for Z = 1	f for Z = 2	f for Z = 3
0	0	1.00	1.00	1.00
0.001	0.03	0.97	0.87	0.73
0.002	0.04	0.95	0.82	0.64
0.005	0.07	0.93	0.74	0.51
0.01	0.09	0.90	0.66	0.40
0.02	0.12	0.87	0.57	0.28
0.05	0.18	0.81	0.43	0.15
0.1	0.24	0.76	0.33	0.10
0.2	0.31	0.70	-	-
0.5	0.41	0.62	-	-

I = ionic strength, Z = charge, f = activity coefficient

Table 6.35 gives the activity coefficients for different charges of the considered ions, calculated from the equation (6.71).

Since calcium hydroxide is the cheapest source of hydroxide ions, it is most often used for the precipitation of metals as hydroxides. In most cases it is necessary to determine the amount of calcium hydroxide required in the laboratory.

Flocculation is carried out after the addition of the chemical and before settling has occured. If the amount of waste water that must be treated per 24 hr is 100 m^3 or less, it is preferable to use a discontinuous treatment. The system in this case consists of two tanks with a stirrer. The waste water is discharged into one tank while the other is used for the treatment process. If chromate is present the treatment processes will follow the scheme below:

1. The concentration of chromate and the amount of acid required to bring the pH down to 2.0 are determined.
2. On the basis of this determination the required amounts of acid and reducing compound are added (about the reduction, see below).
3. Stirring 10-30 minutes.
4. The concentration of chromate remaining is determined and the pH checked. If chromate is present further acid or reducing agent is added. In this case the stirring (10-30 minutes) must be repeated.
5. The amount of calcium hydroxide necessary to obtain the right pH for precipiation is measured. If Cr^{3+} is present a pH value of 8-9.5 is normally required.
6. Flocculation 10-30 minutes.
7. Settling 3-8 hours.
8. The clear phase is discharged into the sewer system.
   The sludge can be concentrated further by filtration or centrifugation.

Stages 1-5 can, of course, be left out if chromate is not present, and the metal ions can be directly precipitated.

P.6.40.   Reduction is a process in which soluble metallic ions are reduced through a redox reaction.

Generally, the process is used in the treatment of plating waste water containing chromate. This water, from chromate acid baths used in electroplating and anodizing processes, contains chromate in the form of CrO_3 or $Na_2Cr_2O_7$, $2H_2O$. The pH of such waste water is low and the Cr(IV) concentration is often very high - up to 20,000 ppm or more. *The most commonly used reducing agents are iron(II) sulphate, sodium meta-hydrogen sulphite or sulphur dioxide.* Since the reduction of chromate is most effective at low pH, it is, of course, an advantage if the waste water itself contains acid, which is often the case. Iron(II) ions react with chromate by reducing the chromium to a trivalent state and the iron(II) ions are oxidized to iron(III) ions.

The reaction occurs rapidly at a pH below 3.0, but since the acidic properties of iron(II) sulphate are low at high dilution, acid must often be added for pH adjustment. The reactions are:

$$CrO_3 + H_2O \; - \; H_2CrO_4 \qquad\qquad\qquad\qquad\qquad (6.72)$$
$$2H_2CrO_4 + 6FeSO_4 + 6H_2SO_4 \; - \; Cr_2(SO_4)_3 + 3Fe_2(SO_4)_3 + 8H_2O \quad (6.73)$$
$$Cr_2O_7^{2-} + 6FeSO_4 + 7H_2SO_4 \; - \; Cr_2(SO_4)_3 + 3Fe_2(SO_2)_3 + 7H_2O + SO_4^{2-}$$
$$\qquad\qquad\qquad\qquad\qquad\qquad\qquad\qquad\qquad\qquad (6.72)$$

It is possible to show that 1 mg of Cr will require 16 mg $FeSO_4$, $7H_2O$ and 6 mg of H_2SO_4, based on stoichiometry.

Reduction of chromium can also be accomplished by the use of meta-hydrogen sulphite or sulphur dioxide. When meta-hydrogen sulphite is used, the salt hydrolyzes to hydrogen sulphite:

$$S_2O_5^{2-} + H_2O - 2HSO_3^- \qquad\qquad (6.75)$$

The hydrogen sulphite reacts to form sulphurous acid:

$$HSO_3^- + H_2O - H_2SO_3 + OH^- \qquad\qquad (6.76)$$

Sulphurous acid is also formed when sulphur dioxide is used, since:

$$SO_2 + H_2O - H_2SO_3 \qquad\qquad (6.77)$$

The reaction is strongly dependent upon pH and temperature.
The redox process is:

$$2H_2CrO_4 + 3H_2SO_3 - Cr_2(SO_4)_3 + 5H_2O \qquad\qquad (6.78)$$

Based on stoichiometry, the following amounts of chemicals are required for 1 mg of chromium to be precipitated: 2.8 mg $Na_2S_2O_5$ or 1.85 mg SO_2.

Since dissolved oxygen reacts with sulphur dioxide, excess SO_2 must be added to account for this oxidation:

$$H_2SO_3 + 1/2\ O_2 - H_2SO_4 \qquad\qquad (6.79)$$

If the amount of waste water exceeds 100 m^3 in 24 hr, the plant shown in Fig. 6.54 can be used.

A potentiometer and pH meter are used to control the addition of acid and reducing agent. These instruments can be coupled to an automatic dosing control. As seen from the figure, the plant consists of a reaction tank, a flocculation tank and a sedimentation tank. Usually a settling time of 24 hr or more is used.

A concentration of 1 1/3 - 3% dry matter for the sludge is usually obtained, which is slightly less than that obtained by discontinuous treatment.

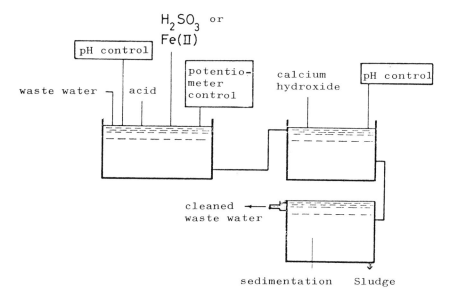

Fig. 6.54. Flowchart of a plant for precipitation of metal ions. This plant includes the reduction of chromate to Cr^{3+}.

Fig. 6.55 shows the so-called Lancy system for treating waste water containing chromate.

This system contains a recirculation tank connected to the reduction tank as well as to the precipitation tank. Filtration of the sludge is continuous and 20-35% dry matter is obtained.

Thomas and Theis (1976) have shown that the coagulation and settling of colloidal chromium(III) hydroxide are functions of both the quantity and type of impurity ions present.

The treatment might not operate effectively because of the complexing or stabilizing effects of high carbonate or pyrophosphate concentrations in the water.

It is possible to eliminate the problem caused by carbonate and phosphate by modifying the treatment scheme for the combined waste. Prime consideration should be given to the modification consisting simply of using lime instead of caustic soda to neutralize the waste after the chemical reduction of chromate to chromium(III) has taken place. The use of lime will cause precipitation and removal of most of the carbonate and pyrophosphate species from the solution while also providing doubly charged counter-ions to aid in coagulating the negatively charged chromium(III) hydroxide colloids that exist at pH values of about 8.

Other treatment modifications, such as the use of polyelectrolytes to flocculate the stabilized chromium(III) hydroxide system, may or may not be effective alternatives, depending on the amount of complexation present in a particular waste system.

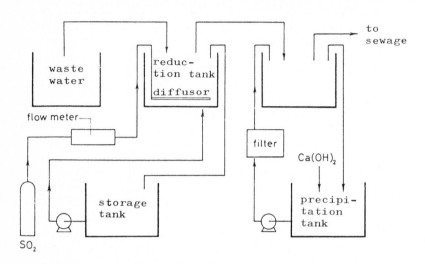

Fig. 6.55. The Lancy system.

Data presented indicate that a hydrogen carbonate alkalinity of 250 mg/l as $CaCO_3$ and a pyrophosphate concentration of 30 mg/l as P together cause appreciable complexation and may make alternatives other than lime neutralization impractical.

The formation of calcium carbonate flocs during lime neutralization could also aid in the removal of chromium(III) hydroxide colloids by the enmeshment mechanism and, at the same time, make the resulting sludge easier to dewater because of the presence of the large volume of calcium carbonate. The increased treatment efficiency and better sludge handling characteristic may make the use of lime a more favourable solution even though the quantity of sludge may be somewhat increased.

The last treatment modification to be considered is the use of Fe^{2+} ions as the reducing agent in the chromate-reduction step. The end product of this reaction is Fe^{3+} ions. In the neutral pH range this is very effective in removing chromium(III) hydroxide colloids.

6.5.3. The application of Ion Exchange for Removal of Heavy Metals.

P.6.41. **A cation exchanger can be used for the removal of metal ions from waste water, such as Fe^{2+}, Fe^{3+}, Cr^{3+}, Al^{3+}, Zn^{2+}, Cu^{2+}, etc.** (Spanier, 1969)

The most common type of cation exchanger consists of a polystyrene matrix, which is a strong acidic ion exchanger containing sulphonic acid groups:

The practical capacity of the cation exchanger is 1-1.5 eqv/l of ion exchang material (Rüb, 1969). To ensure high efficiency, a relatively low flow rate is recommended, often below 5 bed volumes/hour.

A recent development has brought a **starch xanthate and a cellulose polyethyleneimine** ion exchanger onto the market (WRL, 1977). They have a lower capacity than the conventional cation exchangers, but since they are specific for heavy metal ions they have a higher capacity when measured as the volume of water treated between two regenerations per volume of ion exchanger.

It is often preferable to separate waste water from the chromate baths and the rinsing water, since this opens up the possibility of reusing the water. *Chromate can be recovered* and the waste water problem can be solved in an effective way (Mohr, 1969).

Fig. 6.56 is a flow diagram of an ion exchange system in accordance with these principles.

Reuse of the elution liquid from the anion exchanger and of the treated waste water might be considered.

The pH of the elution liquid from the cation exchanger must be adjusted and the metal ions precipitated for removal by the addition of calcium hydroxide (see Fig. 6.56).

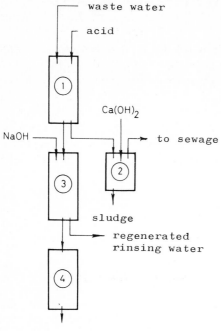

Fig. 6.56. Ion exchange of waste water containing chromate.
(1) Cation exchanger; (2) Precipitation tank; (3) Anion exchanger; (4) Exchange of Na^+ with H^+.

In some instances it is too costly to seperate the different types of waste water and rinsing water. In such cases a simpler ion exchange system is used (Schaufler, 1969), although it is, of course, possible to recover the chromate from the elution liquid.

Since, as well as inorganic impurities, the waste water contains organic material, such as oil, fat, dust, etc., the ion exchanger often becomes clogged if the waste water is not pretreated before being passed into the ion exchange system. *Macroporous ion exchangers* have only partly solved this problem since suspended matter, emulsions and high molecular weight organic compounds will also affect this type of ion exchanger.

Consequently, it is an advantage to pretreat the waste water on sand filters or activated carbon. The sand filter will remove the suspended particles and the activated carbon most of the organic impurities.

Since most of the ions are eluted in the first 60-75% of the eluting volume, it is often of advantage to use the last portion for the subsequent regeneration.

6.5.4. The application of extraction for removal of heavy metals.

P.6.42. **Most heavy metal ions can be recovered from aqueous solutions by extraction. Complexes of the metal ions are formed, for example by reaction with Cl⁻, and these metal complexes are then extracted by means of organic solvents.**

By using different concentrations of the ligand it is possible to separate different metals by extraction, again opening up the possibility of recovering the metals. The calculation of the equilibrium is based on the following reactions:

Metal + ligand — metal complexes   (6.80)

(Metal complex)$_{water}$ — (Metal complex)$_{organic\ solvent}$ (6.81)

From reaction (6.80) and by using the mass action law it is possible to set up the following expression:

$$\frac{[Me\ L]}{[L]\ [Me]} = K \tag{6.82}$$

Reaction (6.81) can be quantitatively expressed by means of the distribution coefficient, D, where:

$$d = \frac{[Me\ L]_{organic\ solvent}}{[Me\ L]_{water}} \tag{6.83}$$

Combining equations (6.82) and (6.83) gives:

$$K * D = \frac{[Me]_{org.}}{[L]\ [Me]} \tag{6.84}$$

where $[Me]_{org.}$ is the total metal ion concentration in the organic phase. By applying this reaction in practice it is possible to vary both the concentration of the ligand, and the volume of the organic phase.

On increasing the concentration of the ligand and increasing the volume of the organic phase, $[Me]_{org}$ will increase and $[Me]$ decrease (see equation (6.83) and (6.84)).

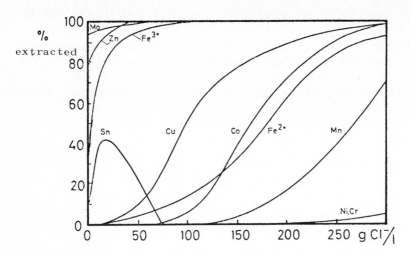

Fig. 6.57. Efficiency of extraction against Cl⁻ concentration.

Fig. 6.57 shows the extraction efficiency of various concentrations of chloride for several metal ions. These calculations are based upon the application of the same volume of organic solvent as the waste water treated. As seen form the diagram, it is possible, for example, to separate Fe^{2+} ions from cobalt by extracting with 60 g Cl⁻/l. The extraction can also be carried out by a liquid ion exchanger of the trialkylamine type R_3NHCl.

The extraction process for $FeCl_4^-$ in aqueous solution will take place according to the following reactions:

$$FeCl_4^-(aq) + R_3NHCl(org) - R_3NHFeCl_4(org) + Cl^-(aq) \qquad (6.85)$$

Recovery of copper from a solution containing Cu^{2+}, SO_4^{2-}, Na^+ and H^+ (pH = approx. 2.0) is possible by extraction with acetone. The efficiency should be 99.9% according to patent DDR 67541. On recovering the acetone by distillation a loss of 0.3 kg acetone/100 kg copper is recorded.

An amine extraction process has also been developed for the recovery of cyanide and metal cyanide from waste streams of plating processes (Chemical Week, 1976).

6.5.5. Application of membrane process for removal of heavy metals.

Membrane separation, electrodialysis, reverse osmosis, ultrafiltration and other such processes are playing an increasingly important role in waste water treatment.

A membrane is defined as a phase which acts as a barrier between other phases. It can be a solid, a solvent-swollen gel or even a liquid. The applicability of a membrane for separation depends on differences in its permeability to different compounds.

Table 6.36 gives a survey of membrane separation processes and their principal driving forces, applications and their useful ranges.

TABLE 6.36
Membrane separation processes

Process	Driving force	Range (μm) particle size	Function of membrane
Electrodialysis	Electrical poten-tial gradient	< 0.1	Selective to certain ions
Dialysis	Concentration	< 0.1	Selective to solute
Reverse osmosis	Pressure	< 0.05	Selective transport of water
Ultrafiltration	Pressure	$5 * 10^{-3} - 10$	Selective to molecular size and shape

Osmosis is defined as a spontaneous transport of a solvent from a dilute solution to a *concentrated solution across a semi-permeable membrane.* At a certain pressure - the so-called osmotic pressure - equilibrium is reached.

The osmotic pressure can vary with concentration and temperature, and depends on the properties of the solution.

For water, the osmotic pressure is given by:

$$\pi = \frac{n}{V} R T \tag{6.86}$$

where
n = the number of moles of solute
V = the volume of water
R = the gas constant
T = the absolute temperature

This equation describes an ideal state and is valid only for dilute solutions. For more concentrated solutions the equation must be modified by **the van Hoff factor** by using an osmotic pressure coefficient:

$$\pi = \emptyset * \frac{n}{V} R T \tag{6.87}$$

For most electrolytes the osmotic pressure coefficient is less than unity and will usually decrease with increasing concentrations. This means that equation (6.86) is usually conservative and predicts a higher pressure than is observed.

P.6.43. **If the pressure is increased above the osmotic pressure on the solution side of the membrane, as shown in Fig. 6.58, the flow is reversed. The solvent will then pass from the solution into the solvent. This is the basic concept of reverse osmosis.**

Reverse osmosis can be compared with filtration, as it also involves the moving of liquid from a mixture by passing it through a filter.

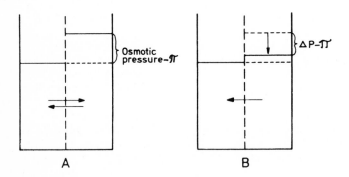

A   B

Fig. 6.58.   A - illustrates equilibrium. An osmotic pressure appears. B - illustrates the principle of reverse osmosis.

However, one important difference is that *the osmotic pressure,* which is very small in ordinary filtration, plays an important role in reverse osmosis. Second, a filter cake *with low moisture content cannot* be obtained in reverse osmosis, because the osmotic pressure of the solution increases with the removal of solvents. Third, the filter separates a mixture on the basis of size, whereas reverse osmosis membranes work on the basis of other factors.   Reverse osmosis has sometimes also been termed hyper-filtration.

The permeate flux, F, through a semipermeable membrane of thickness, d, is given by:

$$F = \frac{D_W * C_W * V}{R\,T\,d} (\Delta P - \pi)$$

(6.88)

where

D_W = the diffusion coefficient

D_W = the concentration of water

V = the molar volume of water

ΔP = the driving pressure (see Fig. 6.58)

The equation (6.88) indicates that the water flux is inversely proportional to the thickness of the membrane. These terms can be combined with the coefficient of water permeation, W_p, and equation (6.88) reduces to:

$$F = W_p * (\Delta P - \pi) \qquad (6.89)$$

where

$$W_p = \frac{D_W * C_W * V}{R T d} \qquad (6.90)$$

For the solute flux, F_s, the driving force is almost entirely due to the concentration gradient across the membrane, which leads to the following equation (Clark, 1962):

$$F_s = D_s \frac{dC'_i}{dx} = D_s \frac{\Delta C'_i}{d} \qquad (6.91)$$

where

C'_i = the concentration of species, i, within the membrane

$\Delta C'_i$ = measured across the membrane

In constructing a system for reverse osmosis many problems have to be solved:

1. The system must be designed to give a **high liquid flux** reducing the concentration potential.
2. **The packaging density must be high** to reduce pressure vessel cost.
3. **Membrane replacement costs must be minimized.**
4. The usually fragile **membranes must be supported** as they have to sustain a pressure of 20-100 atmospheres.

Four different system designs have been developed to meet the solution to the latter problem. *The plate and frame technique, large tube technique, spiral wound technique and the hollow fine fibre technique.*

The various techniques are compared in Table 6.37.

TABLE 6.37
Comparison of the various techniques

Module concept	Packing density (m^2/m^3)	Useful pH range	Ease of cleaning	NaCl rejection	Water flux at 40 atm. $(m^3/m^2/day)$
Plate and frame	450	2 - 8	fair	very good	0.5
Large tubes	150	2 - 8	very good	very good	0.5
Spiral	750	2 - 8	good	very good	0.5
Hollow fine fibres	7500 - 15000	0 - 12[*]	fair	good/fair	0.05 - 0.2

[*]) Polyamide

The most widely used membrane is the *cellulose acetate membrane* in the Loeb-Sourirajan technique. This membrane is asymmetrical and consists of a thin dense skin of approximately 0.2 μm on an approximately 100 μm thick porous support.

As shown in Fig. 6.59, these membranes are not resistant to high or low pH, and a temperature range of 0-30°C must be recommended.

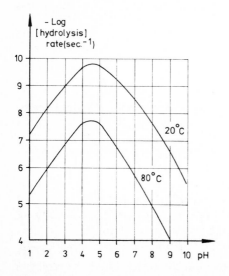

Fig. 6.59. Hydrolysis rate of cellulose acetate membrane as function of pH at 20°C and 80°C. (Vos et al., 1966).

Polyamide membranes have also been developed. They are considerably more resistant to temperature and pH, but give a smaller flux. During the laste decade there has been intense research activity in the development of membranes, resulting in several new types. *Cellulose acetate-butyrate resin, cellulose acetat-methacrylate, polyacrylacid and cellulose nitrate-acetate,* are among the recently developed membrane materials, which are more resistant to pH and temperature, but do not reduce the initial fluxes. Several natural materials could also be of use as membranes and extensive laboratory investigations may hold promise for the application of such natural membranes in the near future (Kraus et al., 1967).

Reverse osmosis is particularly well suited for the treatment of *nickel-plating rinsing water.*

Cellulose acetate membranes are recommended (Hauck et al., 1972). The treatment can contribute significantly to both *water pollution control and recovery of nickel.*

Push et al. (1975) have demonstrated that it is possible to fractionate metal salts by reverse osmosis.

By tailoring membranes for specific applications and by optimizing the process parameters, such as pressure applied, solute concentrations, and added polyelectrolyte, it is possible to fractionate metals from waste water and recover valuable materials.

By using polyamide membranes for reverse osmosis experiments at 100 atmospheres with an equimolar solution of $AgNO_3$ and $Al(NO_3)_3$, the membrane was able to reject 99.98% of aluminium salt, while the silver salt was enriched in the permeate.

6.6. WATER RESOURCES.

6.6.1. Introduction.

P.6.44. **Man cannot exist without water, so there has always been a demand for water and all the nearest and more obvious sources have already been exploited.**
Any future demand must inevitably be met from more remote and increasingly less attractive sites, and for this reason alone development costs must rise continuously in the future.

The relationship between the hydraulic cycle and the water demand now and in the near future has already been touched on in section 2.11.

Most waters have to be purified before they can be used for human consumption. Raw water is so infinitely variable in quality that there is no fixed starting point in the treatment process. Many countries have their own standards of acceptable purity for potable water and these vary. **WHO lays down two standards** (see Appendix 6), which are widely applied in developing countries.

P.6.45. **Virtually no water is impossible to purify to potable standards, but some raw waters are so bad as to merit rejection, because of the expense involved.**
Water from underground sources is generally of better quality than surface water, but it may be excessively hard and/or contain iron and manganese. Full treatment of water may comprise pretreatment, mixing, coagulation, flocculation, settling, filtration and sterilization. Not all waters require full treatment, however.

The difference of the two sources are summarized in Table 6.38.
Pretreatment includes screening, raw water storage, prechlorination, aeration, algal control and straining.

As coagulants aluminium sulphate, sodium aluminate and iron salts are used. They are usually applied in combination with coagulant acids, which include lime, sodium carbonate, activated silica and polyelectrolytes. After coagulation, mixing, flocculation, and sedimentation, take place, as de-scribed in 6.2.2 and 6.3.2 respectively.

For filtration sand filters are widely used either after coagulation, mixing, flocculation and sedimentation of surface water or after aeration of ground water for removal of iron and manganese compounds.

TABLE 6.38
Characteristics of ground water and surface water

Properties	Ground water	Surface water
Salt concentration	high	low
Iron concentration	high	low
Manganese concentration	high	low
$KMnO_4$ number	low	high
Hardness	high	low
pH	6-8	7-9
Turbidity	low	high
Temperature	low	high
Number of E. coli	0	>0
Colour	none	yellowish

Two typical flow diagrams for treatment of ground and surface water are shown in Figs. 6.60 and 6.61. Most of the processes used in water treatment have already been mentioned (see Table 6.2).

However, softening and disinfection are treated in the following paragraphs.

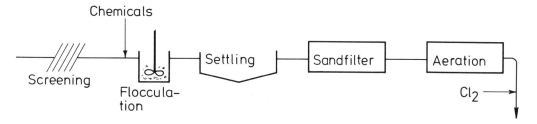

Fig. 6.60.   Typical treatment of surface water.

Fig. 6.61.   Typical treatment of ground water.

P.6.46. Water management is closely linked to waste water management (see section 6.1). Insufficient treatment of waste water can have several consequences, of which two will be mentioned:

1.   low quality surface water that **requires a more advanced treatment of the raw water** for production of potable water,

2.   the treated waste water is discharged to the sea to avoid a deterioration of the quality of the surface water. Consequently, it is necessary **to use supplementary sources of raw water,** which may be difficult close to cities, or areas suffering from water shortage.

6.6.2.   Softening.

The presence of polyvalent cations in water causes hard water, so called because it is hard to form a lather with soap. Polyvalent salts of the long-chain fatty acids present in soap are insolubel. In the days when all washing was done with soaps, the problem of hard water was more of a niusance than it is today, when soaps have been replaced largely with synthetic detergents, whose polyvalent metal salts are relatively soluble. The predominant metal ions in waters used for potable supplies are calcium and magnesium.

P.6.47.   Hardness can be either calcium hardness (including magnesium hardness), **temporary hardness, which is carbonate and hydrogen carbonate hardness, or permanente hardness, which is the difference between the calcium hardness and the temporary hardness,** i.e. calcium hardness that has no carbonate or hydrogen carbonate as counter ions.

P.6.48.   Temporary hardness can be removed by boiling the water, since:

$$Ca(HCO_3)_2 \; - \; CaCO_3 \; + \; CO_2 \; + \; H_2O \tag{6.92}$$

Hardness is often quantified as calcium carbonate equivalents in mg per litre or as **hardness degrees, which uses 10 mg l^{-1} CaO as its base unit.** The content of magnesium ions is taken into consideration by calculating th equivalent amount of calcium carbonate and calcium oxide, respectively. It turns out that 83 mg of magnesium carbonate correspont to 100 mg calcium carbonate and 39 mg of magnesium oxide give the same number of hardness degrees as 56 mg of calcium oxide = 5.6 degrees of hardness.

P.6.49.   There are two principal methods of softening water for municipal purposes; by lime and lime-soda and ion exchange.

The first method is based upon precipitation of calcium as calcium carbonate and magnesium as magnesium hydroxide.
A classification of water in terms of hardness is given in Table 6.39.
The lower limits of softening by this process are based on the solubilities of these precipitates, see Table 6.40.
When lime is used, only the carbonate hardness is reduced. The additional use of soda can reduce the permanent hardness as well.

TABLE 6.39
Classification of water in terms of hardness

mg/l calcium carbonate	Hardness degrees	Specification
0 - 80	0 - 4	Very soft
80 - 160	4 - 8	Soft
160 - 240	8 - 12	Medium hard
240 - 600	12 - 30	Hard
>600	>30	Very hard

TABLE 6.40
Solubilities of $CaCO_3$ and $Mg(OH)_2$ as functions of temperature and pH

pH	$CaCO_3$ solubility mg/l 25°C	60°C	$Mg(OH)_2$ solubility mg/l 25°C	60°C
7	970	240	-	-
8	79	20	-	-
9	8.0	2.1	13,000	5500
10	1.0	0.3	130	55
11	0.4	0.13	1.3	0.6
12	0.3	0.11	0.01	very low

The process can be described in five steps:

1. The reaction between free carbon dioxide and added lime:

$$CO_2 + Ca(OH)_2 \rightarrow CaCO_3 + H_2O \qquad (6.93)$$

2. The reaction of calcium carbonate hardness with lime:

$$Ca(HCO_3)_2 + Ca(OH)_2 \rightarrow 2CaCO_3 + 2H_2O \qquad (6.94)$$

3. The reaction of magnesium carbonate hardness with lime:

$$Mg(HCO_3)_2 + Ca(OH)_2 \rightarrow CaCO_3 + MgCO_3 + 2H_2O \qquad (6.95)$$

$$MgCO_3 + Ca(OH)_2 \rightarrow CaCO_3 + Mg(OH)_2 \qquad (6.96)$$

Combining the above two equations gives:

$$Mg(HCO_3)_2 + Ca(OH)_2 \rightarrow 2CaCO_3 + Mg(OH)_2 + 2H_2O \qquad (6.97)$$

From these equations it can be concluded that 2 moles of lime are required to remove 1 g atom of magnesium, or twice as much as is required for calcium removal.

4. Non-carbonate calcium hardness is removed by soda. The reaction is:

$$CaSO_4 + Na_2CO_3 \rightarrow CaCO_3 + Na_2SO_4 \tag{6.98}$$

The non-carbonate hardness is represented here as sulphate.

5. The reaction of non-carbonate magnesium hardness with lime and soda is:

$$MgSO_4 + Ca(OH)_2 \rightarrow Mg(OH)_2 + CaSO_4 \tag{6.99}$$

$$CaSO_4 + Na_2CO_3 \rightarrow CaCO_3 + Na_2SO_4 \tag{6.100}$$

From these reactions it can be seen that the addition of lime always serves three purposes and may serve a fourth. It removes, in order, carbon dioxide, calcium carbonate hardness and magnesium carbonate hardness. Furthermore, when magnesium non-carbonate hardness must be reduced, lime converts the magnesium hardness to calcium hardness. Soda then removes the non-carbonate hardness in accordance with equations (6.98) and (6.99).

Since softening is usually accomplished at high pH values and the reactions do not go to completion, the effluent from the treating unit is usually supersaturated with calcium carbonate. This would cement the filter medium and coat the distribution system. In order to avoid these problems, the pH must be reduced, so that insoluble calcium carbonate can be converted to the soluble hydrogen carbonate. This is accomplished in practice by recarbonation, i.e. the addition of carbon dioxide. The equations may be as follows:

$$CaCO_3 + CO_2 + H_2O \rightarrow Ca(HCO_3)_2 \tag{6.101}$$

$$CO_3^{2-} + CO_2 + H_2O \rightarrow 2HCO_3^- \tag{6.102}$$

The large amount of sludge containing calcium carbonate and magnesium hydroxide, produced by softening plants, presents a disposal problem, since it can no longer be discharged into the nearest stream or sewer. Some of this sludge can be recycled to improve the completeness of reaction, but a large quantity must be disposed of. The methods used are lagooning, drying for land fill, agricultural liming and lime recovery by recalcination.

In ion exchange softening, the calcium and magnesium ions are exchanged for monovalent ions, usually sodium and hydrogen. This is termed "cation exchange softening", and current practice uses ion exchange resins based on highly cross-linked synthetic polymers with a high capacity for exchangeable cations. In cases where a complete demineralization is required anion exchangers are also employed. In this case, chloride, hydrogen carbonate and other anions are replaced by hydroxide ions, which together with the hydrogen ions released by the cation exchanger, form water:

$$OH^- + H^+ \rightarrow H_2O \tag{6.103}$$

A comparison of the various softening processes is illustrated in Table 6.41.

TABLE 6.41
Comparison of various softening processes

	Lime - soda Cold Hot		Ion exchange Na^+-form	H^+-form
Min. hardness attainable	30 mg/l		10 mg/l	0	0
Total dissolved solids	Decreased		Decreased	Increased slightly	Decreased
Na-content	?		?	Increased	Decreased significantly
Operation cost	low		low	slightly higher	
Capital	high		high	low	low

6.6.3. Disinfection processes.

Micro-organisms are destroyed or removed by a number of physico-chemical waste water treatment operations, such as coagulation, sedimentation, filtration and adsorption.

However, inclusion of a disinfection step has become common practice in water and waste water treatment to ensure against transmission of water-borne diseases. **The disinfection process must be distinguished from sterili- zation. Sterilization involves complete destruction of all micro-organisms including bacteria, algae, spores and viruses, while disinfection does not provide for the destruction of all micro-organisms,** e.g. the hepatitis virus and polio virus are generally not inactivated by most disinfection processes.

P.6.50. The mechanism of disinfection involves at least two steps:
1. Penetration of the disinfectant through the cell wall.
2. Reaction with enzymes within the cell (Fair et al., 1968)

Chemical agents, such as ozone, chlorine dioxide and chlorine, probably cause disinfection by direct chemical degradation of the cell matter, including the enzymes, while application of thermal methods or degradation accomplish essentially physical destruction of the micro-organisms.

The large number of organic and inorganic chemicals exert a poisoning effect on the micro-organisms by an interaction with enzymatic proteins or by disruptive structural changes within the cells.

P.6.51. The rate of destruction of micro-organisms has been expressed by a first order reaction referred to as Chick's law:

$$\frac{dN}{dt} = -k * N \tag{6.104}$$

where N is the number of organisms per volume and k is a rate constant.
By intetration between the limit $t = 0$ and t

$$\int_{N_0}^{N(t)} \frac{1}{N} * \frac{dN}{dt} * dt = -k \int_0^t dt \tag{6.105}$$

or

$$\ln \frac{N(t)}{N_0} = -kt \tag{6.106}$$

$$N = N_0 * e^{-kt} \tag{6.107}$$

Rearrangement of this equation and conversion into common logarithms gives:

$$t = \frac{2.3}{k} * \log \frac{N_0}{N(t)} \tag{6.108}$$

As seen, Chick's law states that the rate of bacterial destruction is directly proportional to the number of organisms remaining at any time. This relationship indicates a uniform susceptibility of all species at a constant concentration of disinfectant, pH, temperature and ionic strength. Many deviations from Chick's law have been described in the literature. In accordance with Fair et al. (1968) chlorination of pure water shows typical deviations from Chick's law, as seen in Fig. 6.62. Often deviation from the first order rate expression is due to autocatalytic reaction. In this case the expression can be transformed to:

$$-\frac{dN}{dt} = k_1 * N(t) + k_2 * N(t) * (N_0 - N(t)) \qquad (6.109)$$

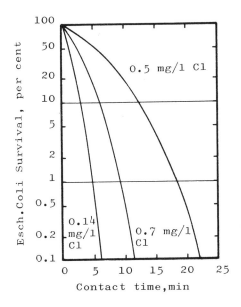

Fig. 6.62. Length of survival of E. coli in pure water at pH 8.5 and 2-5°C.

As pointed out the disinfection rate expression (Chick's law) does not include the effect of disinfectant concentration. The relationship between disinfectant concentration and the time required to kill a given percentage of organisms is commonly given by the following expression:

$$C^n * t = \text{constant} \qquad (6.110)$$

Berg (1964) has shown that the concentration/time relationship for HOCl at 0-6°C, to give a 99% kill of Esch. coli is expressed as:

$$C^{0.86} * t = 0.24 \quad (mg \ l^{-1} \ min) \tag{6.111}$$

Temperature influences the disinfection rate first, by its direct effect on the bacterial action and second, by its effect on the reaction rate. Often an empirical temperature expression is used, such as:

$$k_t = k_{20} * d^{(t-20)} \tag{6.112}$$

where

k_t = the rate constant at t°C

k_{20} = the rate constant at 20°C

d = an empirical constant

P.6.52. Most micro-organisms are effectively killed by extreme pH conditions, i.e. at pH below 3.0 and above 11.0.

The effect of disinfection is also strongly dependent on the coexistence of other matter in the waste water, e.g. organic matter. The disinfectant may (1) react with other species to form compounds which are less effective than the parent compounds, or (2) chemically oxidize other impurities present in the water, reducing the concentration of the disinfectants.

The application of heat is one of the oldest, and at the same time most certain, methods of water disinfection. In addition, *freezing and freeze-drying are effective methods for the preservation of bacteria.* However, these techniques are of little practical significance for waste water treatment, as they are too costly. Disinfecting large volumes of water by heating is clearly not suitable for economic reasons.

The wave length region from 250-265 nm, beyond the visual spectrum, has bactericidal effects.

Mercury vapour lamps emit a narrow band at 254 nm and can be used for small-scale disinfection. It is assumed that the nucleic acids in bacterial cells absorb the ultraviolet energy and are consequently destroyed. These nucleic acids include deoxyribonucleic acid (DNA) and ribonucleic acid (RNA). The main problem in the application of ultraviolet irridation for disinfection is to ensure that the energy is delivered to the entire volume of the water. Even distilled water will absorb only 8% of the applied energy to a depth of 3 cm, and turbidity, dyes and other impurities constitute barriers to the penetration of ultraviolet radiation.

This means that only a thin layer of clear water without impurities, able to absorb the ultraviolet light, can be treated. For a 99% level of kill, the use of a 30 W lamp would allow flows of from 2 to 20 m³/hr to be disinfected. The use of ultraviolet lamps for disinfection has some important advantages. As nothing is added to the water no desirable qualities will be changed. No tastes or odours result from the treatment. The disadvantage of ultraviolet irradiation is that it provides no residual protection against recontamination as the application of chlorine does.

Gamma and X rays are electromagnetic radiations of very short wavelength and have an excellent capacity for destroying micro-organisms. However, their use is relatively expensive. The application of the method requires care, and this will restrict its use considerably.

Chlorine is produced exclusively by electrolytic oxidation of sodium chloride in aqueous solutions:

$$2Cl^- \rightarrow Cl_2 + 2e^- \tag{6.113}$$

After generation, the chlorine gas is purified by washing in sulphuric acid and the product usually has a purity of more than 99%. The gas is liquefied by compression to 1.7 atmospheres between -30°C and -5°C, and stored in steel cylinders or tanks. Chlorine should be handled with caution, as the gas is toxic and has a high chemical activity, with danger of fire and explosion. In the presence of water chlorine is highly corrosive.

When chlorine is added to an aqueous solution it hydrolyzes to yield Cl^- and OCl^-:

$$Cl_2 + 2H_2O \rightarrow H_3O^+ + Cl^- + HOCl \tag{6.114}$$

As can be seen, the process is a disproportionation, since chlorine in zero oxidation state turns into oxidation states -1 and +1.

Hypochlorous acid is a weak acid:

$$HOCl + H_2O \rightarrow H_3O^+ + OCl^- \tag{6.115}$$

The acidity constant K_a:

$$K_a = \frac{[H_3O^+][OCl^-]}{[HOCl]} \tag{6.116}$$

is dependent on the temperature as illustrated in Table 6.42.

HOCl is a stronger disinfectant than OCl^- ions, which explains why the disinfection is strongly dependent on pH.

Fig. 6.63 shows the time/concentration relationship in disinfection with chlorine (after Fair et al., 1968).

TABLE 6.42
The acidity constant for HOCl

Temperature	K_a
0	$1.5 * 10^{-8}$
5	$1.7 * 10^{-8}$
10	$2.0 * 10^{-8}$
15	$2.2 * 10^{-8}$
20	$2.5 * 10^{-8}$
25	$2.7 * 10^{-8}$

The ammonium present in the water is able to react with the chlorine or hypochlorous acid (see 6.3.5).

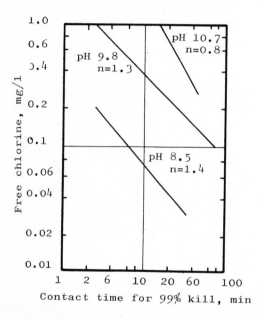

Fig. 6.63. Concentration of free available chlorine required for 99% kill of E. coli at 2-5°C.

The rates of chlorine formation depend mainly on pH and the ratio of the reactants employed. Moore (1951) infers that the distribution of chlorine is based on the equation:

$$2NH_2Cl + H_3O^+ \rightarrow NH_4^+ + NHCl_2 + H_2O \qquad (6.117)$$

for which:

$$K = \frac{[NH_4^+] [NHCl_2]}{[H_3O^+] [NH_2Cl]^2} = 6.7 * 10^5 \quad (25°C) \qquad (6.118)$$

The disinfection power of chloramines measured in terms of contact time for a given percentage kill, is less than that of chlorine. This is seen by comparing Fig. 6.63 with Fig. 6.64.

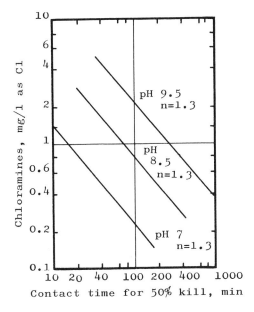

Fig. 6.64. Concentration of combined available chlorine required for 50% kill of E. coli at 2-5°C.

The bacterial properties of chlorine are probably based on the formation of free hypochlorous acid:

$$NH_2Cl + H_2O \rightarrow HOCl + NH_3 \qquad (6.119)$$

$$NH_2Cl + H_3O^+ \rightarrow HOCl + NH_4^- \qquad (6.120)$$

However, the reaction of chlorine with ammonia or amino compounds presents a problem in the practice of chlorination of waste water containing such nitrogen compounds.

Residual
chlorine

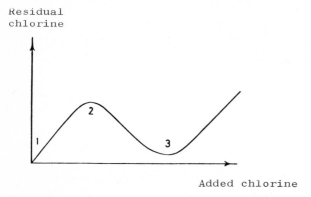

Added chlorine

Fig. 6.65. Breakpoint chlorination.

Fig. 6.65 shows the residual chlorine as a function of the chlorine applied. Between points 1 and 2 in the figure, mono- and di-chloramine are formed. The oxidation processes with chlorine occurring between points 2 and 3 give a decline in residual chlorine. Point 3 is called the breakpoint. Addition of chlorine in this interval probably produces free nitrogen gas as the predominant product of oxidation. Fair et al. (1968) even propose that the reaction involving the formation of NOH as an intermediate, followed by the formation of nitric oxide, NO, could explain the observations between points 2 and 3:

$$2NHCl_2 + 6H_2O \rightarrow 2NOH + 4H_3O^+ + 4Cl^- \qquad (6.121)$$

$$2NOH + HOCl \rightarrow 2NO + H_3O^+ + Cl^- \qquad (6.122)$$

In total:

$$2NHCl_2 + HOCl + 6H_2O \rightarrow 2NO + 5H_3O^+ + 5Cl^- \qquad (6.123)$$

Further addition of chlorine beyond the breakpoint gives an increasing residue of free chlorine.

Chlorine doses below the breakpoint requirement can be used to oxidize ammonia if chlorination is followed by contact with activated carbon (Bauer et al., 1973).

Dichloramine has been shown to be rapidly converted to the end product, the most likely reaction being:

$$C + 2NHCl_2 + H_2O \rightarrow N_2 + 4H^+ + 4Cl^- + CO \qquad (6.124)$$

Further study is, however, needed to show conclusively that surface oxidation results from this reaction. Furthermore, it is important to know that the Cl_2/NH_4-N oxidized mole ratio is 2:1, which is required for ammonium oxidation by this pathway.

The monochloramine reaction with carbon appears more complex. On fresh carbon the reaction is most probably:

$$NH_2Cl + H_2O + C \rightarrow NH_3 + H^+ + Cl^- + CO \qquad (6.125)$$

After this reaction has proceeded to a certain extent, partial oxidation of monochloramine is observed. Possibily according to the reaction:

$$2NH_2Cl + CO \rightarrow N_2 + H_2O + 2H^+ + 2Cl^- + C \qquad (6.126)$$

It has been observed that acclimation of fresh carbon is necessary before monochloramine can be oxidized.

In the removal of ammonia with a dose of chlorine less than the breakpoint followed by contact with activated carbon, pH control can be used to determine the major chlorine species. The studies reported here indicate that a pH value near 4.5 should be avoided, because $NHCl_2$ predominates and thus 10 parts by weight or chlorine are required for each part of NH_3-N oxidized to N_2. At a slightly higher pH and acclimated carbon, the portion of monochloramine increases and the chlorine required per unit weight of NH_3-N oxidized should approach 7.6 parts, ignoring the chlorine demand resulting from other substances. However, further testing should be used to verify this conclusion in each individual case.

When accidental overdosing of chlorine has occurred or after an intentional addition of large quantities of chlorine to accelerate disinfection, it will be desirable to remove the excess chlorine. This is possible with a reducing agent, such as sulphur dioxide, sodium hydrogen sulphite or sodium thiosulphate:

$$SO_2 + Cl_2 + 2H_2O \rightarrow H_2SO_4 + 2HCl \qquad (6.127)$$

$$NaHSO_3 + Cl_2 + H_2O \rightarrow NaHSO_4 + 2HCl \qquad (6.128)$$

$$2Na_2S_2O_3 + Cl_2 \rightarrow Na_2S_4O_6 + 2NaCl \qquad (6.129)$$

Oxidative degradation by chlorine is limited to a small number of compounds. Nevertheless, oxidation of these compounds contributes to overall reduction of BOD_5 in wastes treated with chlorine. A disadvantage is that chlorinated organic compounds may be formed in large quantities. A variety of chlorine compounds is applied in waste water treatments. For these compounds the available chlorine can be calculated. Generally this is expressed as percentage chlorine having the same oxidation ability. Data for the different chlorine-containing compounds are given in Table 6.43.

Table 6.43
Actual and available chlorine in pure chlorine-containing compounds

Compound	Mol. weight	Chlorine equiv. (moles of Cl_2)	Actual chlorine (%)	Available chlorine (%)
Cl_2	71	1	100	100
Cl_2O	87	2	81.7	163.4
ClO_2	67.5	2.5	52.5	260
$NaOCl$	74.5	1	47.7	95.4
$CaClOCl$	127	1	56	56
$Ca(OCl)_2$	143	2	49.6	99.2
$HOCl$	52.5	1	67.7	135.4
$NHCl_2$	86	2	82.5	165
NH_2Cl	51.5	1	69	138

It can be seen that the actual chlorine percentage in chlorine dioxide is 52.5, but the available chlorine is 260%. This is , of course, due to the fact that the oxidation state of chlorine in chlorine dioxide is +4 which means that five electrons are transferred per chlorine atom, while Cl_2 only transfers one electron per chlorine atom.

Hypochlorite can be obtained by the reaction of chlorine with hydroxide in aqueous solution:

$$Cl_2 + 2NaOH \rightarrow NaCl + NaOCl + H_2O \qquad (6.130)$$

Chlorinated lime, also called bleaching powder is formed by reaction of chlorine with lime:

$$Ca(OH)_2 + Cl_2 \rightarrow CaCl(OCl) + H_2O \qquad (6.131)$$

A higher content of available chlorine is present in calcium hypochlorite, $Ca(OCl)_2$. Chlorine dioxide is generated in situ by the reaction of chlorine with sodium chlorite:

$$2NaClO_2 + Cl_2 \rightarrow 2ClO_2 + 2NaCl \qquad (6.132)$$

Theoretically, fluorine could be used for disinfection, but nothing is known about the bactericidal effectiveness of this element at low concentrations. However, bromine is used mainly for disinfection of swimming pools, the reason being **that monobromamine, unlike chloramine, is a strong bactericide.** There is therefore no need to proceed to breakpoint bromination. Bromine has a tendency to form compounds with organic matter, resulting in a high bromine demand. This and the higher cost are the major factors limiting the use of bromine for treatment of waste water.

Iodine can also be used as a disinfectant. It dissolves sparingly in water unless iodide is present:

$$I^- + I_2 \rightarrow I_3^- \qquad (6.133)$$

It reacts similarly with water in accordance with the scheme for chlorine and bromine:

$$I_2 + H_2O \rightarrow HOI + HI \qquad (6.134)$$

It has a number of advantages over chlorination. Iodine does not combine with the ammonium to form iodomines, but rather oxidizes the ammonia. Also it does not combine with organic matter very easily, e.g. it oxidizes phenol rather than forming iodo-phenols. However, iodine is costly and it has, up till now, found a use only for swimming pool disinfection.

Ozone is produced by passing compressed air through a commercial electric discharge ozone generator. From the generator the ozone travels through a gas washer and a coarse centred filter. A dispersion apparatus produces small bubbles with a large surface area exposed to the solution.

Ozone is used extensively in water treatment for disinfection and for removal of taste, odour, colour, iron and manganese, see also 6.4.4.

Ingols and Fetner (1957) have shown that the destruction of Escherichia coli cells with ozone is considerable more rapid than with chlorine when the initial ozone demand of water has been satisfied (see Fig. 6.66).

The activity of ozone is a problem in the disinfection of water containing high concentrations of organic matter or other oxidizable compounds. A further problem arises from the fact that the decomposition of ozone in water does not permit long-term protection against pathogenic regrowth. *Hovever, ozone has the advantage of being effective against some chlorine-resistant pathogens, like certain virus forms* (Stumm, 1958).

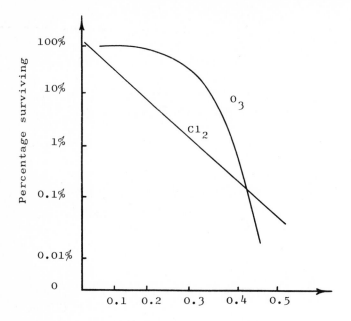

Fig. 6.66.   Disinfection of E. coli by chlorine and ozone (dosage mg/l).

The simultaneous removal of other compounds makes ozonation an advantageous water treatment process.

Ozone can be used to alleviate the toxic and oxygen demanding characteristics of waste water containing ammonia by converting the ammonia to nitrate.

The oxidation is a first order reaction with respect to the concentration of ammonia and is catalyzed by OH^- over the pH range 7-9. The average value of the reaction rate constant at pH 9.0 is $5.2 \pm 0.3 * 10^{-2}$ min^{-1}. Ammonia competes for ozone with the dissolved organic constituents comprising the BOD and is oxidized preferentialy relative to the refractory organic compounds, provided alkaline pH values can be maintained. Due to the elevated pH required, ammonia oxidation by ozone is attractive for the process of lime clarification and precipitation of phosphate.

The reaction of ozone with simple organic molecules has been extensively studied in recent years.

The reactions are usually complex, subject to general and specific catalysts and yield a multitude of partially degraded products.

QUESTIONS.

1. Find the approximate BOD_5 for a waste water containing 120 mg l^{-1} carbohydrates, 80 mg l^{-1} proteins and 200 mg l^{-1} fats.

2. Calculate the total oxygen demand (BOD_5 + nitrification) for domestic waste water with BOD_5 = 180 mg l^{-1} and 38 mg l^{-1} ammonium.

3. Write the chemical reaction for a denitrification of nitrate using acetic acid as a carbon source.

4. Calculate BOD_1, BOD_2, BOD_{10}, and BOD_{20} when BOD_5 = 200 mg l^{-1} and the biological decomposition is considered to be a first order reaction.

5. Nitrification (99%) in a biological plant requires 6 hours retention time at 20°C. What retention time is necessary at 0°C?

6. Calculate the area necessary for spray irrigation of 100,000 m^3 of waste water per year with a BOD_5 of 1000 mg l^{-1}. Maximum is 20,000 kg BOD_5 per ha per year and maximum application rate is 200-300 cm y^{-1}.

7. What disadvantages has the use of iron(III) chloride as a precipitant compared with calcium hydroxide and aluminium sulphate?

8. A waste water has a alkalinity corresponding to its hardness. It contains 120 mg l^{-1} Ca^{2+} and 28 mg l^{-1} Mg^{2+}. How much calcium hydroxide should be used to adjust the pH to 11.0?

9. A waste water contains 6.5 mg P l^{-1}. How much 1) $FeCl_3$, $6H_2O$, 2) $Al_2(SO_4)_3$, $18H_2O$, must be used for chemical precipitation, when 90% P-removal is required?

10. A biological plant is designed to give 90% nitrification. The retention time and the aeration are sufficient, but the observed nitrification is only 80%. What should be done?

11. Calculate the inhibition effect (as %) on the biological treatment of municipal waste water containing 2 mg l^{-1} Hg^{2+} and 10 mg l^{-1} Cu^{2+}. The effects are considered to be additive.

12. Calculate the BOD_5 of a waste water that contains 25 mg l^{-1} acetone, 20 mg l^{-1} acetic acid, 10 mg l^{-1} citric acid and 100 mg l^{-1} glucose. What is the number of person eqv. of 150 m^3 24 hr^{-1} is discharged?

13. Which of the following components can be treated on a mechanical-biological treatment plant (no adaptation is foreseen):
1) Butyric acid, 2) 2,4-dichlorophenol, 3) polyvinyl-chloride, 4) pentanone, 5) stearic acid, 6) butadiene. Suggest a treatment method for each of the six components in the concentration range 10-100 mg l^{-1}.

14. What is the minimum cost of activated carbon (US$ 0.40 per kg) for the treatment of 1 m^3 of waste water containing: a) 25 mg l^{-1} chloro-benzene, b) 10 mg l^{-1} toluene, c) 52 mg l^{-1} dodecylbenzene sulphonate?

15. Suggest a treatment method for waste water containing diethyl-disulphide.

16. How much Ni^{2+} remains in industrial waste water after precipitation with calcium hydroxide at pH = 9.5? What methods are available if a concentration of 0.005 mg l^{-1} or less is required?

17. Calculate the stoichiometrical consumption of chemicals for treatment of 100 m^3 24 hr^{-1} of waste water containing 120 mg l^{-1} $Cr_2O_7^{2-}$. A reduction to Cr^{3+} using Na_2SO_3 + HCl precipitation with calcium hydroxide is suggested.

18. Indicate how 2 g l^{-1} Cu^{2+}, 4 g l^{-1} Zn^{2+} and 1 g l^{-1} Mn^{2+} can be separated by extraction.

19. How much 1) chlorine, 2) ozone, 3) $NHCl_2$, 4) ClO_2 is required to give a 99.9% desinfection of E. coli, provided that no oxidation of organic matter takes place? (Contact time 15 min. at pH 8.5).

20. Which has the highest disinfection effect: 1 mg l^{-1} chlorine at pH 7.2 at 10°C, or 1.2 mg l^{-1} chlorine at pH 7.8 at 20°C?

21. Design an aerated pond (5 steps are considered) for treatment of 250 m^3

d^{-1} waste water from an aquaculture plant. The ponds should bring the BOD$_5$ from 80 mg l^{-1} to 10 mg l^{-1}. How much oxygen must be supplied to the ponds?

22. A lake has a catchment area of 50 ha. It covers an area of 12 km^2 and has an average depth of 15 m. Annual precipitation is 600 mm.
Set up an approximate N- and P-balance for the lake, when it is known that a waste-water plant with mechanical-biological treatment discharges 2000 m^3 d^{-1} to the lake.
Characterize the eutrophication of the lake and consider what improvement one should expect after 90% removal of P or N from the waste water.

23. Design an activated-sludge reactor and determine the weight of waste sludge per unit of time for treatment of 25,000 m^3 d^{-1}. k$_d$ and a are determined by pilot-plant experiments to be 2.5 and 0.8 d^{-1}. BOD$_5$ of the influent is 180 mg l^{-1} and an effluent of BOD$_5$ = 12-15 mg l^{-1} should be obtained.

24. Design a plastic trickling filter for the same problem as given in 23. Use 3 parallel filters with a depth of 7 m. The temperature varies from 14°C to 24°C. Try to recycle 2 and 3 times and comment on a comparison of the 2 results. k = 0.08 min^{-1} at 20°C.

CHAPTER 7

THE SOLID WASTE PROBLEMS

7.1. SOURCES, MANAGEMENT AND METHODS.

7.1.1. Classification of solid waste.

The disposal of solid waste has become a galloping problem in all highly developed countries due to a number of factors:

1. Increased urbanization has increased the concentration of solid waste.
2. Increased use of toxic and refractory material.
3. Increased use of "throw away" mass produced items.

P.7.1. **Solid wastes include an incredible miscellany of items and materials, which makes it impossible to indicate one simple solution to the problem. It is necessary to apply a wide spectrum of solutions to the problems according to the source and nature of the waste.**

Table 7.1 shows a classification of solid waste. Typical quantities in the technological society per inhabitant are included.

TABLE 7.1
Classification of solid waste

Type of waste	kg/inhabitant/day (approx.)
Domestic garbage	4
Agricultural waste (1)	12
Mining waste (2)	18
Wastes from construction (3)	0.1
Industrial waste	1.5
Junked automobiles	0.1

(1) Not mentioned further in this context. Recycling is recommended and also widely in use.
(2) A substantial part is used for landfilling.
(3) Not mentioned further in this context. Recycling is possible and recommended, but a part is also used for landfilling.

7.1.2. Examination of mass flows.

A universal method for treatment of solid waste does not exist, and it is necessary to analyze each individual case to find a relevant solution to the problem.

The analysis begins with an examination of mass flows (principle 2.1) to ascertain whether there is an economical basis for **reuse** (for example, bottles), **recovery of valuable raw materials** (paper and metals) or **utilization of organic matter**, as a soil conditioner or for the production of energy. These considerations are illustrated in Fig. 7.1. Mass balances for important materials, as demonstrated in section 2.8, must therefore be set up. Fig. 7.2 provides another very relevant example.

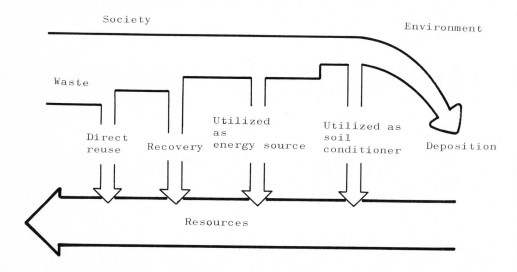

Fig. 7.1. Principles of recycling of solid waste.

The analysis of mass flows is the framework for a feasible solution, which takes economy as well as environmental issues into consideration. However, legislation and economical means are required to achieve the management goals. For instance, the use of returnable bottles can be realized **either by banning the use of throw-away bottles or by placing a purchase tax on throw-away bottles and not on returnable bottles.**

The management problems are highly dependent on the technological methods used for solid waste treatment. The classification mentioned in the introduction to Part B of this book (p. 286) might be used in this context.

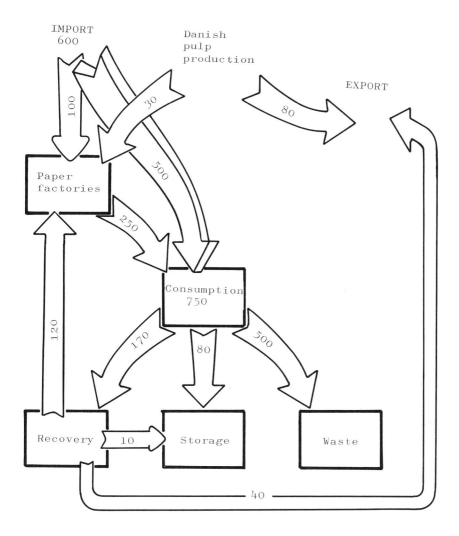

Fig. 7.2. Recovery of paper in Denmark. The mass flows are shown. All number 1000 t/y.

P.7.2. Methods based on alternative technology and recycling, to reduce the amount of waste produced, often require environmental legislation or economical means to guarantee success.
Methods based on deposition will require a comprehensive

knowledge of the environmental effects on the ecosystems involved.

Complete decomposition to harmless components is not possible for most solid waste.

For instance incineration will produce a slag, which might cause a deposition problem, and smoke, which involves air pollution.

7.1.3.   Methods for treatment of solid waste.

P.7.3.   Table 7.2 gives an overview of the methods applicable to solid waste. The table indicates on which principle the method is based - recycling, deposition or decomposition to harmless components. It also shows the type of solid waste the method is able to treat.

TABLE 7.2
Methods for treatment of solid waste

Method	Principle	Applicable to
Conditioning and composting	Decomposition before deposition	Sludge, domestic garbage, agricultural waste
Anaerobic treatment	Decomposition, deposition, utilization of biogas	Sludge, agricultural waste
Thickening, filtration, centrifugation and drying	Dewatering before further treatment and deposition	Sludge
Combustion and incineration	Decomposition and deposition. Utilization of energy	All types of waste
Separation	Recycling	Domestic garbage
Dumping ground	Decomposition and deposition	Domestic garbage
Pyrolysis	Recycling, utilization of energy, decomposition	Domestic garbage
Precipitation and filtration	Deposition or recycling	Industrial waste
Landfilling	Deposition	Industrial-, mining- and agricultrual waste
Aerobic treatment	Decomposition and deposition	Sludge, agricultural waste

In the following sections the problems of solid waste will be discussed in four classes: A. Sludge, B. Domestic garbage, C. Industrial, Mining and Hospital waste, D. Agricultural waste. The methods used for the four classes are, to a certain extent, different, although there is some overlap, between them.

7.2. TREATMENT OF SLUDGE.

7.2.1. Sludge handling.

P.7.4. In most waste water treatments the impurities are not actually removed, but rather concentrated in the form of solutions or a sludge. Only when a chemical reaction takes place does real removal of the impurities occur, e.g. by chemical or biochemical oxidation of organics to CO_2 and H_2O, or denitrification of nitrate to nitrogen gas.
Compare also with principles 2.1 and 2.6.

Sludge from industrial waste water treatment units in most cases requires further concentration before its ultimate disposal. In many cases two- or even three-step processes are used to concentrate the sludge. It is often an advantage to use further thickening by gravity, followed by such treatments as filtration or centrifugation. There are a number of ways of reducing the water content of the sludge that might be used to provide the most suitable solution of how to handle the sludge in any particular case. The final arrangement must be selected not only from consideration of the cost, but also by taking into account that the method used must not cause pollution of air, water or soil.

7.2.2. Characteristics of sludge.

The characteristics of a sludge are among the factors that influence the selection of the best sludge-treatment method. The sludge characteristics vary with the waste water and the waste water treatment methods used.
One of the important factors is **the concentration of the sludge**. Table 7.3 lists some typical concentrations of various types of sludges.
The specific gravity of the sludge is another important factor, since the effect of gravity is utilized in the thickening process. The specific gravity of activated sludge increases linearly with the sludge concentration. This corresponds to a specific gravity of 1.08 g/ml for actual solid. How-

ever, sludge is normally in sufficiently high concentration to exhibit zone-settling characteristics, which means that laboratory measurement of the settling rate must be carried out in most cases before it is possible to design a thickener.

TABLE 7.3
Typical concentations of different types of sludge

Type of sludge	Concentration of suspended matter (w/w%)
Primary sludge (fresh)	2.5 - 5.0
Primary sludge (thickened)	7.5 - 10.0
Primary sludge (digested)	9.0 - 15.0
Trickling filter humus (fresh)	5.0 - 10.0
Trickling filter humus (thickened)	7.0 - 10.0
Activated sludge (fresh)	0.5 - 1.2
Activated sludge (thickened)	2.5 - 3.5
Activated sludge (digested)	2.0 - 4.0
Chemical precipitation sludge (fresh)	1.5 - 5.0
Chemical precipitation sludge (digested)	7.0 - 10.0

The rate with which water can be removed from a sludge by such processes as vacuum filtration, centrifugation and sand-bed drying is an important factor (Nordforsk, 1972), and is expressed by means of the **specific resistance**, R_s, which is calculated from laboratory observations of filtrate production per unit time:

$$R_s = \frac{2b * \Delta P * A^2}{\mu * W} \qquad (7.1)$$

where b = the slope of a plot t/v versus V, t = time, V = filtrate volume, ΔP = the pressure difference across the sludge cake, A = the filter area, μ = viscosity, W = weight of solids deposited per unit filtrate volume.

However, the specific resistance can change during filtration due to compression of the sludge. This is expressed by means of the **coefficient of compressibility**, s, using the following relationship:

$$R_s = R_0 * \Delta P^s \qquad (7.2)$$

where R_0 = the cake constant. When s = 0, sludge is incompressible and R_s = R_0 = a constant.

Table 7.4 gives the dewatering characteristics of variuos sludges.

Studies by Parker et al. (1972) have shown that the filtration time increases with the time of anaerobic storage and with the chloride concentration. Furthermore, it was shown that the filtration time is at a minimum after 5 to 8 days' aeration. The filtration time increases (12-15°C) after aeration for more than 8 days. At higher temperatures the minimum filtration time is reached after a shorter aeration time.

TABLE 7.4
Dewatering characteristics of various sludges

Type of sludge	Specific resistance (sec^2/g)	Pressure (atm)	Compressibility coefficient	Reference
Activated sludge	$2.88 * 10^{10}$	0.5	0.81	Coackley, 1960
Conditioned digested primary and activated sludge	$1.46 * 10^8$	0.5	1.10	Trubnick and Mueller, 1958
Conditioned digested sludge	$1.05 * 10^8$	0.5	1.19	Trubnick and Mueller, 1958
Conditioned raw domestic sludge	$3.1 * 10^7$	0.5	1.00	Trubnick and Mueller, 1958
Thixotropic mud	$1.5 * 10^{10}$	12	-	Gale, 1968
Digested domestic sludge	$1.42 * 10^{10}$	0.5	0.74	Coackley, 1960
Raw domestic sludge	$4.7 * 10^9$	0.5	0.54	Coackley, 1960
Alum coagulation sludge	$5.3 * 10^9$	1.0	-	Gale, 1968
Gelationous $Al(OH)_3$	$2.2 * 10^9$	3.5	-	Gale, 1968
Gelatinous $Fe(OH)_3$	$1.5 * 10^9$	3.5	-	Gale, 1968
Water coagulation sludge	$5.1 * 10^8$	0.7	-	Neubauer, 1966
Colloidal clay	$5 * 10^8$	3.5	-	Gale, 1968
Lime neutralized mine drainage	$3 * 10^8$	1.0	-	Gale, 1968
Conditioned activated sludge	$1.65 * 10^8$	0.5	0.80	Eckenfelder and O'Connor, 1961
Vegetable tanning	$1.5 * 10^8$	1.0	-	Gale, 1968
Ferric oxide	$8 * 10^7$	3.5	-	Gale, 1968
Calcium carbonate	$2 * 10^7$	3.5	-	gale, 1968

The heat value of the sludge is of importance for combustion processes. Fair et al. (1968) have developed the following empirical equation for the heat value of sludge; Q_B (kJ/kg dry solid):

$$Q_B = E \left(\frac{100P_V}{100 - P_c} - B \right) \left(\frac{100 - P_c}{100} \right) \qquad (7.3)$$

where

E and B = empirical constants

P_V = % volatile solid

P_c = dose of conditioning chemical used in dewatering as a percentage of the weight of sludge solid

B = in general 5-10

E = in the range 500-600

When sludge is being considered for use as a soil conditioner, its **chemical properties** are of prime importance. The nutrient content (nitrogen, phosphorus and potassium), in particular, is of interest. Furthermore, knowledge of the heavy metals in sludge is important because of their toxicity.

As shown in Table 7.5, even domestic waste contains certain amounts of heavy metals, and municipal sludge from industrial areas contains a higher concentration of heavy metals. The upper allowable limit for heavy-metal concentration in sludge to be used as a soil conditioner is dependent on the amount of sludge used per ha and on the properties of the soil (Jørgensen, 1975 and 1976).

TABLE 7.5
Characteristic concentration of metals in g per 1000 kg of sludge (dry matter)

	Cr	Ni	Co	Zn	Cd	Cu	Pb	Hg	Ag	Bi
Typical domestic sewage	42	20	6	1380	7	123	218	5.2	13	<25
Mixed domestic and industrial sewage	163	33	10	3665	10	514	317	33	100	<25

Hansen and Tjell (1978) present guidelines used in Scandinavia on sludge application to land; they have taken the present knowledge on this field into consideration.

Finally, **the concentration of pathogenic organisms** in the sludge must be considered. Normal waste water treatment processes, such as sedimentation, chemical precipitation and biological treatments, remove considerable amounts of pathogens which are concentrated in the sludge. A significant reduction in the number of pathogenic organisms has been found to occur during anaerobic digestion, but they are not destroyed entirely.

Combustion, intensive heat treatment of sludge or treatment with calcium hydroxide would eliminate the hazard of pathogenic micro-organisms.

7.2.3. Conditioning of sludge.

P.7.5. Sludge conditioning is a process which alters the properties of the sludge to allow the water to be removed more easily. The aim is to transform the amorphous gel-like sludge into a porous material which will release water. Conditioning of the sludge can be accomplished by either chemical or physical means.

Chemical treatment usually involves the addition of coagulants or flocculants to the sludge. Inorganic as well as organic coagulants can be used, the difference between typical conditioning by polymers or inorganic chemicals being in the amounts of the chemicals used.

Typical doses of inorganic coagulants, such as alunimium sulphate, ferric chloride and calcium hydroxide, are as much as 20% of the weight of the solid, while a typical dose of organic polymer is less than 1% of the weight of the solid. This does not necessarily mean that the cost of using synthetic polymers is lower, since the polymers cost considerably more per kg than the inorganic chemicals used as conditioners.

The amorphous gel-like structure of the sludge is destroyed by heating. Lumb (1951) indicates that the filtration rate of activated sludge is increased by more than a thousand-fold after heat treatment. Typically, the heat-treatment conditions are 30-minutes' treatment at 150-200°C under a pressure of 10-15 atmospheres. A great advantage of heat conditioning is, of course, that the pathogens are destroyed.

Conditioning by freezing has also been reported by Klein (1966) and by Burd (1968), but the process seem to be uneconomic.

7.2.4. Thickening of sludge.

Sludge thickeners are designed on the basis of surface area, which is determined from the material balance:

$$\frac{A}{Q_0 * C_0} = \frac{1/C_1 - 1/C_u}{u}$$
(7.4)

where

A = area of the surface

Q_0 = flow of sludge with the concentration C_0

C_u = the underflow concentration

C_1 and u = the concentration and velocity of any interfacial layer of the settling sludge

The depth of the sludge in the thickener is also a significant design parameters. Roberts (1949) has expressed the rate of sludge thickening by means of the following equation:

$$\log \frac{H - H_\infty}{H_c - H_\infty} = K * (t - t_c)$$
(7.5)

where

H_∞ = the minimum height after infinite time

H = the depth of the sludge after time t

H_c = the depth after time t_c

K = a constant, which must be found experimentally

P.7.6. **Vacuum filtration is used to remove water from a sludge by applying a vacuum across a porous medium.**

The vacuum filter is shown in Fig. 7.3.

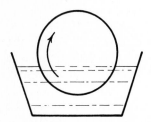

Fig. 7.3. Vacuum filtration.

As the rotary drum passes through the slurry in the slurry tank, a cake of solid is built up on the drum surface and the water is removed by vacuum

-428 -

filtration through the porous medium on the drum surface. As the drum emerges from the slurry the deposited cake is dried further. The cake is removed from the drum by a knife edge. Often the porous filter is washed with water before it is reimmersed in the slurry tank.

Since the specific resistance varies widely with the type of sludge and the waste water treatment used, *it is often best to find the filtration characteristics of the sludge in the laboratory by the Büchner funnel test.*

7.2.5. Centrifugation of sludge.

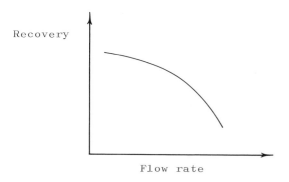

Fig. 7.4. Recovery versus flow rate for centrifugation of sludge.

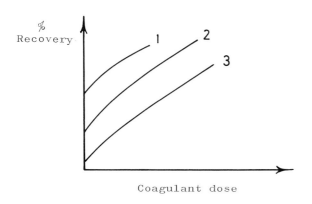

Fig. 7.5. % recovery versus coagulant dose for three flow rates: 1: 1x; 2: 2x; 3: 3x. x = a given amount of electrolytes.

Centrifugation is one of the more recent methods used in the removal of water from waste water sludge.

Of the various types of centrifuge, the solid-bowl centrifuge is considered to offer the best clarification and water-removal properties. It is an important advantage of the centrifugation process that the centrifuge conditions can be adjusted to the concentration of the volatile material (Albertson and Sherwood, 1968). The disadvantage of using a centrifuge is that the cake concentration is generally slightly less than that obtained by vacuum filtration.

The prediction of the behaviour of sludge in a centrifuge is largely a matter of experience. However, some general trends can be noted. If the mass flow rate is increased, recovery is reduced (see Fig. 7.4). The use of electrolytes will increase the recovery at a given flow rate or increase the flow rate for a given recovery. This is illustrated in Fig. 7.5.

7.2.6.   Digestion of sludge.

P.7.7.   If the sludge contains biodegradable organic material it may be advantageous to treat the sludge by aerobic or anaerobic digestion.

Anaerobic digestion is by far the most common method of treating municipal sludge. It creates good conditions for the growth of micro-organisms. The end products of anaerobic digestion are carbon dioxide and methane. The temperature is commonly set at about 35°C, in order to maintain optimum conditions in the digestor. Unfortunately, anaerobic digestion results in considerable quantities of nutrients going into solution (Dalton et al., 1968), which means that a significant amount of nutrient material will be returned to the treatment plant if the supernatant is separated from the sludge.

The principal function of anaerobic digestion is to convert as much as possible of the sludge to end products: liquids and gases. Anaerobic decompositions generally produce less biomass than aerobic processes.

The microorganisms can be divided into 2 broad groups: **the acid formers and the methane formers.**

The acid formers consist of facultative and anaerobic bacteria and soluble products are formed through hydrolysis. The soluble products are then fermented to acids and alcohols of lower molecular weight.

The methane formers are strictly anaerobic bacteria that convert the acids and alcohols, along with hydrogen and carbon dioxide, to methane. A COD-balance for a complete conversion is shown in Fig. 7.6.

Reactors for anaerobic digesters consist of closed tanks with airtight covers. A typical anaerobic digester for a single-stage operation is shown in Fig. 7.7. The digested sludge accumulates in the bottom.

Design parameters for anaerobic digesters are given in Table 7.6.

High-rate digesters are more efficient. The contents are mechanically mixed to ensure better contact between substrate and microorganisms, thus accelerating the digestion process.

TABLE 7.6
Design parameters for anaerobic digesters

Parameter	Normal rate	High rate
Solid retention time	30 - 60	10 - 20
Volatile solid loading (kg/m³ d)	0.5 - 1.5	1.6 - 6.0
Digested solids conc.%	4 - 6	4 - 6
Volatile solid reduction (%)	35 - 40	45 - 55
Gas production m³/kg	0.5 - 0.6	0.6 - 0.65
Methane content	65%	65%

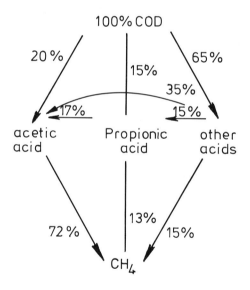

Fig. 7.6. COD-balance for a complete conversion of organic matter. Usually 60-65% of the organic fraction is converted in accordance with the scheme shown.

The properties of aerobically digested sludge are similar to those of an-aerobically digested sludge. An advantage is that some of the operational problems attending anaerobic digestors are avoided, but the disadvantage compared with anaerobic digestion is that the process is more expensive since oxygen must be provided and energy recovery from methane is not possible. Since aerobic digestion is less used in industrial waste water processes than in treatment of municipal waste water, it is not appropriate here to go into further details. For more extensive coverage of these processes, readers are referred to McCarthy (1964) and to Walker and Drier (1966).

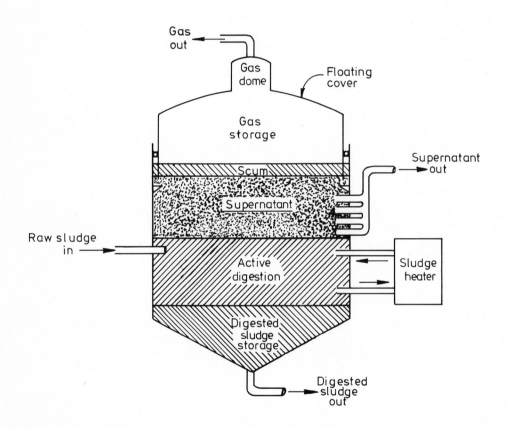

Fig. 7.7.   Diagram of standard-rate anaerobic digester.

7.2.7.   Drying and combustion.

P.7.8.   **The purpose of drying sludge is to prepare it for use as a soil conditioner or for incineration.**

Air drying of the sludge on sand beds is often used to reach a moisture content of about 90%.

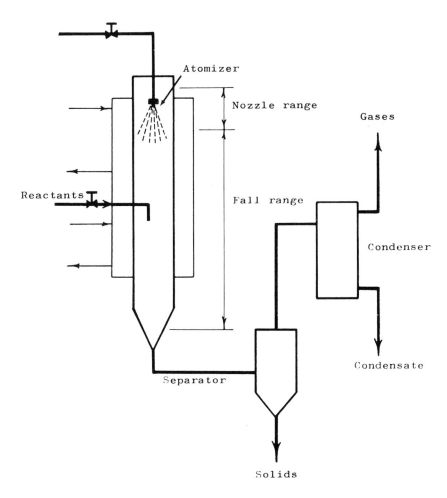

Fig. 7.8.   Flow diagram of atomized suspension technique.

Also, such drying techniques as flash drying and rotary drying are used to remove water from sludge. Often waste heat from the incineration process itself is used in drying. However, it has been reported by Quirk (1964) that the cost of combined drying and combustion is higher than the cost of incineration alone. The economy of sludge drying has recently been reviewed by Burd (1969). He reports that at the present cost of heat drying, it should only be considered if the product (soil conditioner) can be sold for at least US$ 20.00 a ton (1986-dollars).

P.7.9. Combustion serves as a means for the ultimate disposal of the sludge.

Two techniques should be mentioned: the atomized suspension technique (Gauvin, 1947), and the Zimmerman process (Zimmerman, 1958).

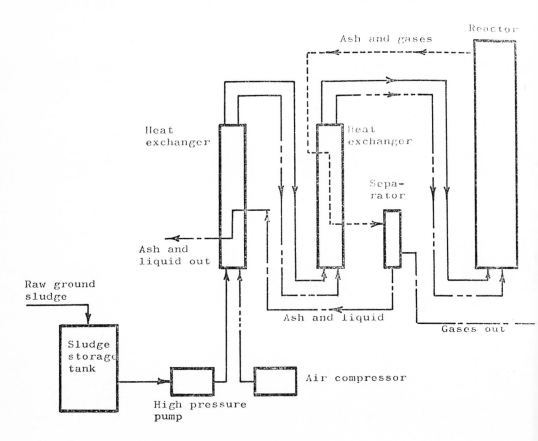

Fig. 7.9. Flow diagram of the Zimmerman process.

In this process the sludge is atomized at the top of the tower, and droplets pass down the tower where the moisture evaporates. The tower walls are maintained at 600-700°C by hot circulating gas. The solid produced is collected in a cyclone and the heat recovered from the stream and gas, as shown in Fig. 7.8.

The Zimmerman process is a wet-air oxidation at high temperature and pressure (Fig. 7.9). Oxidation of organics occurs at 200-300°C and the high pressure is used to prevent evaporation of the water.

The degree of oxidation at various temperatures is plotted in Fig. 7.10. As the oxidation process is an exothermic reaction, *heat is produced,* and it has been calculated that the system is self-sustaining at 4.5% solid of which *70% is volatile matter.* By means of a heat exchanger the heat developed is used to raise the temperature of the incoming sludge.

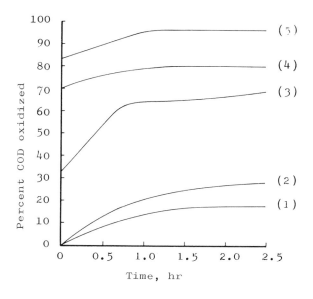

Fig. 7.10. The degree of oxidation at various temperatures plotted against treatment time. (1) 100°C, (2) 150°C, (3) 200°C, (4) 250°C, (5) 300°C.

TABLE 7.7
Composition of domestic garbage

Country	Ash %	Paper %	Org. matter %	Metals %	Glass %	Sundries %	kg/m^3	kg/inh./ year
Belgium (Brussels)	48	20.5	23	2.5	5	3		
Canada	5	70	10	5	3	5	115	380
Czechoslovakia (Prague)	6-65	14-7	39-22	2-1	11-3	18-2	200-400	190-430
Finland (Helsinki)	-	65	10	5	5	15	100-150	310
France (Paris)	24.3	29.6	24	4.2	3.9	14	120-180	300-360
Israel	1.9	24	71.2	1.1	1.0	1.8	255	190
Holland (Hague)	9.1	45.2	14	4.8	4.9	22	160-250	165-190
Norway	0-12	56-24	35-56	3.2-2.6	2.5-5.1	8.4-0	100-280	200
Poland	10-21	2.7-6.2	35-44	0.8-0.9	0.8-2.4	-	250-390	180-240
Spain (Madrid)	22	21	45	3	4	5	330	200
Sweden	0	55	12	6	15	12	140	210
Switzerland	20	40-50	15-25	5	5	-	120-200	150
England	30-40	25-30	10-15	5-8	5-8	5-10	150-250	240-300
USA	10	42	22.5	8	6	11.5	280	520-690
F.R. Germany (West Berlin)	30	18.7	21.2	5.1	9.8	15.2	330-380	210-230
Denmark	10	45	13	4	8	20	150-250	210-310

7.3. DOMESTIC GARBAGE.

7.3.1. Characteristics of domestic garbage.

P.7.10. The composition of domestic garbage might vary from country to country, as illustrated in Table 7.7. To a certain extent the environmental legislation and the economic level of the nation is reflected in this composition.

The amount of domestic garbage per inhabitant is increasing. In most developed countries the growth has been 2-4% from the mid 1950s to 1973, while it has been lower since then (1-2%).

It is often advantageous to carry out a more comprehensive analysis than the one presented in Table 7.7. The analysis will in this case include the following items: *(1) food waste, (2) paper, (3) textiles, (4) leather and rubber, (5) plastics, (6) wood, (7) iron, (8) aluminium, (9) other metals, (10) glass and ceramic products, (11) ash and dust, (12) stone, (13) garden waste, (14) other types of waste.* The knowledge obtained by this analysis can be used to select the most relevant treatment methods.

7.3.2. Separation methods.

P.7.11. Separation of solid waste can be achieved either in a central plant or by the organization of a separate collection of paper, glass, metals and other types of domestic waste.

TABLE 7.8
Collection of paper in some contries

Country		G_p (%)	A_p (%)	I_p (kg/inh./year)
USA	1971	22.7	21.6	57
Canada	1971	21.0	5.9	33
U.K.	1971	27.7	42.3	35
F.R. Germany	1971	30.1	26.4	40
France	1971	27.7	35.1	27
Holland	1971	42.1	40.0	56
Japan	1971	35.9	33.8	44
Sweden	1971	23	6.4	44
Norway	1971	17	7.1	abt. 20
Finland	1971	23	3.3	abt. 30
Denmark	1970	20	41	30
	1971	27	57.7	33
	1987	30	65	50

G_p = percentage recovery, A_p = percentage of returnal paper relatively to the total paper production, I_p = amount of returned paper per inhabitant and year.

This latter method is widely in use for paper and pulp, as demonstrated in Table 7.8, where the percentage recovery, G_p, the percentage of returned paper relative to the total paper production, A_p, and the amount of returned paper per inhabitant and year, I_p, are shown.

A mass flow diagram for a separation plant in Franklin, Ohio, is shown in Fig. 7.11. **After wet grinding, separation of paper fibres, glass,**

iron and other metals occurs. The remaining part of the solid waste is used for heat production.

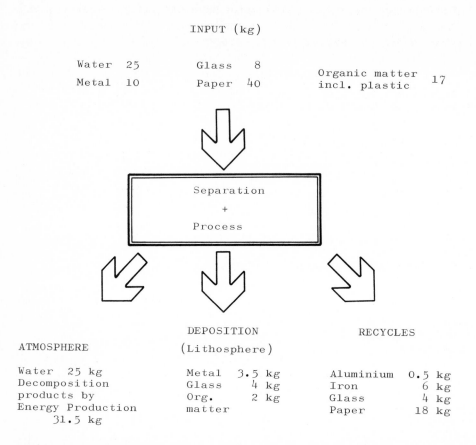

INPUT (kg)

Water 25	Glass 8	Organic matter
Metal 10	Paper 40	incl. plastic 17

Separation
+
Process

ATMOSPHERE

DEPOSITION
(Lithosphere)

RECYCLES

Water   25 kg
Decomposition
products by
Energy Production
        31.5 kg

Metal   3.5 kg
Glass   4 kg
Org.   2 kg
matter

Aluminium   0.5 kg
Iron   6 kg
Glass   4 kg
Paper   18 kg

Fig. 7.11. Separation plant, Franklin, Ohio. 150 t/d. Mass flow diagram (Basis 100 kg solid waste).

7.3.3.   Dumping ground (landfills).

This was previously the most common handling method for solid waste. Today it is mainly in use in smaller towns, often after grinding or compression, which reduce the volume 60-80%.

Deposition of solid waste on dumping ground is an inexpensive method, but it has a number of disadvantages:

1. Possibilities for **contamination of ground water.**
2. Causes inconveniences due to **the smell.**

3. **Attracts noxioux animals,** such as flies and rats.

During the deposition several processes take place:
1. *Decomposition* of biodegradable material.
2. *Chemical oxidation* of inorganic compounds.
3. *Dissolution and wash out* of material.
4. *Diffusion processes.*

Where the decomposition takes place in the aerobic layers carbon dioxide, water, nitrates and sulphates are the major products liberated, while decomposition in anaerobic layers leads to the formation of carbon dioxide, methane, ammonia, hydrogen sulphide and organic acids.

Water percolation from dumping grounds has a very high concentration of BOD and nutrients (see Table 7.9) and *can therefore not be discharged into receiving waters.* If it cannot be recycled on the dumping ground, theis waste water must be subject to some form of treatment. For further details see Persson and Nylander (1974).

7.3.4. Composting.

Composting has been applied as a treatment method in agriculture for thousands of years. The method is still widely applied for treatment of agricultural waste.

P.7.12. Organic matter from untreated solid waste cannot be utilized by plants, but it is necessary to let it undergo a certain biological decomposition, by the action of micro-organisms.

Again we can distinguish between **aerobic and anaerobic processes.** A number of factors control these processes:
1. **The ratio of aerobic to anaerobic processes,** which, of course, is determined by the available oxygen (diffusion process).
2. **Temperature.** Different classes of micro-organisms are active within different temperature ranges (see Table 7.10). Heat is produced by the decomposition processes. The thickness of the layer determines to what extent this heat can be utilized to maintain a temperature of 60-65°C, which is considered to be the optimum. Fig. 7.12 shows a typical course of the temperature by composting in stacks. In this context it is also important to mention that composting at a relatively high temperature means a substantial reduction in the number of pathogenic micro-organisms and parasites.

TABLE 7.10
Classification of micro-organisms

	Temperature optimum	Temperature range
Psychrophile	15-20°C	0-30°C
Mesophile	25-35°C	10-40°C
Thermophile	50-55°C	25-80°C

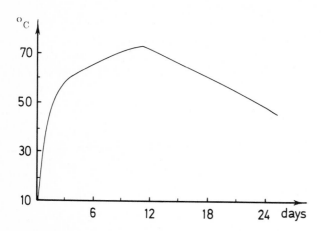

Fig. 7.12. Temperature versus composting time.

3. **The water content** should be 40-60%, as this gives the optimum conditions for the processes of decomposition.
4. **The C/N ratio** should be in accordance with the optimum required by the micro-organisms. Domestic garbage has a C/N ratio of 80 or more due to the high content of paper, while the optimum for composting is 30. It is therefore advantageous to add sludge from the municipal sewage plant to adjust the ratio to about 30. Sludge usually has a C/N ratio of 10 or even less. During the composting the C/N ratio is decreased as a result of respiration, which converts a part of the organic matter to carbon dioxide and water.
5. The optimum conditions for the micro-organisms include a **pH around 7 (6-8)**. Usually pH increases as a result of the decomposition processes. If pH is too low calcium hydroxide should be added and if pH is too high sulphur should be added. Sulphur activates sulphur bacteria, which produce sulphuric acid.

Fig. 7.13 shows a flow chart for a composting plant. Composting with addition of *air accelerates* the decomposition processes, but this process might be excluded in smaller plants.

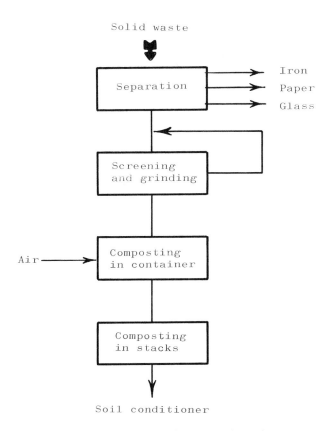

Fig. 7.13. Flow chart of composting plant.

7.3.5. Incineration of domestic garbage.

Incineration is very attractive from a sanitary point of view, but it is a very *expensive method,* which has some environmental disadvantages. Valuable material such as *paper is not recycled, a slag, which must be deposited,* is produced and air pollution problems are involved. Nevertheless, incineration has been the preferred method in many cities and larger towns. This development might be explained by the increasing amount of combustible material in domestic garbage and the possibilities of combining incineration plants with district heating.

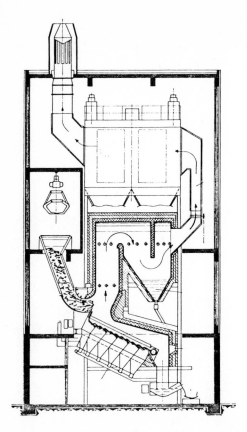

Fig. 7.14. Incineration plant. Type Martin, capacity 1 t hr⁻¹

The optimum combustion temperature is 800-1050°C. If the temperature is below this range incomplete combustion will result (dioxines may be produced), and a higher temperature the slags will melt and prevent an even air distribution. The composition of the solid waste determines whether this combustion temperature can be achieved without using additional fossil fuel. If the ash content is less than 60%, the water content is less than 50% and the combustible material more than 25%, no additional fossil fuel is required.

The heating value can be calculated from the following equation:

$$H_u = H^1 \frac{100 - W}{100} - 600 \frac{W}{100} \qquad\qquad (7.6)$$

where
H¹ = the heat value of dry matter
H_u = the heat value
W = the water content (weight %)

Figs. 7.14 and 7.15 show two typical incineration plants.

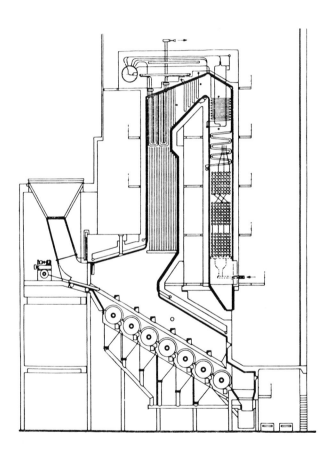

Fig. 7.15. Incineration plant with boiler and screen. System Düsseldorf, capacity 12.5 t hr⁻¹.

The composition of the slag and the flyash, which is collected in the filter, is, of course, dependent on the composition of the solid waste. Table 7.12 gives an analysis of an average sample and also shows the chemical composition of flyash alone. It is notable that a minor amount of the slag and the flyash is unburned material. Slags and flyash comprise 25-40% of the weight of the solid waste.

Although all modern incineration plants have filters, the air pollution problem is not completely solved. Table 7.11 gives a typical analysis of the smoke from an incineration plant. Hydrogen chloride, in particular, can cause difficulties, because it is toxic and highly corrosive.

TABLE 7.11
Typical analysis of the smoke from an incineration plant

Component	Dried slag (65°C) ppm * 10^{-4}/w-%	Gas Before washer ppm * 10^{-4}	Gas After washer vol.-%	Smoke After gas burning ppm
C	25			
CO		5	11	
CO_2	(2.9)	8	18	7 * 10^4
CH_4		5	10	
C_nH_m		1	2	
H_2O	abt. 40	55	4	
H_2		25	55	
H_2S		0.06		
HCl		0.01		6-40
HF		0.0001		0.1-0.8
HCN		0.001		
NH_3		1		
$SO_2 + SO_3$				2-10
NO_x				100-150
N	1.0			
Cl^-	1.1			
S	0.2			
SO_4^{2-}	0.22			
P	0.7			
Ca	4.5			
Fe	2.0			
Si	15			
Al	4			
Mg	0.3			
K	0.8			
Pb	0.1			
Cu	0.03			
Co	0.006			
Cr	0.0026			
Ni	0.0025			
Cd	0.0011			
As	0.0002			
Hg	0.0001			

7.3.6. Pyrolysis.

P.7.13. Pyrolysis is a decomposition of organic matter at elevated temperature without the presence of oxygen.

For pyrolysis of solid waste a temperature of 850-1000°C is generally used. The process produces gas and a slag, from which metals easily, due to the anoxic atmosphere, can be seperated.

Pyrolysis is a relatively expensive process, although less expensive than incineration. The smoke problems are the same as those of incineration, but the easy recovery of metals from the slag is, of course, an advantage.

The gas produced has a composition close to coalgas and in most cases can be used directly in the gas distribution system without further treatment. Approximately 500 m^3 of gas are produced per ton of solid waste, but 300-400 m^3 are used in the pyrolysis process to maintain the temperature.

TABLE 7.12

Chemical composition of dried slag and ash from incineration		Chemical composition of flyash from electrofilter	
Component	w % slag	Component	w % flyash
C	8.79	C	9.8
SiO_2	48.53	SiO_2	40.5
Al_2O_3 + TiO_2	11.59	Al_2O_3	10.0
Fe_2O_3	16.08	SiO_2	1.6
CaO	6.90	Fe_2O_3	13.1
MgO	1.36	CuO	trace
K_2O	5.01	MnO	0.2
SO_3	0.82	CaO	10.4
S	0.16	Mg	0.1
Cl	0.32	BaO	0.8
P_2O_5	0.56	K_2O	3.3
		Na_2O	2.4
		SO_3	6.9
		S	trace
		P_2O_5	0.9

7.4. INDUSTRIAL, MINING AND HOSPITAL WASTE.

7.4.1. Characteristics of the waste.

This types of waste causes particular problems because it may contain toxic matters in relatively high concentrations. Hospital wastes are especially suspect because of contamination by pathogens and the special waste products, such as disposable needles and syringes and radioisotopes used for detection and therapy.

The waste from industries and mining varies considerably from place to place, and it is not possible to provide a general picture of its composition. It the composition permits, it can be used for landfilling, but if it contains toxic matter specia treatment is prequired. The composition may be close to domestic garbage in which case the treatment methods mentioned in section 7.3 can be applied.

Hospital solid waste is being studied in only a few locations to provide data on current practices and their implications. Most hospital waste is now incinerated and this might, in many cases, be an acceptable solution. However, if the solid waste contains toxic matter it should be treated along the lines given for industrial waste in the next paragraph. At least, waste from hospital laboratories, should be treated as other types of chemical waste.

7.4.2. Treatment methods.

Industrial and mining waste containing toxic substances, such as heavy metals or toxic organic compounds, should be treated separately from other types of solid waste, which means that it cannot be treated by the methods mentioned in section 7.2 and section 7.3. A number of countries have built special plants to handle this type of waste, which could be called chemical waste. Such plants may include the following treatment lines:

1. Combustion of toxic organic compounds. The heat produced by this process might be used for district heating. Organic solvents, which are not toxic, should be collected for combustion, because discharge to the sewer mght overload the municipal treatment plant. For example, acetone is not toxic to biological treatement plants, but 1 kg of acetone uses 2.2 kg of oxygen in accordance with the following process:

$$(CH_3)_2CO + 4O_2 \rightarrow 3CO_2 + 3H_2O \qquad (7.7)$$

   A different combustion system might be used for pumpable and non-pumpable waste.
2. Compounds containing halogens should be treated only in a system which washes the smoke to remove the formed hydrogen halogenids.
3. Waste oil can often be purified and the oil reused.
4. Waste containing heavy metals requires deposition under safe conditions after a suitable pretreatment. Recovery of precious metals is often economically viable and is essentioal for mercury because of its high

toxicity. The pretreatment consists of a conversion to the most relevant oxidation state for deposition, e.g. chromate should be reduced to chromium in oxidation state 3. Furthermore metals should be precipitated as the very insoluble hydroxides, (see also 6.5.2) before deposition.

5. Recovery of solvents by distillation becomes increasingly attractive from an economical view-point due to the growing costs of oil products, as most solvents are produced from mineral oil.

7.5. AGRICULTURAL WASTE

7.5.1. Characteristics of agricultural waste

Particular animal waste causes great problems by intensive farming. The productions of chickens, pigs and cattles are in many industrialized countries concentrated in rather large units, which implies that the waste from such production units requires hundreds of hectars for a suitable distribution and feasible use of its value as fertilizer.

Animal waste has a high nitrogen concentration 5-10% based on dry matter (2-4% dry matter).

If the nitrogen is not used as fertilizer it may
1) either evaporate as ammonia
2) be lost to deeper layers, where it contaminates the ground water or
3) be lost by surface run off to lakes and streams.

Agricultural waste has therefore become a crucial pollution problem in many countries with intensive agriculture.

7.5.2. Treatment methods

Many of the methods described above may be used for treatment of animal waste.

The following possibilities give a summary of the available methods to reduce the pollution originated from agricultural waste.

1. Storage capacity for animal waste to avoid spreading on bare fields.

2. Green fields in winter to assure the use of the fertilizing value of agricultural waste.

3. Composting of agricultural waste assure conditioning before it is used. The composting heat may be utilized.

4. Anaerobic treatment of agricultural waste for production of biogas and conditioning before use as fertilizer.

5. Chemical precipitation of agricultural (animal) waste with activated benthonite is applied to bind ammonium and obtain a solid concentration of 6-10%. This process gives 2 advantages:
   A. The needed storage capacity is reduced by a factor 2-4 (the solid concentration is increased from 2-4% to 6-10%)
   B. The loss of nitrogen by evaporation of ammonia is reduced by a factor 3-8 corresponding to an adsorption of about 60-90% of the ammonium on the added benthonite.

6. In China animal waste is applied in fish ponds. Zooplankton eats the detritus and fish feed on the zooplankton.

QUESTIONS

1. Set up a mass flow diagram for iron and glass in a selected district.

2. Discuss the analytical data in Table 7.5 from an ecological view-point.

3. A selected town or district is considered. Compare from both an ecological and an economical point of view the following solutions to the solid waste problem of domestic garbage and sludge: A. Separation of paper, metals and glass followed by either 1) incineration, 2) composting, og 3) pyrolysis.   B. Incineration (no preseparation).   C. Pyrolysis followed by metal separation from the slag.

4. Estimate (roughly) the loss of nitrogen from agricultural waste used
   a) on a bare field
   b) on a green winterfield
   c) after composting
   d) after chemical precipitation by use of activated benthonite.

CHAPTER 8

AIR POLLUTION PROBLEMS

8.1. THE PROBLEMS OF AIR POLLUTION - AN OVERVIEW.

Air pollution control is applying a wide range of remedies, and alternative technology is playing a more important role in air pollution control than in water pollution control. Alternative technology is often employed as a result of increasingly stringent legislation, e.g. the setting of lower threshold levels for lead in gasoline, sulphur in fuel and carbon monoxide in exhaust gases.

Air pollution problems can be considered in terms of the effect on climate, or local or regional effects caused by toxicity of particular pollutants. The principles of air pollution problems were discussed in Part A; this chapter outlines the technology now available for air pollution control. The methods employed are classified according to the problems they solve.

This chapter covers the control of particulate pollution, carbon dioxide, carbon hydride and carbon monoxide and sulphur dioxide problems, nitrogenous gas pollution and industrial gaseous pollution.

P.8.1. **The methods used in air pollution control can also be classified according to the technology applied:**
1. **By changing the distribution pattern of the pollution.** This method could, in principle, be used to solve all local and regional air pollution problems, but it has found its widest application in particulate pollution.
2. **By using alternative technology** to eliminate the problem.
3. **By removing the pollutants.** Distinction should be made between particulate control technology and gas and vapour control technology.

Emission is the output from a source of pollution. It might be indicated as mass or volume per unit of time or per unit of production. If it is given as a concentration unit, e.g. mg per m^3, it is also necessary to know the number of m^3 discharged per unit of time, or the number of m^3 polluted air produced per unit of production.

Imission is the input of pollutants to a given area.. It might be indicated as mg per m^2 and time for particulate matter or as a concentration unit for gaseous pollutants.

Correspondingly, **legislation will distinguish between emission standards and air quality standards respectively** (see 8.6.1. for further details).

Particulate control technology is discussed in sections 8.2.5 - 8.2.10 in context with particulate pollution, although this technology has also found application for the control of other pollutants.

Gas and vapour technology is dealt with under problems of industrial air pollution, but it has also found application in other areas.

8.2. PARTICULATE POLLUTION.

8.2.1. Sources of particulate pollution.

When considering particulate pollution, the source should be categorized with regard to contaminant type. **Inert particulates** are distinctly different from **active solids** in the nature and type of their potentially harmful human health effects. *Inert particulates comprise solid* airborne material, which does not react readily with the environment and does not exhibit any morphological changes as a result of combustion or any other process. Active solid matter is defined as particulate material which can be further oxidized or *which reacts chemically* with the environment or the receptor. Any solid material in this category can, depending on its composition and size, be more harmful than inert matter of similar size.

A closely related group of emissions are from **aerosols**, which are droplets of liquids, generally below 5 μm. They can be oil or other liquid pollutants (e.g. freon) or may be formed by condensation in the atmosphere.

Fumes are condensed metals, metal oxides or metal halides, formed by industrial activities, predominantly as a result of pyrometallurgical processes; melting, casting or extruding operations.

Products of incomplete combustion are often emitted in the form of particulate matter. The most harmful components in this group are often those of **particulate polycyclic organic matter** (PPOM). These materials are homologues and derivatives of benz-a-pyrene.

Natural sources of particulate pollution are *sandstorms, forest fires and vulcanic activity.* The major sources in towns are *vehicles, combustion of fossil fuel for heating and production of electricity, and industrial activity.*

The total global emission of particulate matter is in the order of **10^7 t per year.**

8.2.2. The particulate pollution problem.

Particulate pollution is important with regard to health. The toxicity and the size distribution are the most crucial factors.

Many particles are highly toxic, such as *asbestos and those of heavy metals such as beryllium, lead, chromium, mercury, nickel and manganese.* In addition, it must be remembered that particulate matter is able to absorb gases, so enhancing the effects of these components. In this context the **particle size distribution** is of particular importance, as particles greater than 10μm are trapped in the human upper respiratory passage and the specific surface (expressed as m^2 per g of particulate matter) increases with l/d, where d is the particle size. The adsorption capacity of particulate matter, expressed as g adsorbed per g of particulate matter, will generally be proportional to the surface area.

Table 8.1 lists some typical particle size ranges.

TABLE 8.1
Typical particle size ranges

	μm
Tobacco smoke	0.01 - 1
Oil smoke	0.05 - 1
Ash	1 - 500
Ammonium chloride smoke	0.1 - 4
Powdered activated carbon	3 - 800
Sulphuric acid aerosols	0.5 - 5

However, *size* as well as *shape* and *density* must be considered. Furthermore, particle size has to be determined by two parameters: the mass median diameter, which is the size that divides the particulate sample into two groups of equal mass, i.e. the 50 percent point on a cumulative frequency versus particle size plot (see the examples in Fig. 8.1); and the geometric standard deviation, which is the slope on the curve in Fig. 8.1. It can be found from:

$$\beta = \frac{Y_{84.1}}{Y_{50}} = \frac{Y_{50}}{Y_{15.9}} \qquad (8.1)$$

where the subscript numbers refer to the particle diameters at that percentage on the distribution plot. From the plot is found, in this case:

$$\beta = \frac{7.2}{1.7} = \frac{1.7}{0.4} = 4.25 \qquad (8.2)$$

The particle size also determines the settling out rate of the particulate matters. Particles with a size of 1 μm have a terminal velocity of a few metres per day, while particles with a size of 1 mm will settle out at a rate of about 40 m per day. Particles that are only a fraction of 1 μm will settle out very slowly and can be in suspension in the atmosphere for a very long time.

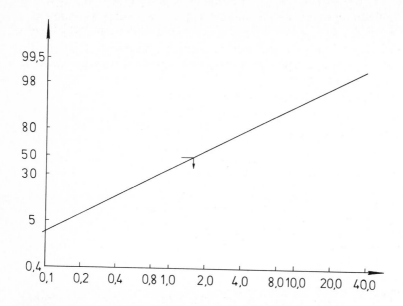

Fig. 8.1. Particle diameter (log scale) plotted against the percentage of particles less than or equal to indicated size.

8.2.3. Control methods applied to particulate pollution.

All three classes of control methods mentioned in section 8.1 are applied to particulate pollution. Legislation has been introduced in an attempt to reduce particulate pollution. In only few cases, however, has the legislation effected a change to alternative technology.

P.8.2. **Particulate control technology can offer a wide range of methods aimed at the removal of particulate matter from gas. These methods are: settling chambers, cyclones, filters, electrostatic precipitators, wet scrubbers and modification of particulate characteristics.**

Table 8.2 summarizes these six technological methods, including their range of application, limitations, particle size range and the efficiencies achieved in general use also.

Particulate pollution is also controlled by modifying the distribution pattern. This method is described in detail in the next paragraph.

TABLE 8.2
Characteristics of particulate pollution control equipment

Device	Optimum particle size (μm)	Optimum concentration (gm^{-3})	Temperature limitations (°C)	Air resistance (mm H_2O)	Efficiency (% by weight)
Settling chambers	> 50	> 100	-30 to 350	< 25	< 50
Centrifuges	> 10	> 30	-30 to 350	< 50-100	< 80
Multiple centrifuges	> 5	> 30	-30 to 350	< 50-100	< 90
Filters	> 0.3	> 3	-30 to 250	> 15-100	> 99
Electrostatic precipitators	> 0.3	> 3	-30 to 500	< 20	< 99
Wet scrubbers	> 2-10	> 3-30	0 to 350	> 5-25	< 95-99

8.2.4. Modifying the distribution patterns.

Although emissions, gaseous or particulate, may be controlled by various sorption processes or mechanical collection, the effluent from the control device must still be dispersed into the atmosphere.

P.8.3. Atmospheric dispersion depends primarily on horizontal and vertical transport.

The horizontal transport depends on the turbulent structure of the wind field. As the wind velocity increases so does the degree of dispersion and there is a correspondingly decrease in the ground level concentration of the contaminant at the receptor site.

The emissions are mixed into larger volumes, of air and the diluted emission is carried out into essentially unoccupied terrain away from any receptors. Depending on the wind direction, the diluted effluent may be funnelled down a river valley or between mountain ranges. Horizontal transport is sometimes prevented by surrounding hills forming a natural pocket for locally generated pollutants. This particular topographical situation occurs in the Los Angeles area, which suffers heavily from air pollution.

The vertical transport depends on the rate of changes of ambient temperature with altitude. The dry adiabatic lapse rate is defined as a decrease in air temperature of 1°C per 100 m. This is the rate at which, under natural conditions, a rising parcel of unpolluted air will decrease in temperature with elevation into the troposphere up to approximately 10,000 m. Under so-called isothermal conditions the temperature does not change with elevation. Vertical transport can be hindered under stable atmospheric conditions, which occur when the actual environmental lapse rate is less than the dry adiabatic lapse rate. A negative lapse rate is an increase in air temperature with latitude. This effectively prevents vertical mixing and is known as inversion.

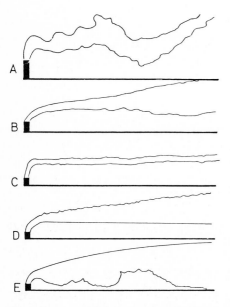

Fig. 8.2. Stack gas behaviour under various conditions. A) Strong lapse (looping), B) Weak lapse (coning), C) Inversion (fanning), D) Inversion below, lapse aloft (lofting), E) Lapse below, inversion aloft (fumigation).

These different atmospheric conditions (U.S.DHEW 1969) are illustrated in Fig. 8.2, where stack gas behaviour under the various conditions is shown. Further explanations are given in Table 8.3.

TABLE 8.3
Various atmospheric conditions

Strong lapse (looping)	Environmental lapse rate > adiabatic lapse rate
Weak lapse (coning)	Environmental lapse rate < adiabatic lapse rate
Inversion (fanning)	Increasing temperature with height
Inversion below lapse aloft (lofting)	Increasing temperature below, app. adiabatic lapse rate aloft
Lapse below, inversion aloft (fumigation)	app. adiabatic lapse rate below, increasing temperature aloft

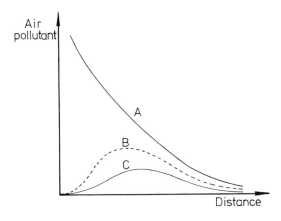

Fig. 8.3. Distribution of emission from stack of different heights under coning conditions. A) height 0 m, B) height 50 m, C) height 75 m.

Fig. 8.3 illustrates the distribution of material emitted from a stack at three different heights under coning conditions. The figure demonstrates that

P.8.4. **the distribution of particulate material is more effective the higher the stack.**

The maximum concentration, C_{max}, at ground level can be shown to be approximately proportional to the emission and to follow approximately this expression:

$$C_{max} = k \frac{Q}{H^2}$$ (8.3)

where Q is the emission (expressed as g particulate matter per unit of time), H is the effective stack height and k is a constant.

The definition of **the effective stack height** is illustrated in Fig. 8.4 and it can be calculated from the following equation:

$$H = h + 0.28 * V_s * D_s \left[1.5 + 2.7 \frac{T_s - 273}{T_s} * D_s \right] \qquad (8.4)$$

where

V_s = stack exit velocity in m per second

D_s = stack exit inside diameter in m

T_s = stack exit temperature in degree Kelvin

h = physical stack height above ground level in m

H = effective stack height in m

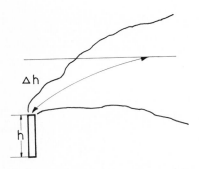

Fig. 8.4. Effective stack height H = h + Δh.

These equations explain why a lower ground-level concentration is obtained when many small stacks are replaced by one very high stack. In addition to this effect, it is always easier to reduce and control one large emission than many small emissions, and it is more feasible to install and apply the necessary environmental technology in one big installation.

Example 8.1.
Calculate the concentration ratio of particulate material in cases A and B, given that the total emission is the same.

A: 100 stacks H = 25 m
B: 1 stack   H = 200 m

Solution:

$$\text{Ratio} = \frac{200^2}{25^2} = 64$$

8.2.5. Settling chambers.

Simple gravity settling chambers, such as the one shown in Fig. 8.5, depend on gravity or inertia for the collection of particles. Both forces increase in direct proportion to the square of the particle diameter, and the performance limit of these devices is strictly governed by the particle settling velocity.

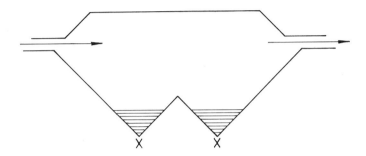

Fig. 8.5. Simple gravity settling chamber.

The pressure drop in mechanical collectors is low to moderate, 1-25 cm water in most cases. Most of these systems operate dry but if water is added it performs a secondary function by keeping the surface of the collector clean and washed free of particles.

The settling or terminal velocity can be described by the following expression, which has general use:

$$V_t = (\partial_p - \partial)\, g\, \frac{d_p{}^2}{18\mu} \tag{8.5}$$

where
V_t = terminal velocity
∂_p = particle density
∂ = gas density
d_p = particle diameter
μ = gas viscosity

This is the equation of Stokes' law, and is applicable to $N_{Re} < 1.9$ where

$$N_{Re} = d_p * V_t * \frac{\partial}{\mu} \tag{8.6}$$

The intermediate equation for settling can be expressed as:

$$V_t = \frac{0.153 * g^{0.71} * d_p^{1.14} (\partial_p - \partial)^{0.71}}{\partial^{0.29} * \mu^{0.43}} \tag{8.7}$$

This equation is valid *for Reynolds numbers between 1.9 and 500,* while the following equation can be applied *above $N_{Re} = 500$ and up to 200,000:*

$$V_t = 1.74 (d_p * g \frac{(\partial_p - \partial)}{\partial})^{1/2} \tag{8.8}$$

The settling velocity in these chambers is often in the range 0.3-3 m per second. This implies that for large volumes of emission the settling velocity chamber must be very large in order to provide an adequate residence time ofr the particles to settle. Therefore, the gravity settling chambers are not generally used to remove particles smaller than 100 μm (= 0.1 mm). For particles measuring 2-5 μm the collection efficiency will most probably be as low as 1-2 percent.

A variation of the simple gravity chamber is the baffled separation chamBer. The baffles produce a shorter settling distance, which means a shorter retention time.

Equations (8.5) - (8.8) can be used to design a settling chamber, and this will be demonstrated by use of equation (8.5).

If it is assumed that (8.5) applies, an equation is available for calculating the minimum diameter of a particle collected at 100% theoretically efficiency in a chamber of length L. In practice, some reentrainment will occur and prevent 100% efficiency. We have:

$$\frac{v_t}{H} = \frac{v_h}{L} \tag{8.9}$$

where H is height of the settling chamber (m), L length of the settling chamber (m) and v_h is the horizontal flow rate (m s^{-1}).

Solving for v_t and substitution into equation (8.5) yields

$$\frac{v_h * H}{L} = \frac{g(\partial_p - \partial) \, d_p^2}{18\mu}$$

$\partial_p \gg \partial$ and this equation gives the largest size particle that can be removed with 100% efficiency in a settling chamber:

$$d_p = (\frac{18\mu v_h * H}{Lg * \partial_p})^{1/2} \qquad (8.10)$$

A correction factor of 1.5 - 3 is often used in equation (8.10).

Example 8.2.
Find the minimum size of particle that can be removed with 100% efficiency from a settling chamber with a length of 10 m and a height of 1.5 m. The horizontal velocity is 1.2 m s⁻¹ and the temperature is 75°C. The specific gravity is 1.5 of the particles. A correction factor of 2 is suggested.

Solution:
At 75°C, see appendix 8, μ is $2.1 * 10^{-5}$ kg m⁻¹ s⁻¹

$$d_p = (2 \frac{18\mu v_h * H}{1 * g * \partial_p})^{1/2} = 2 (\frac{18 * 2.1 * 10^{-5} * 1.2 * 1.5}{9.81 * 10 * 1500})^{1/2}$$

$$= \underline{9.62 * 10^{-5} \ m.}$$

as ∂_p = 1.5 g ml⁻¹ = 1500 kg m⁻³

d_p = 96.2 μ

8.2.6.   Cyclones.

P.8.5.   Cyclones separate particulate matter from a gas stream by transforming the inlet gas stream into a confined vortex. The mechanism involved in cyclones is the continuous use of inertia to produce a tangential motion of the particles towards the collector walls.

The particles enter the boundary layer close to the cyclone wall and loose kinetic energy by mechanical friction, see Fig. 8.6. The forces are involved: the centrifugal force imparted by the rotation of the gas stream and a drag force, which is dependent on the particle density, diameter, shape, etc. A hopper is built at the bottom. If the cyclone is too short, the maximum force will not be exerted on some of the particles, depending on their size and corresponding drag forces (Leith and Licht, 1975). If, however, the cyclone is too long, the gas stream might reverse its direction and spiral up the centre.

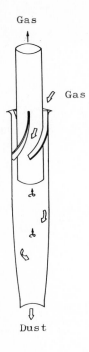
Gas

Gas

Fig 8.6.   Principle of a cyclone.

Dust

It is therefore important to design the cyclone properly. The hopper must be deep enough to keep the dust level low.

The efficiency of a cyclone is described by a graph similar to Fig. 8.7, which shows the efficiency versus the relative particle diameter, i.e. the actual particle diameter divided by **D_{50}, which is defined as the diameter corresponding to 50 percent efficiency. D_{50} can be found from the following equation:**

$$D_{50} = K * (\frac{\mu D_c}{V_c * \partial_p})^{1/2} \tag{8.11}$$

where
D_c = diameter of cyclone
V_c = inlet velocity

∂_p = density of particles

μ = gas viscosity

K = a constant dependent on cyclone performance

Efficiency

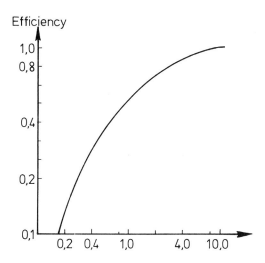

Fig. 8.7.   Efficiency plotted against relative particle diameter.

If the distribution of the particle diameter is known, it is possible from such a graph as in Fig. 8.7 to calculate the total efficiency:

$$\text{eff}_T = \Sigma \, m_i * \text{eff}_i \qquad (8.12)$$

where

m_i = the weight fraction in the i.th particle size range

eff_i = the corresponding efficiency

The pressure drop for cyclones can be found from:

$$\Delta p = N * \frac{V_c^2}{2g} \qquad (8.13)$$

From equations (8.11) and (8.13) it can be concluded tabt higher efficiency is obtained without increased pressure drop if D_c can be decreased with velocity V_c maintained. This implies that *a battery of parallel coupled small cyclones will work more effectively than one big cyclone.* Such cyclones batteries are available as blocks, and are known as multiple cyclones.

Compared with settling chambers, cyclones offer a higher efficiency for particles below 50μm and above 2-10 μm, but involve a greater drop.

Example 8.3.
Determine D_{50} for a flow stream with a flow rate of 7 m s^{-1}, when a cyclone with a diameter of 2 m is used and a battery of cyclones with diameters of 0.24 m are used. Air temperature is 75°C and the particle density is 1.5 g ml^{-1}. K can be set to 0.2. Find also the efficiencies for particles with a diameter of 5μm.

Solution:

1) $\quad D_{50} = K \dfrac{\mu * D_c}{v_c * \partial_p} = 0.2 \dfrac{2.1 * 10^{-5} * 2}{7 * 1500} = 12.7 \ \mu m$

2) $\quad D_{50} = K \dfrac{\mu * D_c}{v_c * \partial_p} = 0.2 \dfrac{2.1 * 10^{-5} * 0.24}{7 * 1500} = 4.4 \ \mu m$

5 μm corresponds to a relative diameter of

1) $\quad \dfrac{5}{12.7} = 0.39$, the efficiency will be about 25% (see Fig. 8.7)

2) $\quad \dfrac{5}{4.4} = 1.14$, the efficiency will be about 55% (see Fig. 8.7)

8.2.7.   Filters.

Particulate materials are collected by filters by three mechanisms (Wong et al., 1956):

Impaction where the particles have so much inertia that they cannot follow the stream line round the fibre and thus impact on its surface (see Fig. 8.8).

Direct interception where the particles have less inertia and can barely follow the stream lines around the obstruction.

Diffusion, where the particles are so small (below 1μm) that their individual motion is affected by collisions on a molecular or atomic level. This implies that the collection of these fine particles is a result of random motion.

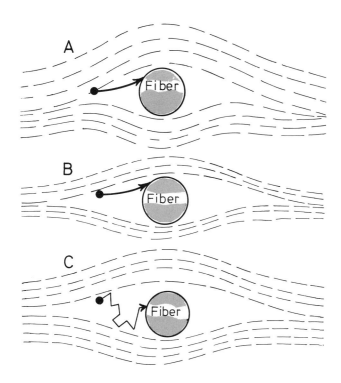

Fig. 8.8.   Particle capture mechanism.
A) Impaction.   B) Direct interception.   C) Diffusion.

Different flow patterns can be used, as demonstrated in Fig. 8.9. The types of fibres used in fabric filters range from natural fibres, such as cotton and wool, to synthetics (mainly polyesters and nylon), glass and stainless steel.

Some properties of common fibres are summarized in Table 8.4. As seen, cotton and wool have a low temperature limit and poor alkali and acid resistance, but they are relatively inexpensive. The selection of filter medium must be based on the answer to several questions (Pring, 1972 and Rullman, 1976):

What is the expected operating temperature?
Is there a humidity problem which necessitates the use of a hydrophobic material, such as, e.g. nylon?

How much tensile strength and fabric permeability are required?
How much abrasion resistance is required?

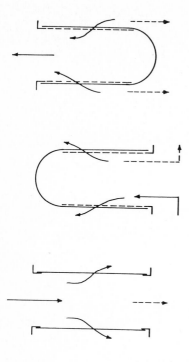

Fig. 8.9.   Flow pattern of filters.

Permeability is defined as the volume of air that can pass through 1 m^2 of the filter medium with a pressure drop of no more than 1 cm of water.

The filter capacity is usually expressed as m^3 air per m^2 filter per minute. A typical capacity ranges between 1 and 5 m^3 per m^2 per minute.

The pressure drop is generally larger than for cyclones and will in most cases, be 10-30 cm of water, depending on the nature of the dust, the cleaning frequency and the type of cloth.

In Fig. 8.10 **the pressure drop is plotted against the mass of the dust deposit** and as can be seen, the maximum pressure drop is strongly dependent on the cleaning frequency.

TABLE 8.4
Properties of fibres

Fabric	Acid	Alkali	Fluoride	Tensile	Abrasion
	r e s i s t a n c e			strength	resistance
Cotton	poor	good	poor	medium	very good
Wool	good	poor	poor	poor	fair
Nylon	poor	good	poor	good	excellent
Acrylic	good	fair	poor	medium	good
Polypropylene	good	fair	poor	very good	good
Orlon	good	good	fair	medium	good
Dacron	good	good	fair	good	very good
Teflon	excellent	excellent	good	good	fair

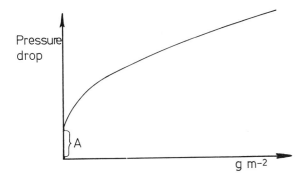

Fig. 8.10. Pressure drop versus dust deposit. A) represents assistance of clean fabric.

There are several specific methods of filter cleaning. The simplest is *backwash,* where dust is removed from the bags merely by allowing them to collapse. This is done by reverting the air flow through the entire compartment. The method is remarkable for its low consumption of energy.

Shaking is another low-energy filter-cleaning process, but it cannot be used for sticky dust. The top of the bag is held still and the entire tube sheath at the bottom is shaken.

The application of *blow rings* involves reversing the air flow without bag collapse. A ring surrounds the bag; it is hollow and supplied with compressed air to direct a constant steam of air into the bag from the outside.

The pulse and improved jet cleaning mechanism involves the use of a high velocity, high pressure air jet to create a low pressure inside the bag and induce an outward air flow and so clean the bag by sudden expansion and reversal of flow.

In some cases as a result of electrostatic forces, moisture on the surface of the bags and a slight degree of hygroscopicity of the dust itself, the material forms cakes that adhere tightly to the bag. In this case the material must be kept drier and a higher temperature on the incoming dirty air stream is required.

Filters are highly effecient even for smaller particles (0.1 - 2 m), which explains their wide use as particle collection devices.

8.2.8.   Electrostatic precipitators.

The electrostatic precipitator consists of four major components:
1. *A gas-tight shell with hoppers* to receive the collected dust, inlet and outlet, and an inlet gas distributor.
2. *Discharge electrodes.*
3. *Collecting electrodes.*
4. *Insulators.*
The principles of electrostatic precipitators are outlined in Fig. 8.11.

P.8.6.   The dirty air stream enters filter, where a high, 20-70 kV, usually negative voltage exists between discharge electrodes. The particles accept a negative charge and migrate towards the collecting electrode.

The efficiency is usually expressed by use of Deutsch's equation (see discussion incl. correction of this equation in Gooch and Francis, 1975):

$$1 - n = e^{\left(\frac{W * l}{\partial * v}\right)} \tag{8.14}$$

where
n = the efficiency
W = velocity of particles (migration velocity)
l = effective length of electrode systems
∂ = distance between electrodes
v = gas velocity

The migration velocity can be found from (see also Rose and Wood, 1966):

$$W = \frac{E_o * E_p * d_p * C}{4\pi\mu} \tag{8.15}$$

where E_o = charging field strength Vm^{-1}

$$C = 1 + \frac{2.5\,l}{d_p} + \frac{0.84}{d_p} \exp\left(\frac{0.435\,d_p}{l}\right)$$

E_p = collecting field Vm^{-1}

l = free path of gas molecules, m

d_p = particle diameter, m

μ = particle gas viscosity, cp

This implies *a relationship between migration velocity and particle diameter similar* to the graph in Fig. 8.12 (see White, 1974, and Dismukes, 1975).

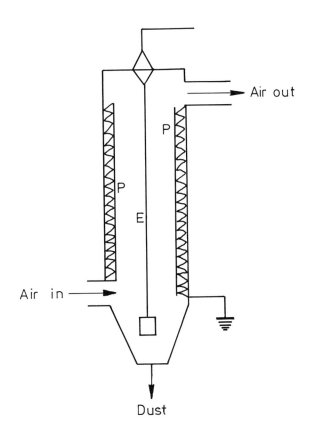

Fig. 8.11. The dust is precipitated on the electrode P. E has a high, usually negative voltage and emits a great number of electrons which give the dust particles a negative charge. The dust particles will therefore be attracted to P.

The operation of an electrostatic precipitator can be divided into three steps:
1. The particles *accept a negative charge.*
2. The charged particles *migrate towards the collecting electrode* due to the electrostatic field.
3. The collected dust *is removed from the collecting electrode* by shaking or vibration, and is collected in the hopper.

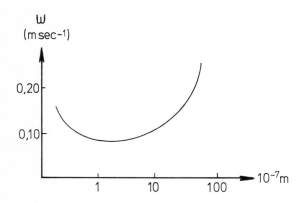

Fig. 8.12.   A typical graph of migration velocity versus particle diameter.

Resistivity, r, is the specific electrical resistance measured in ohm m. It determines the ability of a particle to accept a charge. The practical resistivity can cover a wide range of about four orders of magnitude, in which varying degrees of collection efficiencies exist for different types of particles. Fig. 8.13 demonstrates the effect of resistivity on migration velocity.

The resistivity depends on the chemical nature of the dust, the temperature and the humidity.

Electrostatic precipitators have found a wide application in industry. As the cost is relatively high, the airflow should be at least 20,000 m^3 h^{-1}; volumes as large as 1,500,000 m^3 h^{-1} have been treated in one electrostatic precipitator.
Very high efficiencies   are generally achieved in electrostatic precipitators and emissions as low as 25 mg m^{-3} are quite common. The pressure drop is usually low compared with other devices - 25 mm water at the most. The energy consumption is generally 0.15-0.45 Wh m^{-3} h^{-1}.

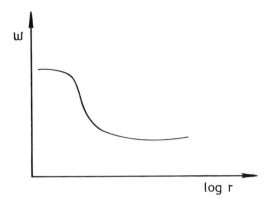

Fig. 8.13.   W versus log resistivity.

8.2.9.   Wet scrubbers.

P.8.7.   **A scrubbing liquid, usually water, is used to assist separation of particles or a liquid aerosol from the gas phase. The operational range for particle removal includes material less than 0.2 μm in diameter to the largest particles that can be suspended in air.**

Four major steps are involved in collecting particles by wet scrubbing. First, *the particles are moved to the vicinity of the water droplets,* which are 10-1,000 times larger. Then the particles *must collide with the droplets.* In this step the relative velocity of the gas and the liquid phases is very important: If the particles have an overhigh velocity in relation to the liquid they have so much inertia that they keep moving, even when they meet the front edge of the shock wave, and either impinge on or graze the droplets. A scrubber is no better than its ability to bring the particles directly into contact with the droplets of the scrubber fluid. The next step is *adhesion, which is directly promoted by surface tension.* Particles cannot be retained by the droplets unless they can be wetted and thus incorporated into the droplets. The last step is *the removal of the droplets containing the dust particles from the bulk gas phase.*

Scubbers are generally very flexible. They are able to operate under peak loads or reduced volumes and within a wide temperature range (Onnen, 1972). They are smaller and less expensive than dry particulate removal devices, but the operating costs are higher. Another disadvantage is that the pollutants are not collected but transferred into water, which means that the related water pollution problem must also be solved (Hanf, 1970).

Several types of wet scrubbers are available (Wicke, 1971), and their principles are outlined below:

1. **Chamber scrubbers** are spray towers and spray chambers which can be either round or rectangular. Water is injected under pressure though nozzles into the gas phase. A simple chamber spray scrubber is shown in Fig. 8.14.

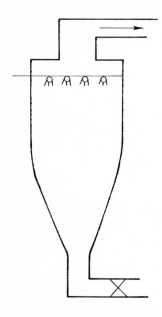

Fig. 8.14.   Chamber scrubber.

2. **Baffle scrubbers** are similar to a spray chamber but have internal baffles that provide additional impingement surfaces. The dirty gas is forced to make many turns to prevent the particles from following the air stream.
3. **Cyclonic scrubbers** are a cross between a spray chamber and a cyclone. The dirty gas enters tangentially to wet the particles by forcing its way through a swirling water film onto the walls. There the particles are captured by impaction and are washed down the walls to the sump. The saturated gas rises through directional vanes, which are used solely to impact rotational motion to the gas phase. As a result of this motion the

gas goes out though a demister for the removal of any included droplets.

4. **Submerged orifice scrubbers** are also called gas-induced scrubbers. The dirty gas is accelerated over an aerodynamic foil to a high velocity and directed into a pool of liquid. The high velocity impact causes the large particles to be removed into the pool and creates a tremendous number of spray droplets with a high amount of turbulence. These effects provide intensive mixing of gas and liquid and thereby a very high interfacial area. As a result reactive gas absorption can be combined with particle removal. The principles of this operation are demonstrated in Fig. 8.15.

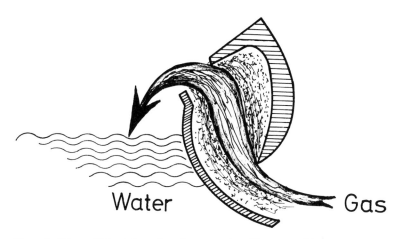

Water — Gas

Fig. 8.15.   Principle of submerged orifice scrubber.

5. The **ejector scrubber** is a water jet pump (see Fig. 8.16). The water is pumped through a uniform nozzle and the dirty gas is accelerated by the action of the jet gas. The result is aspiration of the gas into the water by the Bernoulli principle and, accordingly, a lowered pressure. The ejector scrubber can be used to collect soluble gases as well as particulates.

6. The **venturi scrubber** involves the acceleration of the dirty gas to 75-300 m min^{-1} through a mechanical constriction. This high velocity causes any water injected just upstream of, or in, the venturi throat to be sheared off the walls or nozzles and atomized. The droplets are usually 5-20 μm in size and form into clouds from 150-300 μm in diameter, depending on the gas velocity. The scrubber construction is similar to that of the ejector scrubber, but the jet pump is replaced by a

venturi constriction.

7. **Mechanical scrubbers** have internal rotating parts, which break up the scrubbing liquid into small droplets and simultaneously create turbulence.

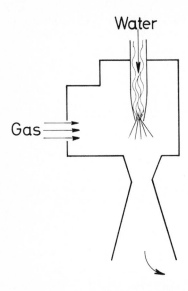

Fig. 8.16.   Principle of ejector scrubber.

8. **Charged-droplet scrubbers** have a high voltage ionization section where the corona discharge produces air ions (as in electrostatic precipitators; see 8.2.8). Water droplets are introduced into the chamber by use of spray nozzles or similar devices. The additional collection mechanism provided by the induction of water droplets increases the collection efficiency.

9. **Packed-bed scrubbers** have a bottom support grid, and an top retaining grid (see Fig. 8.17). The fluid (often water or a solution of alkali or acid) is distributed as shown in the figure over the top of the packed section, while the gas enters below the packing.

The flow is normally counter current. Packed-bed scrubbers offer the possibility of combining gas absorption with removal of particulate material. The pressure drop is often in the order of 3 cm water per m of packing. If the packing consists of expanded fibre, the bed scrubber is known as a fibre-bed scrubber.

The packed-bed scrubber has a tendency to clog under high particulate

loading, which is its major disadvantage.
Common packings used are saddles, rings, etc., like those used in absorption towers.

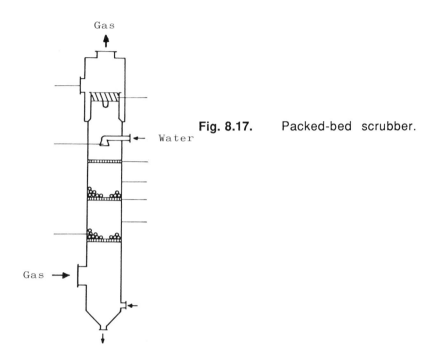

Fig. 8.17.   Packed-bed scrubber.

Some important parameters for various scrubbers are plotted in Fig. 8.18, which demonstrates the relationship between pressure drop, energy consumption and D_{50} (the diameter of the particles removed at 50 percent efficiency).

8.2.10.   Modification of particulate characteristics.

P.8.8.   The human health hazards associated with particulate emission are directly related to the mass median diameter, the influence of which increases with the solubility and toxicity of the pollutant. Particle shape is another significant characteristic of particulate emission.

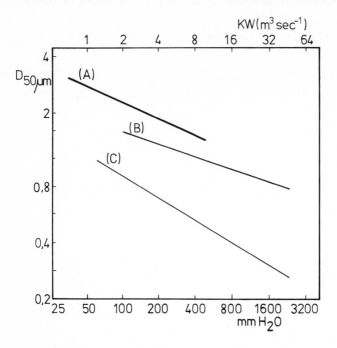

Fig. 8.18. Relationship between D_{50}, pressure drop (mm H_2O) and energy consumption. A: Packed-bed scrubber.   B: Baffled scrubber. C: Venturi scrubber.

Although it is not important in electrostatic precipitators and has very little effecto on cyclone performance, it has a considerable effect on the choise of fabric filter - at least for some applications.

P.8.9. **Particles are conditioned merely to increase their size and thereby increase their ease of collection. Particle conditioning involves two principle mechanisms: agglomeration and condensation.**

P.8.10. **Condensation can be achieved by adjusting the temperature of the water content of the gas phase.**

The degree of particle growth obtained by condensation is dependent on four factors:
1. *The number of nucleation centres.*
2. *The kinetics of particle nucleation.* The surface forces and the character

of the material itself is of importance here. Such properties as surface tension, wetability, hydroscopicity and hydrophobicity effect the rate of particle growth.

3. *The enlargement of the particles into droplets.*
4. *The degree of mixing between water and particles.*

Condensation is carried out by bringing the gas close to the dew point. Saturation can be attained by spraying water into the gas by adding steam. Another possibility is to cool the gas to the dew point either by external cooling or by adiabatic expansion.

Prakash and Murray (1975) have reported on conditioning of process effluent containing talcum powder, under 5 μm in diameter. By adding steam it was possible to reduce the emission by four to five times. Without conditioning the collection efficiency of particles smaller than 1 μm was nil, while with steam injection a 99 percent collection efficiency of particles above 0.3 μm was obtained.

P.8.11. **Agglomeration of particles occurs in nature due to brownian motion.**

The natural rate of agglomeration is, however, too slow under static or laminar flow conditions. If coagulation could be enhanced it would be possible to use a relative inexpensive control element for their removal. At present two techniques are used: **sonic agglomeration** (Dibbs and Marier, 1975) and **coagulation of charged particles**. The former method is more often used in practice. With this technique, particles need to be in insolation chambers for only a few seconds at intensities at 160-170 dB. Sonic aggregation can be *applied to any aerosol or solid particles,* but the energy requirement is relatively high and the high noise level must be eliminated by suitable insulation.

8.3. THE AIR POLLUTION PROBLEMS OF CARBON DIOXIDE, CARBON HYDRIDES AND CARBON MONOXIDE.

8.3.1. Sources of pollutants.

All types of fossil fuel will produce carbon dioxide on combustion, which is used in the photosynthetic production of carbon-hydrates. As such, carbon dioxide is harmless and has no toxic effect, whatever the concentration levels. However, since an increased carbon dioxide concentration in the atmosphere will increase absorption of infrared radiation, the heat balance of the earth will be changed (see Part A, section 2.4 for a detailed discussion).

P.8.12. **Carbon hydrides are the major components of oil and gas, and incomplete combustion will always involve their emission. Partly oxidized carbon hydrides, such as aldehydes and organic acids, might also be present.**

The major source of carbon hydride pollution is *motor vehicles*.

P.8.13. **In reaction with nitrogen oxides and ozone they form the so-called photochemical smog, which consists of several rather oxidative compounds, such as peroxyacyl nitrates and aldehydes.**

In areas where solar radiation is strong and the atmospheric circulation small, the possibility of smog formation increases, as the processes are *initiated by ultraviolet radiation*.

Typical concentration in American cities are shown in Table 8.5. Values from unpolluted areas are included for comparison.

TABLE 8.5
Typical concentrations in American cities (ppm) compared with values from a rural area

	Washing-ton	St. Louis	Phila-delphia	Denver	Chicago	Cincin-nati	Rural area
NO							
Annual average	0.04	0.03	0.06	0.04	0.10	0.004	0.1
Ma. 5 min. value	1.15	0.61	1.98	0.89	0.74	1.18	0.05
NO_2							
Annual average	0.04	0.03	0.04	0.03	0.06	0.04	0.005
Max. 5 min. value	0.19	0.21	0.29	0.35	0.35	0.30	0.01
Carbon hydrides							
Annual average	2.4	3.0	2.5	2.4	2.8	0.6	0.05
Max. 5 mun. value	14.5	14.3	14.4	19.1	14.9	12.8	0.6
CO							
Annual avearge	3	6	7	8	13	5	0.2
Max. 5 min. value	47	68	47	63	66	32	2

P.8.14. **Incomplete combustion produces carbon monoxide. By regulation of the ratio oxygen to fuel a more complete combustion can be obtained, but the emission of carbon monoxide cannot be totally avoided.**

Motor vehicles are also the major source of carbon monoxide pollution. On average, 1 litre of gasoline (petrol) will produce 200 litres of carbon monoxide, while it is possible to minimize the production of this pollutant by using diesel instead of gasoline.

The annual production of carbon monoxide is more then 200 million tons, of which 50 percent is produced by the U.S.A. alone.

In most industrial countries more than 85 percent of this pollutant originates from motor vehicles.

8.3.2. The pollution problem of carbon dioxide, carbon hydrides and carbon monoxide.

As mentioned carbon dioxide is not toxic, but its problem as a pollutant is related solely to its influence on the global energy balance. As this problem is rather complex it will not be dealt with here. For a comprehensive discussion, see Part A.

Carbon hydrides, partly oxidized carbon hydrides and the compounds of the photochemical smog are all *more or less toxic* to man, animals and plants. The photochemical smog *reduces visibility, irritates the eyes and causes damage to plants* with immense economical consequences, for example for fruit and tobacco plantations. It is also able *to decompose rubber and textiles.*

Carbon monoxide is strongly toxic as it reacts with haemoglobin and thereby reduces the blood's capacity to take up and transport oxygen. Ten percent of the haemoglobin occupied by carbon monoxide will produce such symptoms as headache and vomitting. A more detailed discussion of the relationship between carbon monoxide concentration, exposure time and effect is given in section 2.12. However, it should be mentioned here that smoking also causes a higher carboxyhaemoglobin concentration. An examination of policemen in Stockholm has shown that non-smokers had 1.2 percent carboxyhaemoglobin, while smokers had 3.5 percent.

8.3.3. Control methods applied to carbon dioxide, carbon hydride and carbon monoxide pollution.

Carbon dioxide pollution is inevitable related to the use of fossil fuels. Therefore, it can only be solved by the use of other sources of energy.

Legislation is playing a major role in controlling the emission of carbon hydrides and carbon monoxide. As motor vehicles are the major source of these pollutants, control methods should obviously focus on the possibilities of reducing vehicle emission. The methods available today are:

1. **Motor technical methods.**
2. **Afterburners.**
3. **Alternative energy sources.**

The first method is based upon the graphs shown in Fig. 8.19, where the relationship between the composition of the exhaust gas and the air/fuel ratio is illustrated. As seen, a higher air/fuel ratio results in a decrease in the carbon hydride and carbon monoxide concentration, but to achieve this better distribution of the fuel in the cylinder is required, which is only possible by construction of other gasification systems.

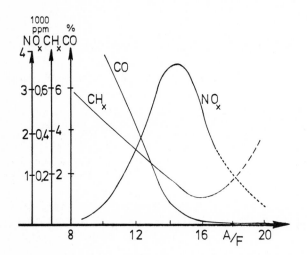

Fig. 8.19. Example of the relationship between the composition of exhaust gas and the air-fuel ratio. CH_x = carbon hydrides.

At present two types of afterburners are in use - **thermal and catalytic afterburners.** In the first type the combustible material is raised above its autoignition temperature and held there long enough for complete oxidation of carbon hydrides and carbon monoxide to occur. This method is used on an industrial scale (Waid, 1972 and 1974) when low-cost purchased or diverted fuel is available; in vehicles a manifold air injection system is used.

Catalytic oxidation occurs when the contaminant-laden gas stream is passed through a catalyst bed, which initiates and promotes oxidation of the combustible matter at lower temperature than would be obtained in thermal oxidation. The method is used on an industrial scale for the destruction of trace solvents in the chemical coating industry. Vegetable and animal oils can be oxidized at 250-380°C by catalytic oxidation. The exhaust fumes from chemical processes, such as ethylene oxide, methyl methacrylate, propylene, formaldehyde and carbon monoxide, can easily be catalytically incinerated at even lower temperatures.

The application of catalytic afterburners in motor vehicles presents some difficulties due to poisoning of the catalyst by lead. With the decreasing lead concentration in gasoline it is becoming easier to solve that problem, and the so-called double catalyst system is now finding a wide application. This system is able to reduce nitrogen oxides and oxidize carbon monoxide and carbon hydrides simultaneously.

Application of alternative energy sources is still at a preliminary stage. The so-called *Sterling motor* is one alternative, as it gives a more complete combustion of the fuel. Most interest has, however, been devoted to *electric vehicles.* Table 8.6 gives a survey of the suggested types of accumulator (battery). In spite of intensive research it is still a problem to produce an accumulator with a sufficiently high energy density to give the vehicle an acceptable radius action.

Fuel cells are able to produce electricity directly form fuel and Table 8.7 mentions the most realistic suggestions so far.

TABLE 8.6
Accumulator types

System	Energy density $(Whkg^{-1})$	Weight/effect ratio $(KG\ kW^{-1})$
Li - Cl	325	1.5
Zn - Air	130	14
Ag - Zn	110	2
Na - S	325	5
Pb-accumulator (conventional)	22	9

TABLE 8.7
Fuel cells

Fuel	Catalyst	Energy density $(Whkg^{-1})$	Weight/effect ratio $(kg\ kW^{-1})$	Volume/effect ratio $(l\ kW^{-1})$
Liquid, O_2 and H_2	Pt	2000	31	51
Air-Ammonium	Ni	1450	40	130
Liquid, air-carbon-hydrides	Pd-Ag	850	70	125

8.4. THE AIR POLLUTION PROBLEM OF SULPHUR DIOXIDE.

8.4.1. The sources of sulphur dioxide pollution.

P.8.15. Fossil fuel contains approximately 2 to 5 per cent sulphur, which is oxidized by combustion to sulphur dioxide. Although fossil fuel is the major source, several industrial processes produce emissions containing sulphur dioxide, for example mining, the treatment of sulphur containing ores and the production of paper from pulp.

The total global emission of sulphur dioxide is approximately *80 million tons per year,* of which U.S.A. and Europe produce more than two-thirds. Emission figures for industrialized areas in Europe are given in Table 8.8.

TABLE 8.8
SO_2-emission in Europe (10^6 t)

	1965	1975	1985
France	2.3	2.7	3.3
Netherlands	0.9	1.3	1.9
Great Britain	6.4	5.5	5.3
Germany (West)	3.2	3.3	3.5
Scandinavia	1.0	1.7	2.5

The concentration of sulphur dioxide in the air is relatively easy to measure, and sulphur dioxide has been used as an indicator component. High values recorded by inversion are typical.

8.4.2. The sulphur dioxide pollution problem.

Sulphur dioxide is oxidized in the atmosphere to sulphur trioxide, which forms sulphuric acid in water. Since sulphuric acid is a strong acid it is easy to understand that sulphur dioxide pollution indirectly causes corrosion of iron and other metals and is able to acidify aquatic ecosystems (these problems have been covered in detail in section 4.7).

The health aspects of sulphur dioxide pollution are closely related to those of particulate pollution. The gas is strongly adsorbed onto particulate matter, which transports the pollutant to the bronchi and lungs. (The relationship between concentration, effect and exposure time has already been discussed in sections 2.12, 2.13 and 4.7).

In Sweden the corrosion problem is estimated to cost more than 200 million dollars per year.

8.4.3. Control methods applied to sulphur dioxide.

Clean Air Acts have been introduced in all industrialized countries during the last decade. Table 8.9 illustrated some typical sulphur dioxide emission standards, although these may vary slightly from country to country.

TABLE 8.9
SO$_2$-emission standards

Duration	Concentration (ppm)
Month	0.05
24 h	0.10 might be exceeded once a month
30 min.	0.25 might be exceeded 15 times per month
	(1% of the time)

The approaches used to meet the requirements of the acts, as embodied in the standards, can be summarized as follow:
1. Fuel switching from *high to low sulphur fuels.*
2. *Modification of the distribution pattern* - use of tall stacks.
3. *Abandonment of very old power plants,*   which have higher emission.
4. *Flue gas cleaning.*

Desulphurization of liquid and gaseous fuel is a well known chemical engineering operation.

In gaseous and liquid fuels sulphur either occurs as hydrogen sulphide or can react with hydrogen to form hydrogen sulphide. The hydrogen sulphide is usually removed by absorption in a solution of alkanolamine and then converted to elemental sulphur. The process in general use for this conversion is the so-called Claus process. The hydrogen sulphide gas is fired in a combustion chamber in such a manner that one-third of the volume of hydrogen sulphide is converted to sulphur dioxide. The products of combustion are cooled and then passed through a catalyst-packed converter, in which the following reaction occurs:

$$2H_2S + SO_2 = 3S + 2H_2O \tag{8.16}$$

The elemental sulphur has commercial value and is mainly used for the production of sulphuric acid.

Sulphur occurs in coal both as pyritic sulphur and as organic sulphur. Pyritic sulphur is found in small discrete particles within the coal and can be removed by mechanical means, e.g. by gravity separation methods. However, 20 to 70 per cent of the sulphur content of coal is present as organic sulphur, which can hardly be removed today on an economical basis. Since sulphur recovery from gaseous and liquid fuels is much easier than from solid fuel, which also has other disadvantages, much research has been and is being devoted to *the gasification or liquefaction of coal.* It is expected that this research will lead to an alternative technology that will solve most of the problems related to the application of coal, including sulphur dioxide emission.

Approach (2) listed above has already been mentioned in 8.2.4, while approach (3) needs no further discussion. The next paragraph is devoted to flue gas cleaning.

8.4.4.   Flue gas cleaning of sulphur dioxide.

When sulphur is not or cannot be economically removed from fuel oil or coal prior to combustion, removal of sulphur oxides from combustion gases will become necessary for compliance with the stricter air pollution-control laws.

The chemistry of sulphur dioxide recovery presents a variety of choices and four methods should be considered:

1. Adsorption of sulphur dioxide on active metal oxides with regeneration to produce sulphur.
2. Catalytic oxidation of sulphur dioxide to produce sulphuric acid.
3. Adsorption of sulphur dioxide on charcoal with regeneration to produce concentrated sulphur dioxide.
4. Reaction of dolomite or limestone with sulphur dioxide by direct injection into the combustion chamber.

8.5.   THE AIR POLLUTION PROBLEM OF NITROGENOUS GASES.

8.5.1.   The sources of nitrogenous gases.

Seven different compounds of oxygen and nitrogen are known: N_2O, NO, NO_2, NO_3, N_2O_3, N_2O_4, N_2O_5 - often summarized as NO_x. From the point of view of air pollution maily NO (nitrogen oxide) and NO_2 (nitrogen dioxide) are of interest.

Nitrogen oxide is colourless and is formed from the elements at high temperatures. It can react further with oxygen to form nitrogen dioxide, which is a brown gas.

P.8.16. The major sources of the two gases are: nitrogen oxide -

combustion of gasoline and oil; nitrogen dioxide - combustion of oil, including diesel oil. In addition, a relatively small emission of nitrogenous gases originates from the chemical industry.

The total global emission is approximately *10 million tons per year.* This pollution has only local or regional interest, as the natural global formation of nitrogenous gases in the upper atmosphere by the influence of solar radiation is far more significant than the anthropogenous emission.

As mentioned above, the nitrogen oxide is oxidized to nitrogen dioxide, although the reaction rate is slow - in the order of 0.007 h^{-1}. However, it can be acclerated by solar radiation.

Nitrogenous gases *take part in the formation of smog,* as the nitrogen in peroxyacyl nitrate originates from nitrogen oxides, see Table 9.14 in 9.4.3. They are highly toxic but as the global pollution problem is insignificant, local and regional problems can partially be solved by changing the distribution pattern (see 8.2.4).

The emission from motor vehicles can be reduced by the same methods as mentioned for carbon hydrides and carbon monoxide. As illustrated in Fig. 8.19, *the air/fuel ratio* determines the concentration of pollutants in the exhaust gas. An increase in the ratio will reduce the emission of carbon hydrides and carbon monoxide, but unfortunately, will increase the concentration of nitrogenous gases. Consequently, the selected air/fuel ratio will be a compromise.

As mentioned in 8.3.3, a *double catalytic afterburner* is available. It is able to reduce nitrogenous gases and simultaneously oxidize carbon hydrides and carbon monoxide. The application of alternative energy sources will, as for carbon hydrides and carbon monoxide, be a very useful control method for nitrogenous gases at a later stage (see 8.3.3).

Between 0.1 and 1.5 ppm of nitrogenous gases, of which 10 to 15 per cent consists of nitrogen dioxide, are measured in urban areas with heavy traffic. As an example, Table 8.5 shows concentrations measured in North American cities. On average the emission of nitrogenous gases is approximately 15 g per litre gasoline and 25 g per litre diesel oil. About toxicity of nitrogenous gases, see section 2.12.

8.5.2. The nitrogenous gas pollution problem.

Nitrogenous gases in reaction with water form nitrates which are washed away by rain water. In some cases this can be a significant source of eutrophication (see 6.3.1). For a shallow lake, for example, the increase in nitrogen concentration due to the nitrogen input from rain water will be rather significant. In a lake with a depth of 1.8 m and an annual precipitation of 600 mm, which is normal in many temperate regions, the annual input will be as much as 0.3 mg per litre.

8.5.3. Control methods applied to nitrogenous gases.

The methods used for control of industrial emission of nitrogenous gases, including ammonia, will be discussed in the next paragraph, but as pointed out above industrial emission is of less importance, although it might play a significant role locally.

The emission of nitrogenous gases by combustion of oil for heating and the production of electricity can hardly be reduced.

8.6. INDUSTRIAL AIR POLLUTION.

8.6.1. Overview.

The rapid growth in industrial production during the last decades has enhanced the industrial air pollution problem, but due to a more pronounced application of continuous processes, recovery methods, air pollution control, use of closed systems and other technological developments, industrial air pollution has, in general, not increased in proportion to production.

P.8.17. Industry displays a wide range of air pollution problems related to a large number of chemical compounds in a wide range of concentrations.

It is not possible in this context to discuss all industrial air pollution problems, but rather to touch on the most important problems and give an overview of the control methods applied today. Only the problems related to the environment will be dealt with in this context.

A distinction should be made between air quality standards, which indicate that the concentration of a pollutant in the atmosphere at the point of measurement shall not be greater than a given amount, and emission standards, which require that the amount of pollutant emitted from a specific source shall not be greater than a specific amount (see also section 8.1).

P.8.18. The standards reflect, to a certain extent, the toxicity of the particular component, but also the possibility for its uptake.

Here the distribution coefficient for air/water (blood) plays a role. The more soluble the component is in water, the greater the possibility for uptake. For example, the air quality standard for acetic acid, which is very soluble in water, is relatively lower than the toxicity of aniline, which is almost insoluble in water.

Table 8.10 gives a survey of the most important air pollutants, the industrial emission source and the average emission per unit of production.

TABLE 8.10
Industrial emission

Production	Pollutant	Emission kg(t product)$^{-1}$
Sulphuric acid	SO_2	10 -30
	$SO_3 + H_2SO_4$	0.6 - 6
Hydrochloric acid	HCl	0.7 - 1.0
	Cl_2	0 - 0.1
	SO_2	2 - 3
	SO_3	0.2 - 0.9
Phosphoric acid	HF	0.1 - 0.2
Ammonium	SO_2	40 - 56
	SO_3	0 - 0.4
	NH_3	0 - 0.06
	NO_x	0 - 0.5
Nitric acid	NO_x	10 - 25
Sodium hydroxide + chlorine	Hg	0.002 - 0.035
	Cl_2	0.005 - 0.02
Cellulose (sulphite)	SO_2	25 - 65
Cellulose (sulphate)	SO_2	10 - 25
Steel and iron	particulate	2 - 20
Cement	particulate	0.5 - 5
Asphalt	particulate	0.1 - 0.5

8.6.2.   Control methods applied to industrial air pollution.

Since industrial air pollution covers a wide range of problems, it is not surprising *that all three classes of pollution control methods* mentioned in section 8.1 have found application: modification of the distribution pattern, alternative production methods and particulate and gas/vapour control technology.

All the methods mentioned in 8.2.4 to 8.2.10 and 8.4 are equally valid for industrial air pollution control.

It is not possible to mention here all alternative production methods that have found application in industry. Generally, it can be stated that a switch from dry to wet methods will often eliminate an air pollution problem, but at the same time create a water pollution problem. In each case it must be estimated which of the two problems it is most difficult or costly to solve properly.

P.8.19. **In gas and vapour technology a distinction has to be made between condensable and non-condensable gaseous pollutants. The latter must usually be destroyed by incineration, while the condensable gases can be removed from industrial effluents by absorption, adsorption, condensation or combustion. Recovery is feasible by the first three methods.**

8.6.3.   Gas absorption.

P.8.20. **Absorption is a diffusional process that involves the mass transfer of molecules from the gas state to the liquid state along a concentration gradient between the two phases.**

Absorption is a unit operation which is enhanced by all the factors generally affecting mass transfer; i.e. *high interfacial area, high solubility, high diffusion coefficient, low liquid viscosity, increased residence time, turbulent contact between the two phases and possibilities for reaction of the gas in the liquid phase.*

This last factor is often very significant and an almost 100 per cent removal of the contaminant is the result of such a reaction. Acidic components can easily be removed from gaseous effluents by absorption in alkaline solutions, and correspondingly alkaline gases can easily be removed from effluent by absorption in acidic solutions. Carbon dioxide, phenol or hydrogen sulphide are readily absorbed in alkaline solutions in accordance with the following processes:

$$CO_2 + 2NaOH - 2Na^+ + CO_3^{2-} \tag{8.17}$$

$$H_2S + 2NaOH - 2Na^+ + S^{2-} + 2H_2O \tag{8.18}$$

$$C_6H_5OH + NaOH - C_6H_5O^- + Na^+ + H_2O \tag{8.19}$$

Ammonia is readily absorbed in acidic solutions:

$$2NH_3 + H_2SO_4 - 2NH_4^+ \ SO_4^{2-} \qquad\qquad (8.20)$$

TABLE 8.11
Absorber reagents

$KMnO_4$	Rendering, polycyclic organic matter
NaOCl	Protein adhesives
Cl_2	Phenolics, rendering
Na_2SO_3	Aldehydes
NaOH	CO_2, H_2S, phenol, Cl_2, pesticides
$Ca(OH)_2$	Paper sizing and finishing
H_2SO_4	NH_3, nitrogen bases

Table 8.11 gives a more comprehensive list of absorber reagents and their application.

Absorption is the inverse process of air stripping and the theory for this unit process (see 6.2.4) is equally valid for the absorption process. By means of air stripping an water pollution problem becomes an air pollution problem, while the absorption transforms an air pollution problem into a water pollution problem. However, both processes open up possibilities for recovery. Gas absorbers and reactive scrubbers are quite flexible in application, dependable and can be highly effective. They have a large capacity at a reasonable pressure drop, are easy to install and are able to control gases and particulates. Their space requirements are usually favourable compared wiht alternative control methods.

8.6.4.   Gas adsorption.

P.8.21.   Adsorption is the capture and retention of a component (adsorbate) from the gas phase by the total surface of the adsorbing solid (adsorbent). In principle the process is the same as that mentioned in 6.3.3, which deals with waste water treatment; the theory is equally valid for gas adsorption.

Adsorption is used to concentrate (often 20 to 100 times) or store contaminants until they can be recovered or destroyed in the most economical way.

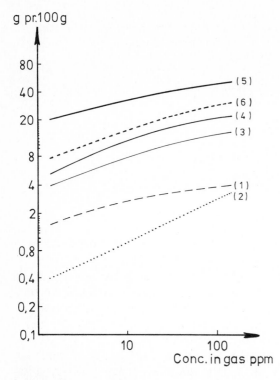

g pr.100g

Fig. 8.20. Adsorption isotherms, at 20°C: (1) Pinene, (2) Methyl mercaptane, (3) Benzene, (4) Pyridine, (5) Isovaleric acid; at 40°C: (6) Isovaleric acid.

Fig. 8.20 illustrates some adsorption isotherms applicable to practical gas adsorption problems. These are (as mentioned in 6.3.3) often described *as either Langmuir's or Freundlich's adsorption isotherms.*

Adsorption is *dependent on temperature,* as illustrated in Fig. 8.21 (in accordance with the theory outlined in 6.3.3). There are four major types of gas adsorbents, the most important of which is activated carbon, but also aluminium oxide (activated aluminia), silica gel and zeolites are used.

The selection of adsorbent is made according to the following criteria:
1. *High selectivity* for the component of interest.
2. *Easy and economical to regenerate.*
3. *Availability of the necessary quantity at a reasonable price.*

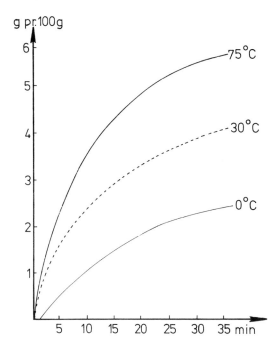

Fig. 8.21. Typical relationship: desorption versus time at three different temperatures for butane on activated carbon. (Adsorption temperature 52°C).

4. *High capacity for the particular application,* so that the unit size will be economical. Factors affecting capacity include total surface area involved, molecular weight, polarity activity, size, shape and concentration.
5. *Pressure drop,* which is dependent on the superficial velocity, as illustrated in Fig. 8.22, where the relationship between the pressure drop and the superficial velocity is shown for different mesh sizes of granular activated carbon.
6. *Mechanical stability in the resistance of the adsorbent particles to attrition.* Any wear and abrasion during use or regeneration will lead to an increase in bed pressure drop.
7. *Microstructure* of the adsorbent should, if at all possible, be *matched to the pollutant* that has to be collected.
8. *the temperature,* which, as illustrated in Figs. 8.20 and 8.21, has a profound influence on the adsorption process.

As already mentioned **regeneration of the adsorbents** is an important part of the total process. A few procedures are available for regeneration:

1. **Stripping** by use of steam or hot air.
2. **thermal desorption** by raising the temperature high enough to boil off all the adsorbed material (see Fig. 8.21).
3. **Vacuum desorption** by reducing the pressure enough to boil off all the adsorbed material.
4. **Purge gas stripping** by using a non-adsorbed gas to reverse the concentration gradient. The purge gas may be condensable or non-condensable. In the latter case it might be recycled, while the use of a condensable gas has the advantage that it can be removed in a liquid state.
5. **In situ oxidation**, based on the oxidation of the adsorbate on the surface of the adsorbent.
6. **Displacement** by use of a preferentially adsorbed gas for the deadsorption of the adsorbate. The component now adsorbed must, of course, also be removed from the adsorbent, but its removal might be easier that that of the originally adsorbed gas, for instance, because it has a lower boiling point.

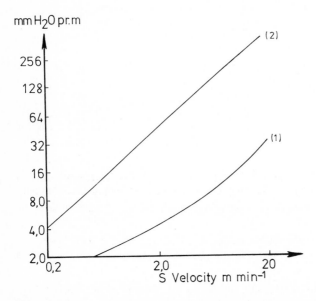

Fig. 8.22.   Relationship between pressure drop (mm H_2O per m) and superficial velocity for two different grades of activated carbon. (1) coarse (2) fine   at 1 atm. and 25°C.

Although the regeneration is 100 per cent the capacity may be reduced 10 to 25 per cent after several regeneration cycles, due to the presence of fine particulates and/or high molecular weight substances, which cannot be removed in the regeneration step.

A flow/chart of solvent recovery using activated carbon as an adsorbent is shown in Fig. 8.23 as an illustration of a plant design.

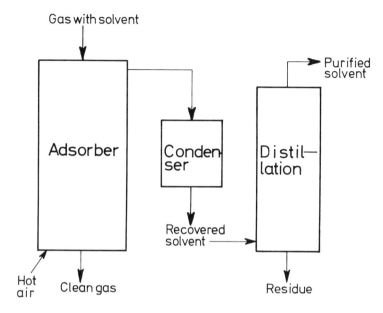

Fig. 8.23. Flow-chart of solvent recovery by the use of activated carbon.

8.6.5. Combustion.

P.8.22. Combustion is defined as rapid, high temperature gas-phase oxidation. The goal is the complete oxidation of the contaminants to carbon dioxide and water, sulphur dioxide and nitrogen dioxide.

The process is often applied to control odours in rendering plants, paint and varnish factories, rubber tyre curing and petro-chemical factories. It is also used to reduce or prevent an explosion hazard by burning any highly flammable gases for which no ultimate use is feasible.

The efficiency of the process is highly dependent on temperature and reaction time, but also on turbulence or the mechanically induced mixing of oxygen and combustible material. The relationship between the reaction rate, r and the temperature can be expressed by Arrhenius' equation:

$$r = A * e^{-E/RT}$$

(8.21)

where A = a constant, E = the activation energy, R = the gas constant and T = the absolute temperature

A distinction is made between combustion, thermal oxidation and catalytic oxidation, the latter two being the same in principle as the vehicles afterburners mentioned in 8.3.3.

QUESTIONS AND PROBLEMS.

1. Calculate the maximum concentration (g m^{-3}) at ground level for the following case:
   Emission: 20 g min^{-1}
   k (constant): 0.2 min m^{-1}
   Stable exit velocity: 2 m sec^{-1}
   Stack exit inside diameter: 1.8 m
   Stack exit temperature: 600°K
   Stack height: 120 m

2. Calculate the ratio in concentration of particulate matter in case A and B provided that the total emission is the same in the two cases:
   A: 2000 stacks   H =   10 m
   B: 1 stack   H = 250 m

3. Indicate a method to remove the following as air pollutants:
   1) ammonia,   2) hydrogen sulphide, 3) phenol, 4) hydrogen cyanide.

4. What is the energy consumption for a packed-bed scrubber treating 650 m^3 min^{-1} with a D_{50} = 2μm?

5. Compare the influence of SO_2 on the sulphur cycle and CO_2 on the carbon cycle.

6. Would the total SO_2 emission on earth be able to change pH of the sea over a period of 100 years, if: a) the total SO_2 emission is constant at the present level, b) the total SO_2 emission increases 2% per year?

7. Calculate the cost of activated carbon per day used for a 90% removal of 10 ppm (v/v) pyridine from 100 m^3 air per hour. The temperature is 20°C. The cost of one kg activated carbon is US$ 0.50.

CHAPTER 9

EXAMINATION OF POLLUTION

9.1 INTRODUCTION.

Quantification of an environmental issue and selection of environmental technology cannot be done only on the basis of theoretical considerations; rather an examination of the pollution situation is required. In the case where an environmental management model is needed, an examination programme must be set up to calibrate and test the model properly.

Current control of the efficiency of installed equipment, whether it solves a waste water, solid waste or air pollution problem, is necessary if the investment is not to be worthless. Where emission standards are set up, they will determine the control programme.

Monitoring on air and water quality is being performed increasingly all over the world. Such programmes are, of course, related to immission standards and can be used to identify and control point sources of pollution and to verify environmental management objectives.

This chapter will not discuss in depth the analytical procedure used in environmental examinations, but will mention only the problems related to such examinations. As these problems are entirely different for water, solid and air pollution they will be dealth with separately in the following three sections.

9.2. EXAMINATION OF WATER AND WASTE WATER.

9.2.1. Collection and preservation of samples.

Samples are collected from aquatic ecosystems and waste water systems for many different reasons. The ultimate aim is to accumulate data, which are used to determine selected properties of water or waste water. In planning any investigation of either, the two most important considerations are the measurement of significant parameters and the collection of representative samples. For most studies it is impossible to measure all variables and it is therefore necessary to limit the analysis to just a few selected parameters. Since time is limited the number of parameters measured must be optimized, so the data recorded must be those which will be of most help in solving the problem under investigation.

The wide variety of conditions under which collections must be made makes it impossible to prescribe a fixed procedure that can be generally used.

P.9.1. **The sampling procedure should also take into account the tests to be performed and the purpose for which the results are needed.**

Thus the problem must be well defined and all facets of the investigation, including economical ones, must be considered prior to sample collection. In many cases a computer programme may be required to carry out this optimization, but the size of the testing programme will dictate whether the application of such techniques is worthwhile

Once the testing programme has been optimized to fit whithin the limits imposed by all the different factors, *specific collection methods* and a *sampling schedule can be selected,* whereupon the size, number, frequency and pattern are determined.

The size of the sample required is related to the concentration of the measured parameter in the sample and to the amount of material required for an accurate analysis. Due to collection, transportation and handling problems, it is desirable to restrict the sample volume to the *minimum* necessary for accurate analysis. The ultimate restraints are the lower limit of precision and the accuracy of the analytical method applied. Often it is necessary *to concentrate* the sample before analysis. Table 9.1 gives a survey of the concentration and separation techniques generally applied in water analysis.

The number of samples and frequency of sampling for a representative value of each sampling site are related to the variability of the parameter being measured and to the level of statistical significance specified. However, the biological significance of the data should also be considered. A good example is the determination of oxygen concentration in bodies of water, since a high daytime concentration of discharged oxygen may be accompanied by very low night values, which might result in fish kills.

P.9.2. **Measures of diurnal or seasonal variations in environmental parameters are often of greater importance than average values, as it is not the average but the extreme conditions that are hazardous to ecosystems** (see e.g. Tarzwell and Gaufin, 1953).

Automatic sampling devices are available but require the preservation of samples as described below. It is often desirable to combine individual samples with sampling proportional to the volume of flow to get the best *overall picture* of the parameter being analysed.

TABLE 9.1
Concentration and separation techniques

Method	Application	References
Precipitation	Trace metals (Mn, Fe, Ni, Co, Zn, Pb, Cu and others	Riley (1965) Joyner et al. (1967) Caitinen (1960)
Liquid-liquid extraction	Pesticides, hydrocarbons, and trace metals (as complexes)	Lamar et al. (1965) Kopp et al. (1967)
Chromatography	Pesticides, detergents, hydrocarbons, phenols, fatty acids and alcohols	Snyder (1961) Stahl (1965) Wren (1960)
Ion exchange	Ca^{2+}, Mg^{2+}, Fe^{3+}, Cl^-, PO_4^{3-}, CrO_4^{2-}, trace metals	Moody and Thomas (1968) Douglas (1967) Harley (1967)
Adsorption	Pesticides, phenols, hydrocarbons, trace metals	Goodenkaul and Erdes (1964) Hassler (1951)
Distillation and sublimation	Trace metals, ammonia, phenols	Kopp and Krover (1965) Mueller et al. (1960)
Membrane techniques	Particulate matter, colloidal matter, Bacteria and viruses	Golterman and Clymo (1969) Clark et al. (1951)
Adsorptive bubble separation	Surface active materials, metals	Rubin (1968)
Freeze concentration	Various organic compounds	Gouw (1968)

Samples taken with an automatic sampling device at a certain time frequency cannot be recommended, unless the parameter's variability alone is of interest. The flow rate for waste water discharge and natural streams is often of varying significance, so flow proportional sampling is required for calculation of total emission or immission values.

The selection of the sampling pattern is of great importance in determining environmental parameters in natural water bodies. The variability of the environment is an important factor that must be considered in formulating the sampling pattern and selecting individual sampling sites. Furthermore, the specificity of the biota for narrow environmental niches and the incomplete mixing of waste effluents with receiving water must also be taken into account (see also Hawkes, 1962 and Hynes, 1960).

As indicated above it is necessary to know the variation over time and place to be able to select the right sampling programme. So, whenever possible, the sampling programme should be based on a previous one or a pilot programme made prior to selection.

Preservation of samples is often difficult because almost all preservation methods interfere with some of the tests. Immediate analysis is ideal but often not possible. In many cases storage at low temperature (4°C) is the best method of preservation for up to 24 hours. No method of preservation is entirely satisfactory and the choice of which to use should be made with due regard for the measurements that will be required.

Some useful methods of preservation and their application are summarized in Table 9.2, but the selection of the right method is dependent on the measuring programme and general guide lines cannot be given.

TABLE 9.2
Methods of preservation

Methods	Application
Storage at low temperature	General
Sulphuric acid pH 2-3	Nutrients, COD
Formaldehyde	Affects many measurements
Sodium benzoate	Sludge and sediment samples (- grease measurement)
Iodine	Algal counts

9.2.2. Insoluble material.

P.9.3. **Insoluble material covers the range from 0.01 - 200 μm in size. Colloidal particles are included as they range in size from about 0.01 - 6 μm. Particles smaller than 0.1 μm are classified as dissolved.**

When waste water is examined, the filterable residue is determined by using different types of filters with a maximum pore size of 5 μm, so that colloidal particles are not included. When examining natural waters, analysis of the colloidal fraction and even determination of particle size distribution are often required. Methods available for such more detailed ecaminations are reviewed in Table 9.3.

An important measure for the determination of bioorganic and inorganic particles is the ability of the particles to accumulate material from the environment by absorption processes. The results of adsorption examinations are often expressed by means of adsorption isotherms. If adsorption is involved the following adsorption isotherms are generally applied.

Freundlich's adsorption isotherm

$$x = a * c^b \qquad\qquad (9.1)$$

Langmuir's adsorption isotherm

$$x = \frac{a * c}{k_m + c} \qquad\qquad (9.2)$$

where
a, k_m and b = constants
c = concentration in liquid
x = concentration of adsorbent

TABLE 9.3
Methods for detailed examination of insoluble material
_____
Centrifugation
X-ray diffusion
Electron microscopy
UV, visible and infrared spectroscopy
Differential thermal analysis
Optical microscopy
Ion exchange
Gel filtration
Electrophoresis
Sedimentation
Dialysis
_____

9.2.3. Biological oxygen demand.

The term BOD refers primarily to the result of a standard laboratory procedure, developed on the premise that the biological reactions of interest can be bottled in the laboratory and that the kinetics and extent of reaction can be observed empirically. The method is, however, *relatively unreliable* as a representation of the system from which the sample was taken. Therefore, the true worth of the standard procedure and the necessity for a more fundamental approach to measurements of biological reactions will be discussed here.

The basis difference between the BOD bottle and a stream is that the BOD bottle *is a static system* which is controlled to provide aerobic conditions, while the stream is a dynamic system which may include anaerobic areas due to formation of bottom deposition. As the synthesized microorganisms age, lysis of the cell walls occurs and the cell contents are released. *The remains of the cell walls* are relatively stable materials, such as polysacharides, which are not degradated during 5 days, which is the normal duration of the BOD test (indicated as BOD_5). Furthermore, *ammonia* may or may not be oxidized to nitrate during this period through the action of nitrifying bacteria.

The analytical procedure of BOD is based on *the assumption* that the biological oxidation is *a first order reaction.* This assumption may, in many cases, be an oversimplification as the biological oxidation consists of independent steps: energy reactions, which support the synthesis of organic matter into new cells, and endogenous reactions.

The rate at which the biological processes occur is also a function *of the food to organism ratio,* so the rate of stabilization of organic material is therefore very much related to upstream activity. In addition the rate at which biological processes take place is strongly dependent on the flow pattern. Shallow streams normally provide an excellent food to organism potential because of their high velocity, which provides turbulence.

In conclusion, the BOD results obtained in a BOD bottle have a limited quantitative correlation with the oxygen demands present in waste water or aquatic ecosystems. However, BOD_5, as presented in many books on standard methods, is still used worldwide for determination of the oxygen consumption ofr biological and chemical oxidation of waterborne substances, despite the fact that the method is very time consuming and it takes 5 days to get the result.

Standard Methods (1980) defines BOD_5 as determined in accordance with the procedure presented in the bood as follows:

The biochemical oxygen demand (BOD) determination described herein constitutes an empirical test, in which standardized laboratory procedures are used to determine the relative oxygen requirements of waste waters, effluents and polluted waters. The test has its wides application in measuring waste loadings to treatment plants and in evaluating the efficiency (BOD removal) of such treatment systems. Comparison of BOD values cannot be made unless the results have been obtained under identical test conditions.

The test is of limited value in measuring the actual oxygen demand of surface waters, and the extrapolation of test results to actual stream oxygen demands is highly questionable, since the laboratory environment does not reproduce stream conditions, particularly as related to temperature, sunlight, biological population, water movement and oxygen concentration.

Nitrification is a confusing factor. Procedures that inhibit nitrification are available, or corrections for nitrification can be carried out through nitrate determinations. The following methods for inhibition have been suggested: pasteurization (Sawyer et al., 1946), acidification (Hurwitz et al., 1947), methylene blue addition (Abbott, 1948), trichloromethyl pyridine addition (Goring, 1962), thiourea and allylthioureas addition (Painter et al.,

1963) and ammonium ion addition (Quastel et al., 1951 and Gaffney et al., 1958).

If BOD_5 determinations are replaced by BOD curves, usefull information about the carbonanceous oxygen demand may be obtained, but this technique requires 10 to 20 days and is complicated by the need for suitable seed organisms. Improved analytical procedures utilizing the dissolved oxygen probe make it possible to determing oxygen required for energy in less than one hour. The synthesis of biological material can be determined through enzyme activity measurements. Such analysis can be carried out in a few minutes.

Oxygen required during stabilization of synthesized cell material can be calculated once synthesis is known. In other words it is possible to arrive at a fundamental measurement of biological oxidation in less than one hour by summation of oxygen needed for energy and endogenous requirements. The result has been termed the stabilization oxygen demand.

Parallel measurements of *COD* (Chemical Oxygen Demand) *and TOC* (Total Organic Carbon) *are good supplements to the BOD-determinations* and thereby might eliminate some of the disadvantages of the BOD-determination.

Often a statistical relationship is worked out for a specific stream between BOD and COD and TOC, to reduce the very time-consuming BOD analysis by translating COD or TOC to BOD.

The above developments have improved the application of oxygen demand measurements although the disadvangages have been only partially eliminated. BOD_5 is and will be for many years the most important measurement of water quality in spite of its disadvantages.

9.2.4. Chemical kinetics and dynamics in water systems.

Kinetic and dynamic studies deal with the changing concentrations of reactants and products is essentially unstable situations. This sets certain requirements for analytical procedures. Spectrophotometric and electrometric methods are widely used in continuous monitoring of reactants or products. Many reactions of interest in water systems are not amenable to continuous monitoring and discrete samples must be taken for analysis. A variety of methods can be used to stop reactions in samples, for example by using inhibitors or poisons like trichloracetic acid, mercuric ions, formaldehyde, heat, or even strong acids or bases.

P.9.4. **The term kinetic analysis can be applied to a rather diverse group of analytical procedures, which determine concentrations of substances by measuring the rate of a**

reaction **rather than the concentration itself.**

The field of kinetic analysis is too broad for more than a cursory coverage here and the interested reader is referred to several review articles (Garrel and Christ, 1965; Stumm, 1967 and Faust and Hunter, 1967).

Table 9.4 gives a survey of some of the most commonly used kinetic analysis of water systems, with reference to the original literature.

TABLE 9.4
Kinetic analysis

Component	Reference
S^{2-}, SO_4^{2-}	Standard Methods (1980 or later editions)
Glucose	Hill and Kessler (1961)
ATP	Patterson (1970)
	Patterson et al. (1970)
	Hamilton and Holm-Larsen (1967)
NADPH	Kratochvil et al. (1967)
Nutrients (low level)	Grouch and Malmstadt (1967)
Ammonia	Weichelbaum et al. (1969)
Mn^{2+}/MnO_2	Morgan and Stumm (1965)

9.2.5. Analysis of organic compounds in aqueous systems.

All waters contain measurable concentrations of organic matter, which has originated from natural processes of biological synthesis and degradation or from the various activities of man.

Distinction can be made between the presence of specific organic compounds, such as alkyl benzene sulphonates, herbicides, insecticides etc., the presence of compounds of a less well defined nature that cause dye, taste and odour problems, and the presence of organic compounds which at present are not associated with known problems. The latter situation requires that the analyst identify and estimate all organic compounds in a dilute aqueous solution and/or suspension, which would, of course, be an impossible task. There are, however, a number of approaches to this problem. One is to guess from knowledge of discharges and previous experience what organic compounds might be present, and then to examine the mixture of these. Another possibility is to detect and estimate classes of materials, rather than individual compounds. Typical examples of classes are carbohydrates, amino acids, ketones, etc.

Several identification methods are available for detecting classes of organic compounds. Among the physical methods should be mentioned determination of the refractive index, magnetic susceptibility, absorption spectra, fluorescense spectra, Raman spectra, mass spectra, NMRS (Nuclear

Magnetic Resonance Spectra), X-ray diffraction, specific dispersion, dielectic constants and solubility. Chemical identification includes the determination of the molecular weight, elemental analysis and estimations of functional groups. For further information, see Hunter (1962); Lee (1965); Mitchell (1966).

Within the analysis of specific organic compounds, that of pesticides and PCB is of great importance, and some guidelines for the estimation of these compounds should be given.

The general procedure consists of three steps:
1. **Concentration** of samples.
2. **Cleaning up.**
3. **Identification** of pesticides and their metabolites.

For the first step adsorption on activated carbon or liquid-liquid extraction can be used. Relatively unpolluted waters may not require an extensive cleanup since pollutants would be of minor importance, but due to the small concentrations present in unpolluted waters the concentration step is required. Adsorption is a very useful method for this first step, as it is able to handle a large volume of water at low concentrations. Its limitation is that it is not quantitative and it is necessary to correct the result by using a recovery percentage. The conditions used for serial liquid-liquid extractions of various pesticides are summarized in Table 9.5, and, as can be seen, a wide range of organic solvents is used for this purpose.

Pollutants may originate from domestic and industrial waste water and a cleanup step is therefore necessary. The following cleanup techniques are in general use:

1. *Partition chromatography on silicid acid columns* for removal of more or less identified organic matter, including organic dyes (see Hinding et al., 1964; Epps et al., 1967; Aly and Faust, 1963).
2. *Thin layer chromatography* for separation of organic pollutants in the analysis of chlorinated hydrocarbon pesticides (Crosby and Laws, 1964).
3. *Adsorption on an aluminia column* for removal of pollutants in identification of chlorinated hydrocarbons (Hamence et al., 1965)
4. *Countercurrent extraction for identification of organic pesticides in food.*
5. *A single-sweep codistillation method* for general removal of organic pollutants (Storherr and Watts, 1965).

For the identification of organic pesticides and their metabolites either thin layer chromatography or gas-liquid chromatography are mainly used.

TABLE 9.5

Conditions used for serial liquid-liquid extraction of pesticides from aqueous samples

Pesticides	Solvents	Sample volume (ml)	Solvent volume (ml)	pH	Mean recovery (%)
DDT	3:1 Ether:n-hexane	1500	250,200,150,150	4.0-5.0	88[a], 61[b]
Dieldrin and metabolites	n-Hexane	100	4 * 50	-	90-95
Five CH	1:1 Ether:pet.ether, CHCl$_3$	1000	100, 4 * 50	-	88
Eight CH	n-Hexane	1000	2 * 25	-	80-115
Aldrin	1:1 Benzene:hexane	850	100, 4 * 50	-	43.4
Aldrin	1:1 Ether:hexane	850	100, 4 * 50	-	69.6
Eleven CH	n-Hexane	12	3 * 2	7.0	2.3-106.0
Heptachlor	2:1 Pet.ether: iso-propyl alcohol	1000	150, 3 * 75	-	80
Five CH	n-Hexane	1000	50	-	95
Nine CH	Benzene	250	25	-	84.6-101.8
Three CH	n-Hexane	1000	30	-	95
Four CH	1:1 n-Hexane:ether	100	3 * 100	-	97-100
Three CH	n-Hexane	1000	100	Acid	> 90
Parathion and diazinon	1:1 Ether:pet.ether or CHCl$_3$	1000	100, 4 * 50	-	90
Dipterex and DDVP	Ethyl acetate	50	50, 25	8.0	94.5
Malathion	Dichloromethane	?	3 * 30	7.0-8.0	100
Diazinon	Benzene	100	100	-	?
Abate	Chloroform	1000	50, 25, 25	1.0	70
Parathion	Benzene	1000	500	Acid	99-100
Methyl parathion, diazinon, malathion, azinphos-methyl	Benzene	250	25	?	95.3-99.7
Parathion, methyl parathion, baytex	n-Hexane	1000	100	Acid	> 90
Parathion, methyl parathion	1:1 Hexane:ether	100	3 * 100	-	98
Dursban	Dichloromethane	50	100, 50	-	92

a) Interferences absent.   b) Interferences present.

References
1. Berck (1953);   2. Cueto et al. (1962);   3. Weatherholtz et al. (1967);   4. LaMar et al. (1965);   5. Kawahara et al. (1967);   6. Cueto et al. (1967);   7. Weatherholtz et al. (1967);   8. Holden et al. (1966); 9. Pionke et al. (1968);   10. Wheatly et al. (1965);   11. Beroza et al. (1966);   12. Warnick et al. (1965); 13. Teasley et al. (1963);   14. El-Refai et al. (1965);   15. Mount et al. (1967);   16. Kawahara et al. (1967);   17. Wright et al. (1967);   18. Mulla (1966);   19. Pionke et al. (1968);   20. Warnick et al. (1965);   21. Teasly et al. (1963);   22. Beroza et al. (1966);   23. Rice et al. (1968).

9.2.6. Analysis of inorganic compounds in aqueous systems.

Water can be considered in three categories from an analytical point of view:

1. *Surface water and ground waters,* which are usually very dilute solutions containing several species of cations, anions and neutral organic and inorganic compounds.
2. *Seawater,* which contains several ions in high but constant concentration.
3. *Waste waters,* which cause special problems because of their variable composition and high concentrations of suspended matter.

The inorganic substances to be considered can be classified into two groups:

A. Ionic species and neutral compounds of more representative elements (see Fig. 9.1), and
B. Other elements, which are present in trace concentrations.

Brief reference is made in Table 9.6 to each of the pertinent methods of water analysis for group A components, and Table 9.7 for group B components. The methods listed are taken mainly from Standard Methods (1980).

I	II	III	IV	V	VI	VII
H						
		B	C	N	O	F
Na	Mg		Si	P	S	Cl
K	Ca			As	Se	Br
						I

Fig. 9.1. The more common elements of the periodic system.

The literature on water analysis is considerable, but it is possible to get a survey of the progress in this field by reading the review every odd numberd year in the April "Annual Reviews" issue of Analytical Chemistry and the annual review of the Jornal of the Water Pollution Control Federation.

TABLE 9.6
Analysis of inorganic compounds (group A)

Element	S.M. = Standard Methods (1980) Reference	Lowest measurable concentration
As	S.M., spectrophotometry	20 ppb
B	S.M., spectrophotometry	0.1 ppm
Br	S.M., spectrophotometry	0.1 ppm
Br	Fishman and Skougstadt (1963) spectrophotometry	5 ppb
Ca	S.M., titration	1 ppm
CO_2, CO_3^{2-}	S.M., titration	1 ppm
CN^-	S.M., spectrophotometry	0.02 ppm
Cl^-	S.M., titration	1 ppm
F^-	S.M., spectrophotometry	0.1 ppm
I^-	S.M., spectrophotometry	1 ppb
Mg	S.M., spectrophotometry	0.5 ppm
NH_4^+	S.M., spectrophotometry	0.01 ppm
NO_2^-	S.M., spectrophotometry	1 ppb
NO_3^-	S.M., spectrophotometry	2 ppb
O_2	S.M., titration	0.1 ppm
PO_4^{3-}	S.M., spectrophotometry	0.01 ppm
K	S.M., spectrophotometry	5 ppm
Si	S.M., spectrophotometry	0.02 ppm
Na	S.M., gravimetric	1 ppm
	S.M., flame photometry	0.01 ppm
S^{2-}	S.M., spectrophotometry	0.05 ppm
SO_4^{2-}	S.M., turbidimetric	0.5 ppm

In view of the present-day emphasis on instrumental methods of analysis based on application of physical properties, it is interesting that the traditional methods still predominate among the standard methods in water analysis of group A compounds. Only about 20 per cent of the methods described in Standard Methods (1980) for these compounds are purely instrumental, without extensive chemical pretreatment. The traditional mehods prevail, including spectrophotometric methods with extensive chemical pretreatment, because they have proven trustworthy and well suited to the sensitivities and precision required.

TABLE 9.7
Analysis of inorganic compounds (group B)

Methods	Applications
Flame photometry	Li, Na, K, Ca, Mg, Sr, Ba
Atomic absorption spectrophotometry (AAS)	Al, Cd, Ca, Cr, Co, Cu, Fe, Pb, Mg, Mn, Hg, Mo, Ni, K, Rb, Sc, Ag, Na, Sr, V, Zn
Atomic fluorescence spectrophotometry	Same as AAS
Emission spectrophotometry	All metals
Adsorption spectrophotometry	Almost all elements
X-ray spectroscopy	All elements from Na up in the periodic table
Activation analysis	As, Ba, Be, Br, Cs, Cr, Co, Cu, Au, Mg, In, Mn, Hg, Ni, P, Ce, Re, R, Ru, Na, Sr, S, Ta, Tl, Th, W, V, Zn

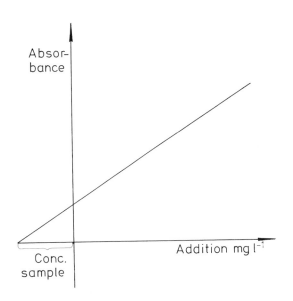

Fig. 9.2. Application of "the addition method".

Group B compounds are more widely analysed by use of instrumental methods, as these are better suited to estimation of minor concentrations. Often the so-called standard addition method must be used to eliminate pollutants. The method is illustrated in Fig. 9.2.

9.2.7.   Bacterial and viral analysis of water.

Natural water free from pollution has a rich microflora. Algae, diatoms, phytoflagellated, fungi, protozoans, metazoans, viruses and many genera of bacteria; including Alcaligenes, Caulobacteria, Chromobacterium, Flavobacterium, Micrococcus, Proteus and Pseudomonas. Sulphur, nitrogen fixing, denitrifying and iron bacteria may be found, depending on the substrate concentration. The microflora are generally strongly affected by solar radiation, temperature, pH, salinity, turbidity, hydrostatic pressure, dissolved oxygen (the redox potential), the presence of organic and inorganic nutrients, growth factors, metabolic regulating substances, and toxic substances.

P.9.5.   Water is one of the most important media for transmission of human microbial diseases and some of the best known examples include typhoid fever, paratyphoid fevers, dysentery, cholera and hepatitis.

It is generally accepted that the addition of pollutants to water may be actually injurious to the health and well being of man, animals and any aquatic life. Intestinal micro-organisms from man and warm-blooded animals enter rivers, streams and large bodies of water. Among these microbes the coliform group are most frequently present in water samples, which explains why they are often used as bacterial indicators, although far from ideal.

However, a wide variety of pathogenic bacteria may also be noted in water samples, such as Brucella, Leptospira, Mycobacterium, Salmonella, Shigella and Vibrio.

A number of biological measurements are made to assess the bacteriological quality of water and thereby indicate the presence or absence of pathogenic enteric bacteria. Standard Methods (1980) contains a number of relevant methods.

The frequency of sampling to determine the bacteriological quality of drinking water *is dependent on the size of the population served;* however, *two samples per month must be considered as the minimum.* A sample frequency of one per day is recommended for a population of 25,000 inhabitants.

Fecal Coliform criteria are widely used for recreational waters and in

Table 9.8 is shown an example (Water Quality Criteria (1968)).

TABLE 9.8
Criteria for recreation waters

Water usage	Fecal Coliform bacteria/100 ml
General recreation	Average < 2000 Maximum < 4000
Designated for recreation	log mean not to exceed 1000 Maximum 10% of samples > 2000
Primary contact recreation	log mean not to exceed 200 Maximum 10% of samples > 400

Viral analysis of water and waste water requires special equipment and techniques, and only a few laboratories are equipped to carry out such analyses. (Further details of these procedures, see Rhodes and van Ruoyen, 1968; Vorthington et al., 1970; England et al., 1967).

More than 100 human enteric viruses, the most important infective viral agents in water, have been described in the literature.

A review of human enteric viruses and their associated diseases is given in Table 9.9.

TABLE 9.9
Human enteric viruses and their associated diseases

	Diseases Number of types	Associated diseases
Polio virus	3	Paralytic polimyelitis, menigitis
Coxsackie virus A Coxsackie virus B	26 6	Herpangina, menigitis Pleurodynia, menigitis and infantile myocarditis
Infectious hepatitis	1	Hepatitis
Adeno virus	30	Eye and respiratory infections
Reo virus	3	Diarrhoea, fever, respiratory infections
Echo virus	29	Menigitis, respiratory infections, fever and rash

9.2.8.   Toxicity of bioassay techniques.

For more than 100 years test organisms have been used to establish the toxic effects of water pollution (Penny and Adams, 1863). Such investigations indicate the degree of dilution and/or treatment that must be provided to prevent damage to the organisms in the receiving waters. Fish, invertebrates and various algae have been variously used for this purpose.

At one time it was hoped that the toxicity of wastes to aquatic organisms could be determined by an analysis of the chemical constituents and comparison with a table listing the toxicities of the various waste substances to fish and other aquatic organisms, and thereby assess the toxicity level. However, this goal has not proved entirely feasible due to the following:

1. Although a great many toxicity data have been recorded (see Jørgensen et al., 1979), *information is still limited* in comparison with the number of toxic substances and species present in aquatic ecosystems.

2. Tests reported in the literature might have been carried out *under conditions different to those in the ecosystem concerned:* pH, dissolved oxygen, salinity, temperature and several other parameters profoundly affect the result of toxicity tests.

3. *The presence or absence of trace components* - often impossible to test in detail - also significantly affect the results.

However, information on the presence of chemical constituents and the related toxicity data found in the literature can be of great value in the interpretation of toxicity bioassay - in practice the two methods of determining the toxicity of waste should be used side by side.

Toxicity bioassay experiments on fish are widely used all over the world, although other vertebrates, invertebrates and algae are also used (see Veger, 1962; Anderson, 1946 and 1950; Gillispie, 1965).

The last five editions of Standard Methods for the examination of Water and Waste Water (WPCF), contain detailed information on toxicity bioassay methods on fish, including selection of test fish (species and size), their preparation (acclimatization 10-30 days, feeding etc.), selection and preparation of experimental water dilutions and sampling and storage of test material. Furthermore, it provides guidelines for general test conditions: temperature, dissolved oxygen content, pH, feeding, test duration, etc.

Procdures for continuous-flow bioassays are also available (see Standard Methods, 1980) and the basic components are shown in Fig. 9.3.

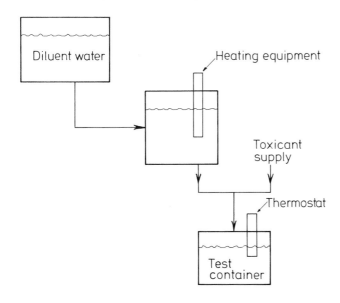

Fig. 9.3.   Continuous-flow bioassay.

P.9.6.   **The ultimate aim of the bioassay test is to find TL_{50} or LC_{50}, the concentration at which 50 per cent of the test fish survive.**

This value is usually interpolated from the percentages of fish surviving at two or more concentrations, at which *less* than half and *more* than half survived. The data are plotted on *semilogarithmic coordinate paper* with concentrations on logarithmic, and percentage survival on arithmetic scales. A straight line is drawn between the two points representing survival at the two concentrations that were lethal to more than half and to less than half the fish. The concentration at which this line crosses the 50 per cent *survival line* (see Fig. 9.4) *is the TL_{50} or LC_{50} value.* (Other methods of determining this value are sometimes more satisfactory).

A precision within about 10 per cent is often attainable, but better precision should not be expected even under favourable circumstances. In most cases 48-hour and 96-hour values are also determined, but these concentrations do not, of course, represent values that are safe in fish habitats. Long-term exposure to much lower concentrations may be lethal to fish and still lower concentrations may cause sublethal effects, such as impairment of swimming ability, appetite, growth, resistance to disease and reproductive capacity. Formulae for the estimation of permissible discharge rates or dilution ratios for water pollutants on basis of acute toxicity evaluation have been tentatively proposed and discussed (see Henderson, 1957; Warren and Doudoroff, 1958).

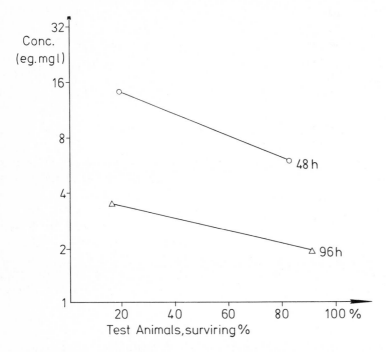

Fig. 9.4.   Interpretation of toxicity bioassay results.

A widely used method employs fractional application factors by which the LC_{50} values can be multiplied until a presumably safe concentration is arrived at. However, it would seem that no single application factor can be applied to all toxic materials. The National Technical Advisory Committee on Water Quality Requirements for Aquatic Life has tentatively grouped toxicants into three main categories according to their persistance (see Federal Water Pollution Control Administration, U.S., 1968). This solution to the problem must only be considered as provisional, however, and possible synergistic and antagonistic factors should always be taken into consideration.

The effects of the presence of two or more toxicants in the water being tested may be calculated by the following equation:

$$ST = \frac{C_1}{TL_{m_1}} + \frac{C_2}{TL_{m_2}} + \dots\dots\dots\dots\dots\dots \qquad (9.3)$$

where
C_1, C_2 .... = the actual concentrations
TL_{m_1}, TL_{m_2} .... = the corresponding threshold limits
ST = the sum of the effects

The value of ST, whether > or < than one, is determined.
This method can be applied, provided that no knowledge on synergism and anatgonism is available.

9.2.9. Measurement of taste and smell.

Direct measurement of material that produces tastes and odours can be made if the causative agents are known, for instance, by gas or liquid chromatography.
Quantitative tests that employ the human senses of taste ans smell can also be used, for instance, the test for the threshold odour number (TON). The amount of odorous water is varied and diluted with enough odour-free distilled water to make a 200 ml mixture. An assembled panel of 5-10 "noses" is used to determine the mixture in which the odour is just barely detectable to the sense of smell. The TON of that sample is then found, using the equation

$$TON = \frac{A + B}{B} \qquad (9.4)$$

where A is the volume of odorous water (ml) and B is the volume of odour-free water required to produce a 200 ml mixture.

9.2.10. Standards in quality control.

P.9.7. A standard is a definite rule or measure established by authority. It is official, but not necessarily rationally based on the best scientific knowledge and engineering practice. Standards can be applied either on receiving waters - rivers, lakes, estuaries, oceans or groundwater - or on the effluent from municipal, industrial and

agricultural pursuits.

Receiving water standards reflect the assimilative capacity of the water body receiving effluents.

Their purpose is to preserve the water body at a certain minimum quality. Effluent or emission standards are easy to administer, but do not take into account the most economic use of the assimilative capacity of receiving waters.

P.9.8. **Use of concentration effluent standards for a substance can easily be misleading since dilution can mask the true pollution level.**
It is more appropriate to limit the amount of pollutant, that can be discharged in a given time, e.g. kg per day, week or month.

A relationship between the two types of standards is difficult to estabilsh. It requires knowledge of the complex relationship between the discharge and the ultimate concentration in the receiving water, and since almost all discharged components are involved in many processes in the aquatic ecosystem, the problem must be solved by systems analysis (see the section on ecological models in 1.5).

Standards for receiving waters attempt *to protect* the water resource for uses such as *drinking, swimming, fishing, irrigation, etc.* However, standards should not be set at minimum levels, lests water quality be destroyed for future generation, but should prevent a polluter from profiting at the expense of neighbouring water users and compensate for differences in summer and winter and dry versus wet years, etc.

The use of water resources directly determines the standards applied to receiving waters and indirectly determinies those applied to effluent.

Table 9.10 illustrates how standards can be set for different water usage. Although it includes only a limited number of items, it may be considered representative of standards applied to fresh surface waters.

A major source of pollution is the municipal waste water. It is the responsibility of the municipality to construct facilities to meet the standards, but design and construction of a waste water treatment plant is only the first part of an extensive pollution abatement programme. The plant must be operated to meet the design objectives and, therefore, proper operation and maintenance will require additional funds. Furthermore, an effluent standard should be set for every plant to ensure it is functioning as represented to the taxpayers, and the key to this will be the monitoring of treated effluent.

TABLE 9.10
Examples of standards

Class	A	B	C	D
Best usage	Drinking water supply	Bathing	Fishing	Agricultural Industrial
Sewage or waste effluent	None, which is not effectively disinfected	None, which is not effectively disinfected		
pH	6.5 - 8.5	6.5 - 8.5	6.5 - 8.5	6.0 - 9.5
Dissolved oxygen	≥5.0 mg l⁻¹	≥5.0 mg l⁻¹	≥5.0 mg l⁻¹	≥3.0 mg l⁻¹

The other major source of pollution, industrial waste water, is another problem. Industries usually have little reason to devote money to waste treatment without tremendous pressure form all levels of government and the general population.

Finally funds must be used for monitoring and surveillance of standards of receiving waters in order to control the water quality and to assess the success of control methods.

9.3. EXAMINATION OF SOLIDS.

9.3.1. Pretreatment and digestion of solid material.

Procedures intended for the physical and chemical examination of waste waters and waters can also be applied in most cases to examination of soil, sludge, sediments and solid industrial waste, after suitable pretreatment. Standard Methods (1980) gives details on pretreatment procedures for sludge and sediments, and these are generally applicable with some minor modification, to all types of solid material. Some problems related to the examination of solid material should be mentioned, however:

1. *representative samples are very difficult to obtain,* but a fixed procedure cannot be laid down because of the great variety of conditions under which collection must be made. In general the sampling procedure must take into account the tests to be performed and the purpose for which the results are required,

2. *determination of components that are subject to significant* and *unavoidable changes on storage* cannot be made on composit samples, but

should be performed on individual samples *as soon as possible* after collection or preferable at the sampling point,

3. *preservation of samples is often difficult because most* preservatives interfere with some of the tests. Chemical preservation should be used *only* when it can be shown not to interfere with the examination being made. Storage at a low temperature (0-4°C) is often the only way of preserving samples overnight.

As mentioned, a pretreatment is usually necessary before the standard procedures used for water and waste water assessment can be used on solid material. Digestion of samples are necessary before determination of total concentration. The digestion procedure is dependent on the component to be analyzed, as it must be certain that the total amount is dissolved.

9.3.2. Examination of sludge.

P.9.9.   **The control of sewage treatment is dependent on tests on sludge. This control is of special importance in the activated sludge process, where the sludge parameters determine the operation of the plant.**

Tests for the various forms of residue determine classes of material with similar physical properties and similar responses to ignition. The following forms of residue are widely used in the examination of sludge, and also for soil samples:

1. **Total residue of evaporation.** The procedure (Standard Methods, 1980) is arbitrary and the result will generally not represent the weight of actual dissolved and suspended solids. A discussion of the temperature for drying residues can be found in Standard Methods (1980), but 103°C is generally applied.
2. **Total volatile and fixed residue** is determined by igniting the residue on evaporation at 550°C in an electric furnace to constant weight.
3. **Total suspended matter** is determined by filtration. The amount of suspended matter removed by a filter varies with the porosity of the filter.
4. **Volatile and fixed suspended matter** are determined by evaporation and ignition of the suspended matter obtained under point 3.
5. **Dissolved matter** is calculated from the difference between the residue on evaporation and the total suspended matter.
6. Matter may be reported on either a **volume or a weight basis.**

Additionally two indices are used to estimate the various forms of residue. The so-called **sludge volume index,** which is the volume in ml occupied by 1 g of activated sludge after the sludge has been allowed to settle for 30 minutes. **The sludge density index**, which is the reciprocal of the sludge volume index multiplied by 100.

9.3.3.   Examination of sediments.

P.9.10.   Analyses of sediments are widely used for environmental control, because sediment is able to accumulate various pollutants. Significantly higher concentrations of toxic organic compounds and heavy metals are found in the sediment than in the overlying water.

Therefore, *more accurate mapping* of the pollution is achieved by examinating the sediment than the water. Sediment is able to bind contaminants by adsorption, biological uptake of benthic organisms and chemical reactions, for each of which it is often important to find the binding capacity and its allocation, in order to predict the ecosystem's future reaction to pollution. However, this involves a rather time-consuming examination of the chemical and biological composition of the sediment. Kamp-Nielsen (1974) has described a quicker but empirical method for determining the nutrient binding capacity of sediment, which can be expanded slightly to cover toxic compounds and heavy metals.

Examination of sediment also has the advantage that sediment has *memory.* Analysis of a sediment core taking samples at various depth can give clues to previous conditions in the ecosystem by determination of the net settling rate expressed, e.g. in cm per annum.

Fig. 9.5 illustrates the results of such an analysis.

The digestion methods used on sludge, can also be applied to sediment. It is often necessary to use the most complicated digestion methods if determination of the total content is required. However, from an environmental management point of view, it is generally sufficient to use simpler digestion methods, as these give only slightly lower results. For example, to determine the heavy metal concentration in sediment it is often sufficient to use wet chemical digestion by nitric acid. The difference between this method and the use of sodium carbonate melting as pretreatment to an extraction with hydrogen fluoride and nitric acid is only minor and will always represent only very non-labile material (see Nordforsk, 1975).

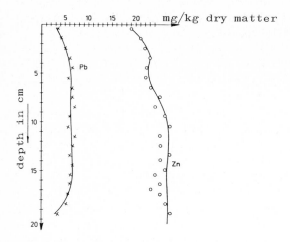

Fig. 9.5. Metal analysis of a core from Lake Glumsoe. Zinc and lead (mg/kg dry matter) are plotted against the depth.

9.4. EXAMINATION OF AIR POLLUTION.

9.4.1. Introduction.

The prevalent air pollution at a given time is a function of a number of factors, which may be categorized as follows:

1. Meteorological conditions
2. Source strength and position
3. Quantitative composition of source emission
4. Deposition and/or decomposition rate

P.9.11. A complete characterization of the total immission is practically impossible, and indeed any given air pollution description is only a reflection of the actual true immission,

i.e. it is limited in several ways with respect to number of analysed compounds, knowledge of statistical variation in time and space ans analytical problems of a prectical and/or theoretical nature. Consequently, it is necessary to define as precisely as possible the aim of the examination of the air. Is the purpose only to monitor the major pollutants in quantity or to gain knowledge about the distribution of these compounds with major adverse effects on human health, plants and animals and/or materials? In other words: choice of immission component and preliminary determinations of concentration range are prerequisite.

Table 9.11 lists the most commonly measured air pollutants and typical ranges in American and British cities. It is of interest that only a very limitied number of these compounds have not been present in the atmosphere during the period of mans presence on earth (compare with Table 9.12).

These few compounds originate mainly from the organic chemical industry and include halogenated aromatics (DDT, PCBs) and polyaromatics (PAH).

P.9.12. The present air pollution is more a matter of quantitative than qualitative change, and a number of living organisms have actually been shown to possess potential adaptivity towards this new situation.

Similarly, it is within reach of modern technology to produce less corrosive materials in order to prevent damage due to air pollution.

Man, however, is subject only to very slow evolutionary change, and will certainly not develop within only a couple of hundred years.

In conclusion, therefore, it makes sense to focus on protecting human health when considering air pollution control and indeed this is common practice.

TABLE 9.11
Most commonly measured Air Pollutants

A. Composition of the atmosphere (Bowen, 1966)

Compounds	$\mu g/m^3$ STP	Residence time (year)
CH_4	850 - 1100	100
CO	1 - 20	0.3
CO_2	600,000	4
H_2	36 - 90	
H_2O	$3 * 10^4 - 3 * 10^7$	0.027
H_2S	3 - 30	0.11
O_2	$2.99 * 10^{10}$	
O_3	0 - 100	2
N_2	$9.73 * 10^{10}$	
NH_3	0 - 15	short
N_2O	500 - 1200	4
NO_2	0 - 6	short
SO_2	0 - 50	0.014

TABLE 9.11 (continued)

B. Elementary composition of smoke in $\mu g/m^3$ polluted air (Bowen, 1966)

Element	N. American cities	British cities
A l	3 - 4	
As	0.01 - 0.02	0.01 - 0.2
Be	0.0001 - 0.0003	0.0001 - 0.001
Ca	2 - 16	
Co		0.0007 - 0.004
Cr		0.002 - 0.02
Cu	0.05 - 0.9	0.02 - 0.25
F	0.01 - 0.4	
Fe	3 - 15	
Mg	1 - 7	
Mn	0.1 - 0.3	0.01 - 0.1
Mo		0.0005 - 0.006
Ni		0.002 - 0.2
Pb	0.5 - 3	0.2 - 1.4
S	1 - 8	
Sb		0.004 - 0.25
Si	4 - 6	
Sn	0.01 - 0.03	
T i	0.04 - 1	0.01 - 0.2
V	0.001 - 0.1	0.001 - 0.04
Zn	0.2 - 2	0.07 - 0.5

TABLE 9.12
Composition of "clean" Air (Fenger, 1979)

Component	Concentration	
N_2	78.084	%
O_2	20.949	%
A r	0.934	%
Ne	18.2	ppm
He	5.2	ppm
K r	1.1	ppm
H_2	0.5	ppm
N_2O	0.3	ppm
Xe	0.09	ppm
CO_2	320	ppm
CH_4	1.5	ppm
CO	0.1	ppm
O_3	0.02	ppm
NH_3	0.01	ppm
NO_2	0.001	ppm
SO_2	0.0002	ppm
H_2S	0.0002	ppm
H_2O	0 - 7	%

Finally, two important aspects of the impact of air pollution on human health must be stressed. First is the traditional separation *between outdoor and indoor ambient air quality,* which has different acceptable limit values for pollutants.

Secondly, a distinction between indirect and direct effects has to be made; the current anxiety about heavy metals is partly due to direct effects (e.g. organic lead compounds in the city atmosphere), partly to accumulation with time in solids, plants and animals, which may result in elevated levels in foodstuffs. Equally, sulphur dioxide in combination with suspended particulate matter is directly toxic to humans at rather high concentrations; but the long-term indirect effects of this compound are also a matter of concern, as decreasing pH and increasing leaching of inorganic components of the soil - some being rather toxic, e.g. Al compounds - are detrimental to fresh- and groundwater systems. Bearing the above introductory statements in mind, the aim of an framework for air pollution examination should be clearer.

9.4.2. Meteorological conditions.

P.9.13. **A detailed knowledge of the meteorological conditions is necessary in order to employ any emission measurement for environmental protection, because transportation from any source to the receptor is a function of meteorological variables. Furthermore, under certain meteorological circumstances, a build-up of pollutants take place, even if the source strength remains constant. (Principles 2.1 and 2.6 are used).**

The dispersion of air pollution from a point source as a function of different tropospheric lapse rate conditions (Fig. 8.2) can be represented by aplume from one chimney, see also section 2.11.

Fig. 9.6 shows the relative frequency of the different conditions in an open country, such as Denmark. Several larger cities of the world apply an early warning system involving meteorological data describing the lapse rat e conditions. In Denmark, this has not been necessary. The meteorological situation giving the highest ground level concentrations is inversion (see also 8.2.4). Inversions may occur for a number of reasons, the most important being:
1. (Fig. 9.7) *strong cooling of the soil/concrete surface overnight* (clear sky), which results in the formation of a very sable layer close to the ground; during the day this inversion is often eliminated as the sun heats up the atmosphere.

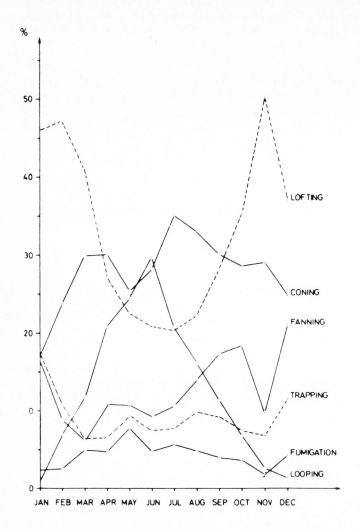

%

50

40

LOFTING

30

CONING

FANNING

20

TRAPPING

0

FUMIGATION

LOOPING

0

JAN FEB MAR APR MAY JUN JUL AUG SEP OCT NOV DEC

Fig. 9.6. Frequency of types of stack-smoke behaviour (1962-1964) for a 60 meter stack. (Perkins, 1974) .

2. (Fig. 9.7) *subsidence inversion.* During high pressure situations, cold air with high density subsides and forms stable layer close to the ground with a rather small mixing height.
3. *Front inversion.* During the pregression of a front, the cold surface atmospheric layer is gradually covered or replaced by warmer air, and consequently the mixing height is reduced.
4. *Sea-land breeze inversions* (Fig. 9.8). Cities close to the sea are subject

to a special phenomenon due to the time lag difference in heating and cooling the concrete of a city compared wiht water. The net results is a seaward breeze at night and a land-ward breeze during day, the effect being most obvious on sunny days in the midafternoon.

These inversion situations may be modified by environmental factors, such as the presence of large amounts of concrete in a city resulting in a local air circulation system as shown in Fig. 9.9.

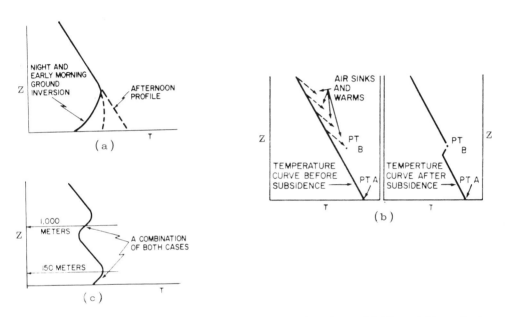

Fig. 9.7. Temperature inversions. (a) During the day. (b) The sinking of air leads to warming aloft, the formations of inversions. (c) A combination of cases. (Perkins, 1974).

As industrial areas are often situated in the city periphery the inflow of air from here may increase the immission of some pollutants in the city centre.

In principle, the immission around a point source may be calculated on the basis of source strength, emission characteristics (such as stack height, smoke temperature, etc.) and meteorological parameters (see section 2.11). Even larger areas, like cities, may be treated in a similar way, but the presence of a large number of sources (point-, line- and area sources) defined by the following examples:

    point source: a chimney, a vehicle,
    line soruce: a highway,
    area source: a city, an industrial area,

make it necessary to rely upon rather more empirical relationships, and the data therefore become less reliable. At present emission-based model calculations have not been able to replace field measurements of immission, and a combination of the two approaches will probably be a must for a long time yet.

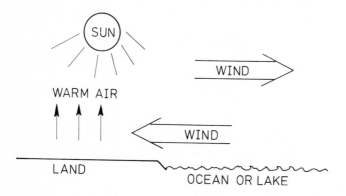

Fig. 9.8.   Sea breeze during the day. (Wright and Sjessing, 1976).

Before turning to a description of the most common air pollutants and the examination of their occurrence, it is relevant to stress a final point related to the aim of monitoring programmes.

P.9.14.   **It is generally not possible to conclude anything definitive with respect to harmful effects on living organisms due to immission exposure. Of course, the bases for limit values are controlled experiments, but their results may only rarely be translated directly to field conditions.**

This is because:
1. *organisms exhibit a complex dose-response relationship* (Fig. 9.10), see also section 2.13,
2. *synergistic/antagonistic effects are common.*

    In conclusion, this points to the necessity of introducing effect monitoring using sensitive indicator organisms as a part of any environmental surveillance programme.

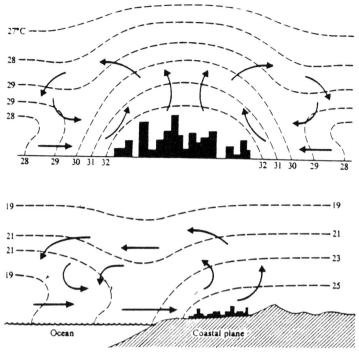

Fig. 9.9. Examples of nonhomogenous = nonstationary wind patterns around cities and near coastlines during the day. (Rohde, 1978).

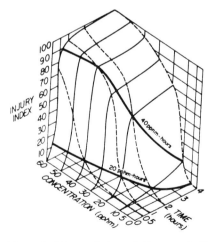

Fig. 9.10. Interrelations of time and concentration on the sensitivity of pinto bean plants to ozone (Stern, 1979).

9.4.3. Gaseous air pollutants.

SO_2 is one of the major gaseous air pollutants emitted from power plants and industries. Some of the strongest air pollutions have been recorded around, e.g. Zn and Cd smelters, when sulphide ores are roasted to produce these metals. The results were heavy emissions of SO_2 that devastated the environment, in particular the growth of trees and herbaceous plants. Fortunately, such strong emissions occur in only a few places; a larger problem occurs with the SO_2 produced during combustion of sulphur-containing fossil fuel, see also 8.4.1 and 4.7. This results in local as well as regional pollution problems.

Very often, the SO_2 is emitted together with particulates, and this combination of suspended matter and SO_2 may cause health problems in cities and industrial areas. SO_2 alone is not believed to be very harmful to man at its actual ambient levels. Regional pollution problems due to SO_2 are mainly a result of its acidifying effects (see section 4.5).

However, the acidification is also due to the presence of NO_2 and acidic rain, which reacts with water as follows

$$2NO_2 + H_2O \rightarrow NO_3^- + NO_2^- + 2H^+.$$

It is believed that the contribution to the total acidification of precipitation from NO_2 will increase during future decades as the immission of NO_2 increases at present, the major part of the pH decrease is attributed to SO_2.

The measurement of SO_2 and SPM (Suspended Particulate Matter) in the atmosphere is often performed with a so-called "OECD-instrument". Several other methods are available, and some of them permit an instantaneous reading, an important facility, when describing the immission in an area subject to very strong fluctuations with problematically high short-time peak values.

The design of the monitoring network and choice of equipment depend on the following factors.

First, effective design can only be made on the basis of some **prior knowledge** about the distribution of the pollutant within the area. The statistician needs to know pilot values about the variation in time and space in order to calculate the necessary number of measuring points. Often, however, practical and economic reasons limit the possibility of setting up an optimal network.

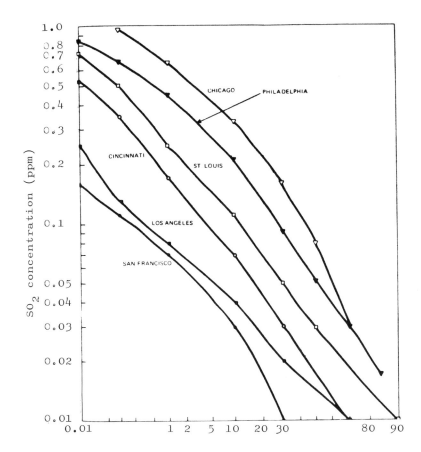

Fig. 9.11. Frequency distribution of sulphur dioxide levels in selected American cities. 1962 to 1967 (1-hour averaging time). The approximate log normality of the distribution of sulphur dioxide concentrations is shown by the rather straight lines of the distribution functions when these are plotted on a logratihmic scale (concentrations) against a normal frequency. (Perkins, 1974).

Second, as regards **choice of equipment**, the kind of data required by the administration limit the number of feasible possibilities. Typically, a set of data for management purposes is defined in relation to the legislation, which may mean definitive averaging times, concentration ranges and analytical variation, the reason being that *peak values* and their frequency of occurrence are extremely important in relation to human health and plant damage. In order to estimate and thus control the peak value frequency, the data distribution must be known - very often it is lognormal (Fig. 9.11) - and a limit guideline may be a statement about a few maximum values not to be exceeded more than a given percentage of the total period in question. Table 9.13 provides an example of limit values for SO_2.

NO/NO_2 has already been dealt with in relation to acid precipitation. The term NO_x should be avoided, as it is often misunderstood.

TABLE 9.13
Limit values for sulphur dioxide and suspended particulate matter (SPM) (Measured by the OECD method).
Limit values of sulphur dioxide expressed as $\mu m \ m^{-3}$ together with the corresponding values for SPM expressed as $\mu m \ m^{-3}$

Period	Limit value for SO_2	Corresponding value for SPM
Year	Median of diurnal averages over one year 80 120	> 40 ≥ 40
Winter Oct.1 - Mar.1	Median of diurnal averages over one winter 130 180	> 60 ≥ 60
Year divided in periods of 24 hours	98-percentile of all diurnal averages over one year 250 350	> 150 ≥ 150

N_2O, NO, NO_2 occur commonly in the atmosphere as they are formed by denitrifying bacteria, but it is inert. NO_2 is a stronger toxic agent than NO, but even NO possisses phytotoxic properties. The analytical procedure for measuring nitrogenous oxides often involves oxidation of NO to NO_2. The NO level may thus only be determined by subtraction. The trend in present administrative practice is specially to writ NO_2 or NO. The formation of NO/NO_2 takes place at high temperatures as a result of the reaction between atmospheric nitrogen and oxygen $N_2 + O_2 - 2NO$.

NO is slowly oxidized to NO_2, and the ratio NO_2/NO is always lower close to the source, e.g. in city centres, than in the rural surroundings. The oxidation of NO involves the release of oxygen ions, O^-, which may react with an oxygen molecule O_2 to form O_3 ($NO + O_2 - NO_2 + O^-$; $O^- + O_2 - O_3$).

TABLE 9.14
A photochemical model

NO$_x$ cycle	1. $NO_2 + h\nu - NO + O$	
	2. $O + O_2 + M - O_3 + M$	
	3. $O_3 + NO \ ^- NO_2 + O_2$	
	4. $O_3 + NO_2 - NO_3 + O_2$	produces
	5. $NO_3 + NO_2 \ \overline{\ H_2O\ } \ 2HNO_3$	nitric acid
	6. $NO + NO_2 \ \overline{\ H_2O\ } \ 2HNO_2$	nitrous acid
	7. $HNO_2 + h\nu - NO + OH$	
	8. $CO + OH \ \overline{\ O_2\ } \ CO_2 + HO_2$	CO produces an OH chain
	9. $HO_2 + NO - NO_2 + OH$	
	10. $HC + O - aRO_2{}^*$	
	11. $HC + O_3 - bRO_2{}^* + cRCHO$	aldehyde and peroxy radicals
	12. $HC + OH - dRO_2{}^* + eRCHO$	
	13. $HC + RO_2 - fRO_2{}^* + gRCHO$	
	14. $RO_2 + NO - NO_2 + hOH$	oxidation of NO to NO$_2$
	15. $RO_2 + NO_2 - PAN$	production of peroxy-acetyl nitrate

HC = hydrocarbon   a - h stoichiometric coefficients
RO$_2{}^*$ = peroxy radical   M = catalyst

Source: Seinfeld, 1971. See also T.A. Hecht adn J.H. Seinfeld, Environ. Sci. Technol. vol. 6, pp. 47-57, 1972; R.G. Lamb and J.H. Seinfeld, Environ. Sci. Technol. vol. 7, pp. 253-261, 1973.

A large number of photochemical reactions may occur (Table 9.14) in the presence of hydrocarbons, which act deirectly and catalytically, resulting in the so-called photochemical complex with O_3 and PAN/PPN (see also section

5.7). So, NO_2/NO pollution should not only be considered in isolation, but also as part of the photochemical smog formation, for which nitric oxides are precursors.The measurement of NO_2 is usually based on chemiluminescence, and may be made with very short averaging times - half an hour generally being pre- ferred. The measurement of O_3 is also based on chemiluminescence and may be done routinely.

The analysis of PAN/PPN, however, is more complicated and involves sophisticated gas chromatographic techniques.

9.4.4. Particulate matter.

Suspended particulate matter (SPM) has been measured for decades in heavily polluted areas of Europe and North America. SPM may be separated in two fractions, *smoke and fly ash,* both comprising paritcles less than 10 μ in aerodynamic diameter and therefore not significantly influenced by gravitational forces. *Soot is black,* but smoke be a variety of colours depending on the industrial processes involved in its formation. The reflectrometric method of analysing SPM only gives a measure of the soot component + the black smoke fraction. It is preferable simply to weigh the collecting filters before and after the passage of a given volume of air.

The filters may later be analysed for their content of, for example, heavy metals, using AAS (Atomic Absorption Spectrophotometry) or PIXE (Pictum Induced X-ray Emission spectrophotometry). The major soruces of SPM in cities are traffic exhaust and to a lesser extent point sources, like power plants. Car exhausts contain particles of very small (< 0.5 μ) size, composed mainly of lead halogenides. Thus a 1:2 mole ratio between Pb and Zn is found close with distance from the source. Only a small proportion less than 1/4 of the total emission of Pb-containing particles is deposited in the vicinity (± 200 m) of highways etc. The remainder is dispersed and contributes to the common rate of ~ 10 mg $m^{-2}y^{-1}$ of Pb in Europe, for example.

The composition of *small particles emitted from oil-fired* power plants is characterized by relatively high concentrations of V and Ni, because these metals accumulate in oil deposits, bound to porphyrin systems from the degraded chlorophyll of carbonaceous plants. The presence of such metal oxides in these particles has a catralytic effect on the oxidation of SO_2 to SO_3, resulting in the production of very acidic particles small enough to be transported deep into the respiratory tract causing health problems.

Soil dust, occasionally transported over large areas, is characterized by a high content of relatively harmless metals, such as Ti, Fe and Mn. Thus, the composition of airborne particles is important when considering emission reduction and immission guidelines.

Sedimentary dust consists of particles larger than 10 μ in diameter that remain in the atmosphere only for short periods, i.e. have residence times of less than 1 hour. Air pollution due to these rather large particles is thus restricted to the immediate surroundings of its factory or power plant source. The adverse health effects from sedimentary dust are negligible, as most of the dust does not reach the respiratory tract and if it does, the particles are retained in the nose or mouth. Only toxic, soluble compounds may then enter the bloodstream and result in health problems.

One of the problems of sedimentary dust concerns its light-inhibiting properies when it covers surfaces, such as glass or plants leaves. Very alkaline or acidic dust particles may have a strong corrosive effect. This type of air pollutant is measured using funnels placed in a network around the factory or in a selected investigation area. It is generally not considered a major air pollution problem, as its control is rather simple and effective.

HF/F^- are mainly emitted from brick, glass, porcelain and fertilizers factories and the impact on the environment is generally restricted to the immediate neighbourhood of the source.

Fluorides are very soluble in water and thus easily washed off and attached to particles with a surface water film. The emission of fluorides is thus controlled by deposition measurements close to the source. Deposition on plants has led to local cattle fluorosis.

9.4.5. Heavy metals.

When dealing with air pollution due to heavy metals, *the complexity* of *the chemistry of these elements must be borne in mind.* Lead provides a good example (see Table 9.15).

P.9.15. **The different chemical compounds have different toxic properties and it makes no sense to reduce the immission exclusively on a quantitative basis simply by reducing the level of the most common compounds in the atmosphere.**

Generally, the heavy metals are attached to particles or form particles with a few anions (SO_4^{2-}, X^-), mercury being an important exception.

Emission and re-emission form soil of gaseous Hg and alkyl-Hg occur, and re-emission probably also occurs to a smaller degree with a few other heavy metals, such as Cd.

Heavy metals immission is determined by *filter analysis* following passage of a known volume of air through the filter. Separation into particle size groups is often made, e.g. between paricles less than 2.5 μ and between

2.5 and 20 μ.

Lead halogenides are confined to small particles, 60-70 per cent being smaller than 2 μ.

TABLE 9.15
Lead compounds as airpollutants

$Pb(C_2H_5)_4$	$PbCl_2$
$Pb(CH_3)_4$	PbO
$Pb(NH_3)_4Cl_2$	$PbCO_3$
$Pb(NH_3)_4Br_2$	$Pb_3(OH)_2(CO_3)_2$
$PbBr_2$	$PbSO_4$

9.4.6. Hydrocarbons and carbon monoxide.

Hydrocarbons are present in high concentrations in city atmospheres, originating mainly from car exhausts. *The complexity of the hydrocarbon immission is illustrated in Fig. 9.12.* By far the most common hydrocarbon is methane (CH_4), which usually constitutes more than 80 per cent of the total hydrocarbon content of the atmosphere. The hydrocarbons are either formed during combustion, being more or less completely oxidized, or simply evaporated into the surrounding air from cars (crank case, carburetor and petrol tank).

P.9.16. The hydrocarbons in city atmosphere represent a rather large health problem, and several are believed to be carcinogenic.

Due to the complexity of hydrocarbon immission, its control has been limited in most areas; instead, efforts have been concentrated on reducing emission, and cars have been fitted with a veriety of devices to improve their combustion efficiency.

This has also resulted in a reduction in CO emission. Carbon monoxide is frequently measured in city atmosphere because of its potentially hazardous effect of inhibiting the blood's oxygen uptake. Carbon monoxide quickly becomes well dispersed in the air because its relative density is very close to the average air density (28 and 29). It has practically no effect on plants, however, partly because of its very low solubility in water.

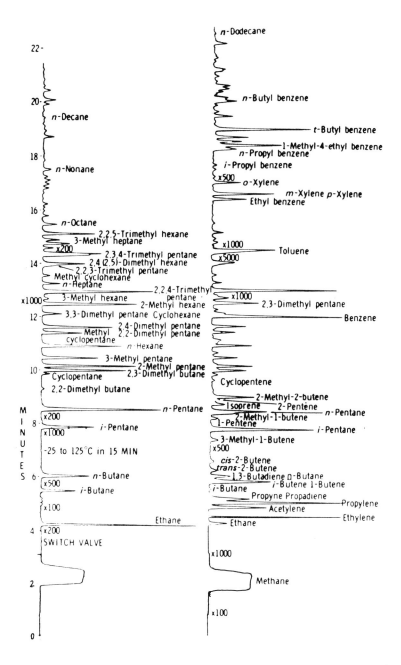

Fig. 9.12. Gas chromatograph showing products of combustion in automobile engine. (Hafstad, 1969).

TABLE 9.16
Ambient air quality standards

TABLE 9.16

Ambient air quality standards

Pollutant	Averaging time	California standards Concentration[g]	Method[a]	Federal standards[d] Primary[b,g]	Secondary[c,g]	Method[e]
Photochemical oxidants (corrected for NO_2)	1 hour	0.10 ppm (200 $\mu g/m^3$)	Neutral buffered KI	160 $\mu g/m^{3h}$	Same as primary standards	Chemiluminescent method
Carbon monoxide	12 hours	10 ppm (11 mg/m^3)	Nondispersive infrared spectroscopy		Same as primary standards	Nondispersive infrared spectroscopy
	8 hours			10 mg/m^3 (9 ppm)		
	1 hour	40 ppm (46 mg/m^3)		40 mg/m^3 (35 ppm)		
Nitrogen dioxide	Annual average	100 $\mu g/m^3$ (0.05 ppm)	Saltzman method	100 $\mu g/m^3$ (0.05 ppm)	Same as primary standards	Colorimetric method using NaOH
	1 hour	0.25 ppm (470 $\mu g/m^3$)				
Sulphur dioxide	Annual average			80 $\mu g/m^3$ (0.3 ppm)	60 $\mu g/m^{3i}$ (0.02 ppm)	
	24 hours	0.04 ppm (105 $\mu g/m^3$)	Conductimetric method	365 $\mu g/m^3$ (0.14 ppm)	260 $\mu g/m^3$ (0.10 ppm)	Pararosaniline method
	3 hours				1300 $\mu g/m^3$ (0.5 ppm)	
	1 hour	0.5 ppm (1310 $\mu g/m^3$)				
Suspended particulate matter	Annual geometric mean	60 $\mu g/m^3$	High volume sampling	75 $\mu g/m^3$	60 $\mu g/m^3$	High volume sampling
	24 hours	100 $\mu g/m^3$		260 $\mu g/m^3$	150 $\mu g/m^3$	
Lead (particulate)	30 day average	1.6 $\mu g/m^3$	High volume sampling, dithizone method			
Hydrogen sulphide	1 hour	0.03 ppm (42 $\mu g/m^3$)	Cadmium hydroxide STRactan method			
Hydrocarbons (corrected for methane)	3 hours (6-9 am)			160 $\mu g/m^3$ (0.24 ppm)	Same as primary standards	Flame ionization detection using gas chromatography
Visibility reducing particles	1 observation	In sufficient amount to reduce the prevailing visibility[f] to 10 miles when the relative humidity is less than 70%				

Notes:
a. Any equivalent procedure which can be shown to the satisfaction of the Air Resources Board to give equivalent results at or near the level of the air quality standard may be used.
b. National Primary Standards: The levels of air quality necessary, with an adequate margin of safety, to protect the public health. Each state must attain the primary standards no later than three years after that state's implementation plan is approved by the Environmental Protection Agency (EPA).
c. National Secondary Standards: The levels of air quality necessary to protect the public wellfare from any known or anticipated adverse effects of pollutant. Each state must attain the secondary standards within a "reasonable time" after implementation plan is approved by the EPA.
d. Federal Standards, other than those based on annual averages or annual geometric means, are not to be exceeded more than once per year.
e. Reference method as described by the EPA. An "equivalent method" of measurement may be used but must have a "consistent relationship to the reference method" to be approved by the EPA.
f. Prevailing visibility is defined as the greatest visibility which is attained or surpassed around at least half of the horizon circle, but not necessarily in continuous sectors.
g. Concentration expressed first in units in which it was promulgated. Equivalent units given in parantheses are based upon a reference temperature of 25°C and a reference pressure of 760 mm of Hg.
h. Corrected for SO_2 in addition to NO_2.
i. Revoked in 1973.

9.4.7. Biological monitoring.

P.9.17. **Biological monitoring involves several different approaches, but aims always to evaluate the impact of
pollutant on living systems by observing changes in the
systems themselves after exposure to the immission.**

The impact recorded is most often *an adverse effect,* but may also be
the *accumulation of heavy metals,* for example. The following paragraphs
discuss some typical examples of biological monitoring.

Epiphytic lichens have been used to reflect levels of a number of particularly gaseous pollutants, such as SO_2 and HF/F^-. A number of cities in
Europe and North America have been extensively studied with respect to
lichen growth on trees. The general pattern is a *decrease in the frequency* of
occurrence of most species with proximity to the city centre, with a few
species showing the opposite trend. The correlation between species
distribution and levels of SO_2 is often very good, and a number of laboratory
experiments have shown *a causal relationship between SO_2 levels and lichen
injury/performance;* the species sensitivity sequence observed in the
laboratory closely follows the field observations.

The presence of suspended particulate matter in the city atmosphere
does not in itself contribute significantly to a reduction in lichen growth. On
the contrary, the predominance of oxides in the particles may result in an
alkaline reaction when suspended in water; the presence of SO_2, however,
more than neutralizes this effect, and the lichen substratum, the tree bark,
becomes more acidic the closer it is to the city centre. Of course, this effect
contributes indirectly to the overall change in population distribution.

In relation to biological monitoring, a number of lichen species can be
used as indicators to estimate SO_2 levels; if a certain species occurs in the
investigation area, the SO_2 immission cannot exceed the critical value for
that species. An example is given in Fig. 9.13 A-E.

Several questions arise when relating species distribution to ambient
SO_2 levels, namely how specific is the reaction; are other pollutants present
that may also adversely affect the lichens? The question of SPM has already
been discussed, and it may safely be anticipated that the above-mentioned
indirect effects of SPM generally correlate closely with SO_2 immission.

With regard to NO_2/NO, these pollutants are known to be much less toxic
to plants in general than SO_2 and at low levels even beneficial. Hydrogen fluoride and fluorides, however, are as or more toxic towards lichens than SO_2.

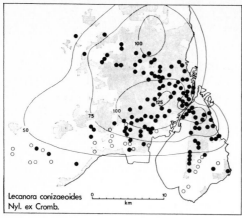

A Lecanora conizaeoides
Nyl. ex Cromb.

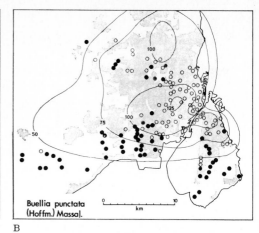

B Buellia punctata
(Hoffm.) Massal.

C Xanthoria parietina
(L.) Th. Fr.

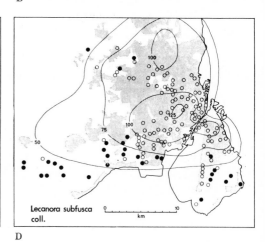

D Lecanora subfusca
coll.

E Physcia pulverulenta
(Schreb.) Hampe

Fig. 9.13. The distribution of the indicator lichen species in the investigation area. o Sampling station. • The species present above 30 cm from the ground. O The species present only below 30 cm from the ground. (Johnsen and Søchting, 1973).

In areas, where HF/F⁻ are major components of the immission, **a biological index** has been developed analogous to the one described for SO_2. Fortunately, however, HF/F⁻ only play a very minor role in the city atmospheres of Europe and North America. In conclusion, epiphytic lichen distributions may be regarded as an overall reaction to the SO_2-immission, and so their reflect to SO_2 levels is fully justified. It must be emphasized, however, *that any biological effect index can never be regarded* as being 100 percent specific, *unlike physicochemical measurements.* Table 9.17 outlines the major advantages and disadvantages connected with the two methods.

Transplantation is another method of biological monitoring. This method has the great advantage that the investigator is not concerned with the occurrence of trees when planning the study.

TABLE 9.17
Comparison of biological and physicochemical monitoring

	Advantages	Disadvantages
Biological monitoring	Biological effects of pollutants in the actual environment are recorded Synergism/antagonism is detectable	Non-specific, essentially Relation to ambient concentration values complicated
Physicochemical monitoring	Specific Pollutant concentrations are measured	No effects recorded Synergism/antagonism not detectable

This means that the injury data resulting from a transplantation experiment are generally easier to subject to statistical analysis, as the uniformity of the stations, their distribution and species composition can be determined beforehand and thus optimized. It is probably in the field of transplantation techniques that biological monitoring is progressing best at present.

Transplantation may comprise very different plant groups, and a few examples are given below:

1. Epiphytic lichens

Hypogymnia physodes has been transplanted to around industrial areas or into cities in order ot evaluate the average SO_2 immission (Fig. 9.14). The plants used have been taken from trees either by cutting out bark discs or using of Abies or Larix species. The injury index is based on the bleaching effect of SO_2, which will kill the lichen algal component before the lichen fungus. Using the method for different periods of time may be complicated by the difference in weather conditions. These differences may interfere strongly with the responese in lichens, and when comparing two time periods precautions have to be taken.

2. Higher plants
a. O_3-monitoring

Tobacco plants of the variety Bec W3 in particular are very sensitive to O_3, and when exposed to levels that, in general, would not result in crop damage, a characteristic pattern of spots may occur on the leaves (Fig. 9.15). These spots are caused by dieback of regions of teh mesophyll adjacent to, and the palisade tissue above, the stomata following diffusion of O_3 into the leaf.

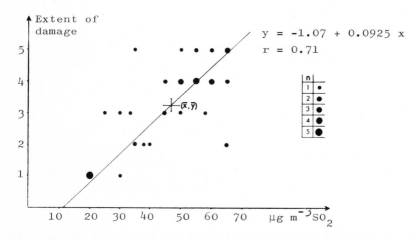

Fig. 9.14. The correlation between SO_2 levels at transplanation sites and the extent of transplant thallus damage. $r \neq 0$ at $p = 0.01$. n = number of coinciding dots.

Recording of the total area of the leaf that has been injured gives a semiquantitative measure of the O_3 immission, because the reaction is rather specific. This method has been used extensively in USA, UK and the

Netherlands. The plants may be grown on waste land or in so-called open-top chambers, the last method giving the most easily compared data, because the stations become more uniform.

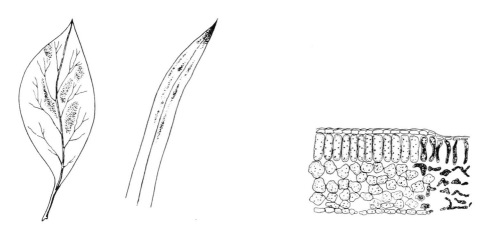

Fig. 9.15A. Sulphur dioxide injury. Leaf necrosis in dicotyledons occur intercostally, in monocotyledons as necrotic stripes, particularly near the tip. Right: Transect of healthy and injuried leaf part. (Garber, 1967).

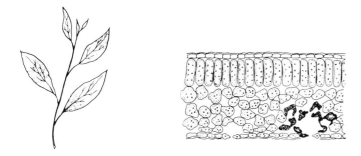

Fig. 9.15B. Smog-injury. In old leaves the leaf base is injured, in young leaves the tip. Transect: The injury begins in the cells around the stomata (Garber, 1967).

b. HF/F⁻

Accumulation of fluorides in beaves of higher plants leads to a typical picture of injury: dead whitish tissue surrounded by black, dead bands (Fig. 9.16); in monocotyledons, the dark bands occur close to the leaf tips.

Gladiolus gandavensis has been used to monitor fluorides, and the degree of damage, e.g. measured as the length of the dead (necrotis) area, correlates well with the deposition of fluorides.

Finally, one biological assessment method, that involves a more sophisticated equipment and a larger time scale, is ecosystem monitoring. The use of whole ecosystems, which are limited in some way, e.g. watersheds and lakes, or form a minor part of large, homogeneous systems have the advantage that accumulation rates may be compared with subtle effects at a very early stage.

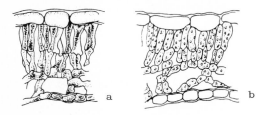

Fig. 9.16A.   a) Transect section of a Prunus leaf injuried by fluorides.   b) Normally developed transect section of a Prunus leaf. (Garber, 1967)

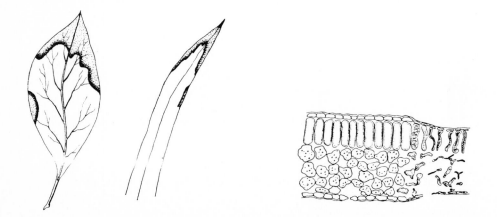

Fig. 9.16B.   Fluoride injury in decotyldons and monocotyledons. Typical leaf edge and "tip burn" necrosis. In the leaf transect the tissue destruction is clearly visible. (Garber, 1967)

A net of such monitor systems have been developed or are under development in a few countries and will meet the requirements for long-term surveillance of trends in the environment that at present may seem insignificant; typical examples are heavy metal deposition and the slowly increasing levels of organic pollutants in the atmosphere.

APPENDIXES

APPENDIX 1

A. Composition of the earth (total)

	Weight (%)
Fe	34.63
O	29.53
Si	15.20
Mg	12.70
Ni	2.39
S	1.93
Ca	1.13
Al	1.09
Na	0.57
Cr	0.29
Mn	0.22
Co	0.13
P	0.10
K	0.07
Ti	0.05
	100.00

B. Background concentration: Abundance in average crustal rock

Item	Value (% by weight)
Ag	0.000007
Al	8.1
Au	0.000004
Co	0.0025
Cr	0.010
Cu	0.0055
Fe	5.0
Hg	0.000008
Mg	2.1
Mn	0.10
Mo	0.00015
Ni	0.0075
Pb	0.0013
Pt	0.000001
Sb	0.00002
Sn	0.00020
Ti	0.44
U	0.00018
V	0.014
W	0.00015
Zn	0.0070

C. Background concentration: Biosphere

Item	Value (moles/hectare)
Al	4.12
C	6560
Ca	18.9
Cl	2.80
Fe	1.38
H	13,100
K	11.7
Mg	8.18
Mn	0.765
N	71.6
Na	1.65
O	6540
P	3.35
S	4.44
Si	8.62

D. Composition of the hydrosphere

element	mg/l	tons	element	mg/l	tons
Cl	19,000.0	$29.3 * 10^{15}$	V	0.002	$3.0 * 10^{9}$
Na	10,500.0	$16.3 * $ -	Mn	0.002	$3.0 * $ -
Mg	1,350.00	$2.1 * $ -	Ti	0.001	$1.5 * $ -
S	885.0	$1.4 * $ -	Sb	0.0005	$0.8 * $ -
Ca	400.0	$0.6 * $ -	Co	0.0005	$0.8 * $ -
K	380.0	$0.6 * $ -	Cs	0.0005	$0.8 * $ -
Br	65.0	$0.1 * $ -	Ce	0.0004	$0.6 * $ -
C	28.0	$0.04 * $ -	Y	0.0003	$0.5 * $ -
Sr	8.0	$12.0 * 10^{12}$	Ag	0.0003	$0.5 * $ -
B	4.6	$7.1 * $ -	La	0.0003	$0.5 * $ -
Si	3.0	$4.7 * $ -	Kr	0.0003	$0.5 * $ -
F	1.3	$2.0 * $ -	No	0.0001	$15.0 * 10^{7}$
A	0.6	$0.93 * $ -	Cd	0.0001	$15.0 * $ -
N	0.5	$0.78 * $ -	W	0.0001	$15.0 * $ -
Li	0.17	$0.26 * $ -	Xe	0.0001	$15.0 * $ -
Rb	0.12	$0.19 * $ -	Ge	0.00007	$11.0 * $ -
P	0.07	$0.11 * $ -	Cr	0.00005	$7.8 * $ -
I	0.06	$93.0 * 10^{9}$	Th	0.00005	$7.8 * $ -
Ba	0.03	$47.0 * $ -	Sc	0.00004	$6.2 * $ -
In	0.02	$31.0 * $ -	Pb	0.00003	$4.6 * $ -
Zn	0.01	$16.0 * $ -	Hg	0.00003	$4.6 * $ -
Fe	0.01	$16.0 * $ -	Ga	0.00003	$4.6 * $ -
Al	0.01	$16.0 * $ -	Bi	0.00002	$3.1 * $ -
Mo	0.01	$16.0 * $ -	Nb	0.00001	$1.5 * $ -
Se	0.004	$6.0 * $ -	Tl	0.00001	$1.5 * $ -
Sn	0.003	$5.0 * $ -	He	0.000005	$0.8 * $ -
Cu	0.003	$5.0 * $ -	Au	0.000004	$0.6 * $ -
As	0.003	$5.0 * $ -	Pa	$2 * 10^{-9}$	3000
U	0.003	$5.0 * $ -	Ra	$1 * 10^{-10}$	150
Ni	0.002	$3.0 * $ -	Rn	$0.6 * 10^{-15}$	$1 * 10^{-3}$

E. Composition of the atmosphere

	Volume (ppm)	Weight (ppm)	Mass $* 10^{20}$ g
N_2	780,900	755,100	38,648
O_2	209,500	231,500	11,841
A	9,300	12,800	0.665
CO_2	300	460	0.0233
Ne	18	12.5	0.000636
He	5.2	0.72	0.000037
CH_4	1.5	0.9	0.000043
Kr	1	2.9	0.000146
N_2O	0.5	0.8	0.000040
H_2	0.5	0.03	0.000002
O_3	0.4	0.6	0.000031
Xe	0.08	0.36	0.000018

F. The atomic composition of the four spheres

Element	Atoms % in		(v.l. = very low)	
Element	Biosphere	Lithosphere	Hydrosphere	Atmosphere
H	49.8	2.92	66.4	v.l.
O	24.9	60.4	33	21
C	24.9	0.16	0.0014	0.03
N	0.27	v.l.	v.l.	
Ca	0.073	1.88	0.006	v.l.
K	0.046	1.37	0.006	v.l.
Si	0.033	20.5	v.l.	v.l.
Mg	0.031	1.77	0.034	v.l.
P	0.030	0.08	v.l.	v.l.
S	0.017	0.04	0.017	v.l.
Al	0.016	6.2	v.l.	v.l.
Na	v.l.	2.49	0.28	v.l.
Fe	v.l.	1.90	v.l.	v.l.
Ti	v.l.	0.27	v.l.	v.l.
Cl	v.l.	v.l.	0.33	v.l.
B	v.l.	v.l.	0.0002	v.l.
Ar	v.l.	v.l.	v.l.	0.93
Ne	v.l.	v.l.	v.l.	0.0018

APPENDIX 2

Pesticides: Degradation rate

Item	Conditions -	in soil, time used to reach 70-100% of control
Heptachlor	90 weeks	
IPA	5	-
Linuron	18	-
MPCA	12	-
Monuron	35	-
Picloram	80	-
Prometryn	10	-
Propazine	80	-
Simazine	40	-
TCA	12	-
Trifluralin	23	-
2,3,5-T	21	-
2,3,6-TBA	47	-
2,4-D	5	-

Pesticides: Half-life time in soil

Item	Conditions	
Akton	1.5 years	10 lbs/acre, spray, corn, sultan silt loam
Akton	1 year	2 lbs/acre, granules, corn, sultan silt loam
Akton	1.2 years	2 lbs/acre, spray, corn, sultan silt loam
Akton	32 weeks	LAB.
Aldicarb	7-17 days	Oxidation, value depending of soil
Aldrin	10 days	In irrigated soil, cotton, Gezira, Sudan
Aldrin	40% remaining after 14 years	Max value
Aldrin	28% remaining after 15 years	Max value
Azinphosmethyl	484 days	Sterile soil, dry, lag period included, 279 K
Azinphosmethyl	135 days	Sterile soil, dry, lag period included, 298 K
Azinphosmethyl	36 days	Sterile soil, dry, lag period included, 313 K

Pesticides: Half-life time in soil   (continued)

Item	Conditions	
Benomyl	3-6 months	On turf, Delaware
Benomyl	6.12 months	Bare soil, Delaware
BHC	10% remaining after 14 years	Max. value
Bux	1-2 weeks	Hydrolysis, lab.
Carbaryl	8 days	Agricultural soil
Carbaryl	64 days	Sandy loam, Application = 25.4 kg/ha
Carbofuran	0.60 years	In soil, better drained, Application = 2.5 ppm, clay-muck, 1973
Carbofuran	0.92 years	In soil, poorly drained, Application = 2.5 ppm, clay, 1973
Carbofuran	0.56 years	In soil, well drained, Application = 2.5 ppm, clay-muck, 1973
Carbofuran	30 days	Hydrolysis, formation of phenol
Chlordane	40% remaining after 14 years	Max. value
Chlorobenzilate	21 days	In Leon and lakeland sands
DDT	8 months	P,P'-DDT, subtropical soil, fall and winter, Application = 5 kg/ha
DDT	14 days	In irrigated soil, cotton, Gezira, Sudan
DDT	39% remaning	Max. value after 17 years
Dicamba	3 weeks	On litter, Texas
Dicamba	2 weeks	On native grasses, Texas
Dieldrin	2.7 days	Loss by volatilization, grass, first 5 d
Dieldrin	7.5 months	P,P'-DDT, subtropical soil, fall and winter, Application = 5 kg/ha
Dieldrin	11 days	In irrigated soil, cotton, Gezira, Sudan
Dieldrin	31% remaining after 15 years	Max. value
Dioxathion	55 days	Soil dust 2(?069), pH = 7.3, org. matter = 2.1%
Dioxathion	45 days	Soil dust 3(?069), pH = 7.3, org. matter = 2.3%
Dioxathion	30 days	Soil dust 4(?069), pH = 7.6, org. matter = 1.8%

Pesticides: Half-life time in soil   (continued)

Item	Conditions	
Endosulfan	7 days	In irrigated soil, cotton, Gezira, Sudan
Endrin	41% remaining after 14 years	Max. value
Ethion	420 days	None
Heptachlor	1.7 days	Loss by volatilization, grass, first 5 d
Heptachlor	16% remaining after 14 years	Max. value
Malathion	3 days	Basic silty loam, Illinois, Application = 10 ppm, pH = 6.2
Malathion	7 days	Basic silty loam, Illinois, Application = 10 ppm, pH = 8.2
Malathion	4.3 days	Basic silty loam, Illinois, Application = 10 ppm, pH = 7.2
Meobal	7 days	Field condition
Methomyl	30-42 days	None
Naproamide	54 days	Soil moisture content = 10.0%, 301 K, initial = 4.5 kg/ha
Naproamide	63 days	Soil moisture content = 7.5%, 301 K, initial = 4.5 kg/ha
Carbaryl	1.0 day	pH = 8.0, 301 K, seawater
Carbaryl	99 min	pH = 10.0, 285 K, init. concentration= $3 * 10^{-3}$ M, dark
Carbaryl	20 min	pH = 10.0, 298 K, init. concentration= $3 * 10^{-3}$ M, dark
Carbaryl	8 min	pH = 10.0, 308 K, init. concentration= $3 * 10^{-3}$ M, dark
Carbaryl	27 min	pH = 9.8, 298 K, init. concentration= $3 * 10^{-3}$ M, dark
Carbaryl	58 min	pH = 9.5, 298 K, init. concentration= $3 * 10^{-3}$ M, dark
Carbaryl	116 min	pH = 9.2, 298 K, init. concentration= $3 * 10^{-3}$ M, dark
Carbaryl	173 min	pH = 9.0, 298 K, init. concentration= $3 * 10^{-3}$ M, dark
Carbaryl	1 month	pH = 8.0, 276.5 K, seawater
Carbaryl	4.8 days	pH = 8.0, 290 K, seawater
Carbaryl	3.5 days	pH = 8.0, 293 K, seawater
Carbaryl	3.2 hours	Hydrolysis, pH = 9.0, 300 K
Carbaryl	1.3 days	Hydrolysis, pH = 8.0, 300 K
Carbaryl	13 days	Hydrolysis, pH = 7.0, 300 K
Carbaryl	4.4 months	Hydrolysis, pH = 6.0, 300 K
Carbaryl	3.6 years	Hydrolysis, pH = 5.0, 300 K

Pesticides: Half-life time in soil   (continued)

Item		Conditions
Diazinon	0.49 days	Hydrolysis, pH = 3.1
Diazinon	31 days	Hydrolysis, pH = 5.0
Diazinon	185 days	Hydrolysis, pH = 7.5
Diazinon	136 days	Hydrolysis, pH = 9.0
Diazinon	6 days	Hydrolysis, pH = 10.4
Diazinon	0.017 days	Hydrolysis, pH = 3.1
Diazinon	1.27 days	Hydrolysis, pH = 5.0
Diazinon	29 days	Hydrolysis, pH = 7.5
Diazinon	18 days	Hydrolysis, pH = 9.0
Diazinon	0.42 days	Hydrolysis, pH = 10.4
Dimilin	22.9 days	pH = 7.7, 283 K, initial = 0.1 ppm
Dimilin	80.5 days	pH = 10.0, 283 K, initial = 0.1 ppm
Dimilin	28.7 days	pH = 7.7, 297 K, initial = 0.1 ppm
Dimilin	8.31 days	pH = 10.0, 297 K, initial = 0.1 ppm
Dimilin	8 days	pH = 7.7, 311 K, initial = 0.09 ppm
Dimilin	3.45 days	pH = 10.0, 311 K, initial = 0.1 ppm
Heptachlor	23.1 hour	299.8 K, for hydrolysis, in distilled water
Methoprene	30 hours	Freshwater pound, initial = 0.01 ppm, February, in sunlight, California
Methoxychlor	58 hours	Conc. in water = 10^{-7} M, NO H_2O_2 added, 338 K
Methoxychlor	58 hours	Hydls. water conc. = 10^{-6} M, 5% acetonitrile, NO H_2O_2 added, 338 K
Methoxychlor	<1 hour	Hydls. water conc. = 10^{-6} M, 5% acetonitrile, H_2O_2-conc. = 0.1 M, 338 K
Methoxychlor	<1.7 hours	Hydls. water conc. = 10^{-6} M, 5% acetonitrile, H_2O_2-conc. = $8 * 10^{-2}$ M, 338 K
Methoxychlor	2 hours	Hydls. water conc. = 10^{-6} M, 5% acetonitrile, H_2O_2-conc. = $8 * 10^{-3}$ M, 338 K

Rate of degradation: biological degradation in water

Item	Conditions	
Diethanolamine	19.5 mg COD/ gram-hour	Init. COD = 200 mg/l, no other source of C, aerob, 293 ± 3 K
Diethylene glycol	13.7 mg COD/ gram-hour	Init. COD = 200 mg/l, no other source of C, aerob, 293 ± 3 K
Dimethyl-cyclohexanol	21.6 mg COD/ gram-hour	Init. COD = 200 mg/l, no other source of C, aerob, 293 ± 3 K
Ethylene diamine	9.8 mg COD/ gram-hour	Init. COD = 200 mg/l, no other source of C, aerob, 293 ± 3 K
Ethylene glycol	41.7 mg COD/ gram-hour	Init. COD = 200 mg/l, no other source of C, aerob, 293 ± 3 K
Furfuryl alcohol	41.1 mg COD/ gram-hour	Init. COD = 200 mg/l, no other source of C, aerob, 293 ± 3 K
Furfuryl-aldehyde	37.0 mg COD/ gram-hour	Init. COD = 200 mg/l, no other source of C, aerob, 293 ± 3 K
Gallic acid	20.0 mg COD/ gram-hour	Init. COD = 200 mg/l, no other source of C, aerob, 293 ± 3 K
Gentisic acid	80.0 mg COD/ gram-hour	Init. COD = 200 mg/l, no other source of C, aerob, 293 ± 3 K
Glucose	180.0 mg COD/ gram-hour	Init. COD = 200 mg/l, no other source of C, aerob, 293 ± 3 K
Glucose	8 to > 26 days	Half-life time, 278 K, 5 stations at Southampton
Glucose	3 to 10 days	Half-life time, 295 K, 5 stations at Southampton
Glucose	21 to > 30 days	Half-life time, 300 K, Porto Novo 2 stations from River Vellar
Glucose	11 days	Half-life time, 300 K, Porto Novo Kille Backwater
Glucose	9 to 10 days	Half-life time, 300 K, 2 stations from Kille Backwater and Mangrove Swamp
Glucose	> 17 days	Half-life time, 300 K, Porto Novo Bay of Bengal
Glucose	45 days	Half-life time, 300 K, destilled water
Glycerol	85.0 mg COD/ gram-hour	Init. COD = 200 mg/l, no other sources of C, aerob, 293 ± 3 K
Hydroquinone	54.2 mg COD/ gram-hour	Init. COD = 200 mg/l, no other sources of C, aerob, 293 ± 3 K

Rate of degradation: biological degradation in water (continued)

Item	Conditions	
Iso-propanol	52.0 mg COD/ gram-hour	Init. COD = 200 mg/l, no other sources of C, aerob, 293 ± 3 K
Isophthalic acid	85.0 mg COD/ gram-hour	Init. COD = 200 mg/l, no other sources of C, aerob, 293 ± 3 K
Lineal alkyl ben- zen sulfonate in presence of:		
Activated sludge	0.79 mg sur- fact./l day	21 days, batch culture, initial conc. = 20 mg surfactant/l
Anabaena cylindrica	0.72 mg sur- fact./l day	21 days, batch culture, initial conc. = 20 mg surfactant/l
Anabaena variabilis	0.92 mg sur- fact./l day	21 days, batch culture, initial conc. = 20 mg surfactant/l
Anacystis nidulans	0.40 mg sur- fact./l day	21 days, batch culture, initial conc. = 20 mg surfactant/l
Ankistrodes- mus braunii	0.35 mg sur- fact./l day	21 days, batch culture, initial conc. = 20 mg surfactant/l
Calothrix parietina	0.93 mg sur- fact./l day	21 days, batch culture, initial conc. = 20 mg surfactant/l
Chlorella pyrenoidosa	0.58 mg sur- fact./l day	21 days, batch culture, initial conc. = 20 mg surfactant/l
Chlorella vulgaris	0.42 mg sur- fact./l day	21 days, batch culture, initial conc. = 20 mg surfactant/l
Cylindro- spernum sp.	0.84 mg sur- fact./l day	21 days, batch culture, initial conc. = 20 mg surfactant/l
Gloeocapsa alpicola	0.92 mg sur- fact./l day	21 days, batch culture, initial conc. = 20 mg surfactant/l
Nostoc Muscorum	0.46 mg sur- fact./l day	21 days, batch culture, initial conc. = 20 mg surfactant/l
Oscillatoria borneti	0.58 mg sur- fact./l day	21 days, batch culture, initial conc. = 20 mg surfactant/l
P-chloroanti- line	5.7 mg COD/ gram-hour	Init. COD = 200 mg/l, no other sources of C, aerob, 293 ± 3 K
P-chlorophenol	11.0 mg COD/ gram-hour	Init. COD = 200 mg/l, no other sources of C, aerob, 293 ± 3 K
P-cresol	55.0 mg COD/ gram-hour	Init. COD = 200 mg/l, no other sources of C, aerob, 293 ± 3 K
P-hydroxyben- zoic acid	100.0 mg COD/ gram-hour	Init. COD = 200 mg/l, no other sources of C, aerob, 293 ± 3 K
P-nitroaceto- phenone	5.2 mg COD/ gram-hour	Init. COD = 200 mg/l, no other sources of C, aerob, 293 ± 3 K
P-nitro aniline	No degradation	Init. COD = 200 mg/l, no other sources of C, aerob, 293 ± 3 K
P-nitrobenz- aldehyde	13.8 mg COD/ gram-hour	Init. COD = 200 mg/l, no other sources of C, aerob, 293 ± 3 K
P-nitrobenz- oic acid	19.7 mg COD/ gram-hour	Init. COD = 200 mg/l, no other sources of C, aerob, 293 ± 3 K

Rate of degradation: biological degradation in water (continued)

Item	Conditions	
P-nitrophenol	17.5 mg COD/ gram-hour	Init. COD = 200 mg/l, no other sources of C, aerob, 293 ± 3 K
P-nitro toluene	32.5 mg COD/ gram-hour	Init. COD = 200 mg/l, no other sources of C, aerob, 293 ± 3 K
P-phenylen-diamine	More degrad-able	Init. COD = 200 mg/l, no other sources of C, aerob, 293 ± 3 K
P-toluenesul-phonic acid	8.4 mg COD/ gram-hour	Init. COD = 200 mg/l, no other sources of C, aerob, 293 ± 3 K
Phenol	80.0 mg COD/ gram-hour	Init. COD = 200 mg/l, no other sources of C, aerob, 293 ± 3 K
Phloroglucinol	22.1 mg COD/ gram-hour	Init. COD = 200 mg/l, no other sources of C, aerob, 293 ± 3 K
Phosphorus	0.14 l/day	Potomac (Estuary), 293 K, org. P
Phosphorus	0.40 l/day	Lake Erie, 293 K, org. P
Phosphorus	0.14 l/day	Lake Ontario, 293 K, org. P
Phthalic acid	78.4 mg COD/ gram-hour	Init. COD = 200 mg/l, no other sources of C, aerob, 293 ± 3 K
Phthalimide	20.8 mg COD/ gram-hour	Init. COD = 200 mg/l, no other sources of C, aerob, 293 ± 3 K
Pyrocatechol	55.5 mg COD/ gram-hour	Init. COD = 200 mg/l, no other sources of C, aerob, 293 ± 3 K
Pyrogallol	No degradation	Init. COD = 200 mg/l, no other sources of C, aerob, 293 ± 3 K
Resorcnol	57.5 mg COD/ gram-hour	Init. COD = 200 mg/l, no other sources of C, aerob, 293 ± 3 K
Ribose	8 to > 30 days	Half-life time, 278 K, 5 stations at Southampton
Ribose	3 to 9 days	Half-life time, 295 K, 5 stations at Southampton
Ribose	> 36 days	Half-life time, 300 K, Porto Novo, River Vellar
Salicyloic acid	95.0 mg COD/ gram-hour	Init. COD = 200 mg/l, no other sources of C, aerob, 293 ± 3 K
Sec. butanol	55.0 mg COD/ gram-hour	Init. COD = 200 mg/l, no other sources of C, aerob, 293 ± 3 K
Si	0.0015 l/day	Detritus Si to dissolved Si
Sulphanilic acid	4.0 mg COD/ gram-hour	Init. COD = 200 mg/l, no other sources of C, aerob, 293 ± 3 K
Sulphosali-cylic acid	11.3 mg COD/ gram-hour	Init. COD = 200 mg/l, no other sources of C, aerob, 293 ± 3 K
Tert. butanol	30.0 mg COD/ gram-hour	Init. COD = 200 mg/l, no other sources of C, aerob, 293 ± 3 K
Tetrahydrofur-furyl alcohol	40.0 mg COD/ gram-hour	Init. COD = 200 mg/l, no other sources of C, aerob, 293 ± 3 K
Tetrahydro-phthalic acid	No degradation	Init. COD = 200 mg/l, no other sources of C, aerob, 293 ± 3 K

Rate of degradation: biological degradation in water (continued)

Item	Conditions			
Tetrahydro-phthalimide	No degradation	Init. COD = 200 mg/l, no other sources of C, aerob, 293 ± 3 K		
Thymol	15.6 mg COD/gram-hour	"	"	"
Triethylene	27.5 mg COD/gram-hour	"	"	"
1-Napthalene-sulfonic acid	18.0 mg COD/gram-hour	"	"	"
1-Naphtol	38.4 mg COD/gram-hour	"	"	"
1-Naphthol-2-sulfonic acid	18.0 mg COD/gram-hour	"	"	"
1-Naphthyl-amine	No degradation	"	"	"
1-Naphthylamine 6-sulfonic acid	No degradation	"	"	"
1,2-Cyclo-hexanediol	66.0 mg COD/gram-hour	"	"	"
1,3-Dinitro-benzene	No degradation	"	"	"
1,4-Butanediol	40.0 mg COD/gram-hour	"	"	"
1,4-Dinitro-benzene	No degradation	"	"	"
2-Chloro-4-nitrophenol	5.3 mg COD/gram-hour	"	"	"
2-Naphthol	39.2 mg COD/gram-hour	"	"	"
2,3-Dimethyl-aniline	12.7 mg COD/gram-hour	"	"	"
2,3-Dimethyl-phenol	35.0 mg COD/gram-hour	"	"	"
2,4-Diamino-phenol	12.0 mg COD/gram-hour	"	"	"
2,4-Dichlor-phenel	10.5 mg COD/gram-hour	"	"	"
2,4-Dimethyl-phenol	28.2 mg COD/gram-hour	"	"	"
2,4-Dinitri-phenol	6.0 mg COD/gram-hour	"	"	"
2,4-Trinitro-phenol	No degradation	"	"	"
2,5-Dimethyl-aniline	3.6 mg COD/gram-hour	"	"	"
2,5-Dinitri-phenol	10.6 mg COD/gram-hour	"	"	"

Rate of degradation: biological degradation in water (continued)

Item	Conditions			
2,5-Dinitro-phenol	No degradation	Init. COD = 200 mg/l, no other sources of C, aerob, 293 ± 3 K		
2,6-Dimethyl-phenol	9.0 mg COD/ gram-hour	"	"	"
2,6-Dinitro-phenol	No degradation	"	"	"
3,4-Dimethyl-aniline	30.0 mg COD/ gram-hour	"	"	"
3,4-Dimethyl-phenol	13.4 mg COD/ gram-hour	"	"	"
3,5-Dimethyl-phenol	11.1 mg COD/ gram-hour	"	"	"
3,5-Dinitro-benzoic acid	No degradation	"	"	"
4-Methylcyclo-hexanol	40.0 mg COD/ gram-hour	"	"	"
4-Methylcyclo-hexanone	62.5 mg COD/ gram-hour	"	"	"

APPENDIX 3

Concentration factors (CF), ww in brackets means that CF is based upon wet weight

Component	Species	CF	Concentration in water	Conditions
Ag	Daphnia magna	26 (ww)	0.5 mg l^{-1}	-
Ag	Phytoplankton	620-15,000 (ww)	wide range	-
Aldrin	Buffalo fish	30,000 (ww)	0.007 µg l^{-1}	-
Aldrin	Catfish	1590 (ww)	0.044 µg l^{-1}	-
Aldirn	Oyster	10 (ww)	0.05 µg l^{-1}	-
As	Salmo gardneri egg	18.5 (ww)	0.05 mg l^{-1}	279.5°K 33 days
Au	Brown algae	270 (ww)	wide range	-
Cd	Brown algae	890 (ww)	wide range	-
Cd	Zooplankton	6000 (ww)	10^{-4} mg m^{-3}	-
Cd	32 Freshwater plant species	1620	wide range	-
Chlordane	Algae	302 (ww)	6.6 ng l^{-1}	-
Chlorinated naphthalene	Chloroccum sp.	120 (ww)	100 µg l^{-1}	24 h
Co	32 Freshwater plant species	4425	wide range	-
Cr	Fish species	10 (ww)	wide range	Freshwater
Cr	Molluscs	21,800 (ww)	wide range	Marine sp.
Cs	Salmo trutta	1020 (ww)	wide range	Soft water 6.6 g fish
Cu	Chorda filum	560	$2.5 \ast 10^{-7}$ g l^{-1}	Seawater
Cu	Ulva sp.	47,000-56,000	low	Seawater
DDT	Algae	500 (ww)	0.016 ng l^{-1}	-
DDT	Crab	144 (ww)	50 µg l^{-1}	Seawater
DDT	Crayfish	97 (ww)	0.1 µg l^{-1}	-
DDT	Oyster	70,000	0.1 µg l^{-1}	-
DDT	Sea squirt	160,000 (ww)	0.1 µg l^{-1}	Seawater
DDT	Snail	480	50 µg l^{-1}	-
DDT	Trout	200 (ww)	20 µg l^{-1}	-
Dieldrin	Algae	4091	0.011 ng l^{-1}	-
Dieldrin	Catfish	4444 (ww)	0.009 µg l^{-1}	-

Concentration factors (CF), ww in brackets means that CF is based upon wet weight   (continued)

Component	Species	CF	Concentration in water	Conditions
Dieldrin	Trout	3300 (ww)	2.3 μg l^{-1}	-
Fe	Brown algae	17,000 (ww)	-	-
Fe	Zooplankton	144,000 (ww)	0.01 mg m^{-3}	Seawater
Heptachlor	Bluegill	1130 (ww)	50 μg l^{-1}	-
Hexabro-mobiphenyl	Salmo salar	1.73 (ww)	all	5.3 g fish 48 h
Hexachlo-robenzene	Salmo salar	690 (ww)	all	288°K 6 g fish
Hg	Daphnia magna	50 (ww)	2 mg m^{-3}	10 weeks
Hg	Zooplankton	650 (ww)	0.02 mg m^{-3}	Freshwater
Trimethyl-naphthalene	Rangia cuneata	26.7 (ww)	0.03 mg l^{-1}	24 h
Pb	Brown algae	70,000 (ww)	all	-
Pb	Fucus vesiculosus	870	Very low	Seawater
Pb	Zooplankton	1500 (ww)	2 mg m^{-3}	-
PCB	Salmo salar	282 (ww)	wide range	5.29 g fish, 24 h
PCB	Yellow perch	17,000 (ww)	1.0 μg l^{-1}	Freshwater
Ra	Brown algae	370 (ww)	wide range	-
Se	Phytoplankton	900-5500 (ww)	wide range	-
W	Brown algae	87 (ww)	wide range	-
Zn	Phytoplankton	8900-75,000 (ww)	all	-
Zn	Pike	1250 (ww)	low	Freshwater

APPENDIX 4

A - mg per kg dry matter, elements in dry plant tissues

Element	Plankton*)	Brown algae	Ferns	Bacteria	Fungi
Ag	0.25	0.28	0.23		0.15
Al	1,000	62		210	29
As		30			
Au		0.012			
B		120	77	5.5	5
Ba	15	31	8		
Be					<0.1
Br		740			20
C	225,000	345,000	450,000	538,000	494,000
Ca	8,000	11,500	3,700	5,100	1,700
Cd	0.4	0.4	0.5		4
Cl		4,700	6,000	2,300	10,000
Co	5	0.7	0.8		0.5
Cr	3.5	1.3	0.8		1.5
Cs		0.067			
Cu	200	11	15	42	15
F		4.5			
Fe	3,500	690	300	250	130
Ga	1.5	0.5	0.23		1.5
H	46,000	41,000	55,000	74,000	55,000
Hg		0.03			
I	300	1,500			
K		52,000	18,000	115,000	22,300
La		10			
Li		5.4			
Mg	3,200	5,200	1,800	7,000	1,500
Mn	75	53	250	30	25
Mo	1	0.45	0.8		1.5
N	38,000	15,000	20,500	96,000	51,000
Na	6,000	33,000	1,400	4,600	1,500
Ni	36	3	1.5		1.5
O	440,000	470,000	430,000	230,000	340,000
P	4,250	2,800	2,000	30,000	14,000
Pb	5	8.4	2.3		50
Ra	$4 * 10^{-7}$	$9 * 10^{-8}$			
Rb		7.4			
Re		0.014			
S	6,000	12,000	1,000	5,300	4,000
Se		0.84			2
Si	200,000	1,500	5,500	180	
Sn	35	1.1	2.3		5
Sr	260	1,400	13		320
Ti	80	12	5.3		
U					0.25
V	5	2	0.13		0.67
W		0.035			
Y			0.77		0.5
Zn	2,600	150	77		150
Zr	20	2.3		5	

*) Mainly diatoms

B - mg per kg dry matter elements in dry animal tissues *)

Element	Coelenterata	Annelida	Mollusca	Crustacea	Insecta	Pisces	Mammalia
Ag	5?				≤0.07	11?	0.006
Al		340	50	15	100	10	<3
As	30	6	0.005	0.08		0.3	0.2
Au	0.007		0.008	0.0005		0.0003	<0.0009
B		2.1?	20	15		20	<2
Ba			3	0.2			2.3
Bi	0.3?					0.04?	
Br	1,000	100?	1,000	400		400	4
C	436,000	402,000	399,000	401,000	446,000	475,000	484,000
Ca	1,300	11,000	1,500	10,000	500	20,000	85,000
Cd	1		3	0.15			3
Cl	90,000		5,000	6,000	12,000	6,000	3,200
Co	4?	5?	2	0.8	<0.7	0.5	0.3
Cr	1.3					0.2	<0.3
Cs							0.06
Cu	50	4?	20	50	50	8	2.4
F			2	2		1,400	500
Fe	400	630	200	20	200	30	160
Ga	0.5?					0.15?	
Ge	1.5?					0.3?	
H	45,000	59,000	60,000	60,000	73,000	68,000	66,000
Hg			1?			0.3?	0.05
I15	160	4	1	0.9	1	0.43	
K	3,000	16,000	19,000	13,000	11,000	12,000	7,500
La							0.09
Li			1?		≤7		<0.02
Mg	5,500	6,000	5,000	2,000	750	1,200	1,000
Mn	30?	0.06?	10	2?	10	0.8	0.2
Mo	0.7		2	0.6	0.6	1	<1
N	63,000	99,000	85,000	84,000	123,000	114,000	87,000
Na	48,000		16,000	4,000	3,000	8,000	7,300
Ni	26?	11?	4	0.4	9	1	<1
O	271,000	340,000	390,000	400,000	323,000	290,000	186,000
P	14,000	8,100	6,000	9,000	17,000	18,000	43,000
Pb	35?		0.7	0.3	≤7	0.5	4
Ra			$1.5*1^{-7}$	$7*10^{-9}$		$1.5*10^{-8}$	$7*10^{-9}$
Rb			20				18
Re			0.006	0.0005		0.0008	
S	19,000	14,000	16,000	7,500	4,400	7,000	5,400
Sb	0.2					0.2	0.14
Sc							0.006
Se							1.7
Si		150	1,000	300	6,000	70	120
Sn	23?		15?	0.2		3?	<0.16
Sr		20	60	500			21
Th	0,03						
Ti	7		20	17	160	0.2	<0.7
U						≤0.06	0.023
V	2.3	1.2	0.7	0.4	0.15	0.14	<0.4
W			0.05	0.0005		0.0014	
Zn	1500?	6?	200	200	400	80	160

*) Most of the figures for marine animals were derived from the compilation by Vinogradov (1953)

C - mg elements per kg dry mammalian tissues

Element	Brain	Heart	Kidney	Liver	Muscle	Skin	Hair
Ag	0.04	0.01	<0.005	0.03	<0.004	0.022	
Al	0.92	0.8	1.1	1.7	0.67	4.4	30
As	0.08	0.01	0.34	0.5	0.16	0.36	1.1
Au	<0.5	0.00013	<0.5	<0.0001	<0.4	<0.2	
B	<0.6	0.2	<0.5	0.48	0.31	<0.2	
Ba	0.012	0.08	0.06	<0.007	0.013	0.15	
Be	<0.002	<0.002	0.002	0.0009	<0.003	<0.04	
Bi	<0.1	<0.08	<0.09	<0.07	<0.08	<0.03	
Br	3	8	16	10	4	10	6
Ca	320	150	390	140	105	360	200
Cd	<3	0.05	130	6.7	<0.06	<1	
Ce		0.0064			0.00003		
Cl	8,000	6,000	9,000	4,800	2,800	11,000	20,000
Co	0.0055	0.05	0.05	0.23	0.016	<0.03	15
Cr	0.12	0.025	0.05	0.026	0.042	0.29	2
Cs	0.03	0.05	0.03	0.05	0.09	<0.04	
Cu	22	14	12	196	3.1	1.7	80
Eu					0.00012		
F	2	2	3.2	4	5		
Fe	200	190	290	520	140	29	130
Ga	<0.04	<0.04	<0.04	<0.04	<0.04	<0.02	
Hf					<0.04		
Hg		0.17	0.25	0.022	0.02		
I	0.4	0.09	0.0015	0.12	1.7		
In					0.016		
Ir					0.00002		
K	11,600	9,200	7,800	7,400	10,500	1,900	
La		0.00012					
Li	<0.03	<0.03	<0.03	<0.02	<0.02	0.084	
Lu					0.00012		
Mg	550	640	550	480	630	150	
Mn	1.1	0.8	3.8	3.7	0.21	0.22	1
Mo	<0.2	0.2	1.4	2.8	<0.2	<0.07	
N	99,000	132,000	115,000	112,000	108,000	161,000	
Na	10,000	4,500	800	5,500	4,000	9,300	
Ni	<0.3	<0.2	<0.2	<0.2	0.008	0.8	6
P	12,200	6,000	6,900	7,400	6,300	680	800
Pb	0.24	0.2	4.5	4.8	<0.2	0.78	35
Pd					0.002		
Pt					0.002		
Ra			$4*10^{-9}$	$8*10^{-9}$	10^{-10}		
Rb	15	13	17	30	24	8	
Ru	<0.5	<0.4	<0.4	<0.4	0.002	<0.2	
S	6,700	9,500	6,600	8,400	6,800	3,200	38,000
Sb		0.006					
Sc		0.00006			0.008		
Se	2.1	0.7	2.1	2.1	2.5		0.3-13
Si	80	100	95	70	130	450	
Sm		0.01					

C - mg elements per kg dry mammalian tissues (continued)

Element	Brain	Heart	Kidney	Liver	Muscle	Skin	Hair
Sn	<2	0.2	0.74	0.85	<0.2	0.36	
Sr	0.085	0.1	0.24	0.06	0.05	0.15	
Te					0.02		
Ti	<0.3	<0.2	<0.2	<0.2	<0.2	0.54	3
Tl	<0.5	<0.4	<0.4	<0.4	<0.4	<0.2	
Tm					0.0004		
U		0.03	0.03	0.04	0.03		0.13
V	<0.3	<0.04	<0.05	<0.04	<0.04		0.02
W		0.005					
Yb					0.00012		
Zn	46	110	210	130	180	13	170
Zr	<5	<4	<4	<4	<0.3	<2	

References for tables A, B and C:

Aten et al., 1961; Arrhenius, 1963; Beharrell, 1942; Baumeister, 1958; Bowen and Dymond, 1955; Bertrand, 1950; Bowen, 1963; Boirie et al., 1962; Brooksbank and Leddicotte, 1953; Black and Mitchell, 1952; Bowen and Cawse, 1963; Bowen, 1956; Bertrand, 1942; H.J.M. Bowen, unpublished; Bowen, 1960; Bertrand and Levy, 1931; Cannon, 1960; Cannon, 1963; Chau and Riley, 1965; Chilean Iodine Educational Bureau, 1952; Dye et al., 1963; Fukai and Meinke, 1959 and 1962; Fore and Morton, 1952; Forbes et al., 1954; Hunter, 1953; Hunter, 1942 and 1953; Ferguson and Armitage, 1944; Moon and Pall, 1944; Hamaguchi et al., 1960; Henderson et al., 1962; Harrison et al., 1963; International Commission on Radiological Protection, 1964; Johnson and Butler, 1957; Jervis et al., 1961; Koczy and Titze, 1958; Kehoe et al., 1940; King, 1957; Koch et al., 1956; King and Belt, 1938; Koch and Roesmer, 1962; Kringsley, 1959; Long, 1961; Lounamaa, 1956; Leddicotte, 1959; Leroy and Koksoy, 1962; Low, 1949; Lux, 1938; Mayer and Gorham, 1957; Matsumura et al., 1955; McCance and Widowson, 1960; McConnell, 1961; Muth et al., 1960; Mitchell, 1944; Moiseenko, 1959; Mackle et al., 1939; Monier-Williams, 1950; Mullin and Riley, 1956; Neufeld, 1936; Newman, 1949; Porter, 1946; Pavlova,1956; Parr and Taylor, 1963, 1964; Smales and Salmon, 1955; Stitch, 1956; Soremark and Bergman, 1962; Schofield and Hackin, 1964; Shacklette, 1965; Schwartz and Foltz, 1958; Shimp et al., 1957; Shibuya and Nakai, 1963; Stock, 1940; Smales and Pate, 1952; Sowden and Stitch, 1957; Spector, 1956; Soremark, 1964; Samsahl and Soremark, 1961; Stamm and Fernandez, 1958; Suzuki and Hamada, 1956; Tipton and Cook, 1963; Turner et al., 1958; Thompson and Chow, 1956; Thomas, Hendricks and Hill, 1950; Tyutina et al., 1959; Vinogradow, 1953; Vinogradova and Kobalsky, 1962; Wester, 1965; Wakita and Kigoshi, 1964; Young and Langille, 1950; Yamagata, 1950, 1962; Yamagata, Murata and Toril, 1962.

Amounts of elements *) in the diet of adult mammals in mg day^{-1}

Species	Man (Homo sapiens)				Rat (Ratus norvegicus)			
Mean weigth	70 kg				0.3 kg			
Wt. of dry diet	750 g/day				10 g/day			
Element/ state	Defi- cient	Normal	Toxic	Lethal	Defi- cient	Normal	Toxic	Lethal
Ag$^+$		0.06-0.08	60	1300				
Al^{3+}		10-100			0.001		200	220
As$^{III \text{ or } V}$		0.1-0.3	5-50	100-300	0.002		0.6	1.3-5
B Borate		10-20	4000		0.0006		0.15	130-270
Ba^{2+} soluble		(1-5)	200					70-100
Bi^{3+}		(0.06)					1.5	
Br$^-$		1-10	3000		0.005			800
Ca^{2+}		400-1500			1	45-60		>400
Cd^{2+}		(0.6)	3				0.5	16
Cl$^-$	70	2400-4000			0.4	5-30		>900
Co^{2+}		0.0002	500				0.7	
Cr^{VI} Chromate		(0.05)	200	3000			5	
Cu^{2+}		2-5	250-500			0.05-0.2		20
F$^-$		0.5	20	2000	0.0007	0.001	0.1	30
Fe$^{II \text{ or } III}$		12-15				0.1-0.5		>60
Ga^{3+}		(0.02)					10	
Hg^{II}		0.005-0.02		150-300				8
I$^-$	0.015	0.2	10,000			0.001-0.002		
In^{3+}		(0.01)					30	200-300
K$^+$		1400-3700	6000		0.3	50		>400
Li$^+$		2	200					
Mg^{2+}		220-400			0.1	2-5		
Mn^{2+}		3-9			0.003	0.03-0.2		
Mo^{VI} Molybdate		(0.7)			0.0005	0.0005-0.001	5	50
N Organic		8000-22,000						
Na$^+$	45	1600-2700			0.2	5-50		
Ni^{2+}		0.3-0.5					50	
P Phosphate		1200-2700				35-45		
Pb^{2+}		0.3-0.4		10,000				270
Rb$^+$		(10)				0.5	10	
S Sulphate, etc.		420-3000						
Sb$^{III \text{ or } V}$		(0.1)	100					11-75
Se^{IV} Selenite		(0.2)	5		0.0007		0.06	1-2
Si Silicate		600						
Sn^{2+}		17-45	2000					
Sr^{2+}		1.5-5					8	900
Ta^V Tantalate		(1)						300
Te^{VI} Tellurate		(0.02)		2000			0.25	1-9
Ti TiO$_2$		(1-10)						
Tl$^+$		(0.1)		600				7.5
U^{VI}UO$_2^{2+}$		(0.05)						36
V^V Vanadate		(0.3)					0.5	1.5
W^{VI} Tungstate		(0.05)						30-50
Zn^{2+}		10-15			0.016	0.02-0.04	50	150
Zr^{IV}		(0.1)						250-700

*) For comparative purpose, and for order of magnitude estimates for other species of mammals, the amounts in mg per kg body weight are more useful than the absolute amounts given above. Figures given in parantheses are provisional.

APPENDIX 5

A: Effects of Trace Amounts of Methyl Mercury Hydroxide (MMH) on the Growth of Tomato Seedlings after Thirteen Days. Six replications

Treatment (ppm) MMH	Mean shoot growth (cm)	% inhibition of mean shoot growth	Mean seedlings wet weight (mg)	% inhibition of mean wet weight	Mean seedlings dry weight (mg)	% inhibition of mean dry weight
0.05	0.6	88.8	42	95.8	5	94.3
0.04	0.8	83.6	61	93.9	8	89.3
0.03	1.3	75.5	141	86.5	15	81
0.02	3.2	37.9	320	68.4	28	64
0.01	3.9	25.3	353	65.1	30	60.6
0	5.2	-	1013	-	77	-

B: Growth response of loblolly Pine (LP) and Red Maple (RM) versus Lead Concentration (pb) in mole/l.

Pb	Height (cm)		Root dry weight (g)		Root/shoot ratio		Anthocyanin (relative)	
	LP	RM	LP	RM	LP	RM	LP	RM
0	7.26	10.02	0.8	1.89	2.8	2.86	0.99	3.18
$2*10^{-4}$	7.23	10.51	0.61	1.61	2.38	2.24	0.99	2.49
10^{-3}	5.03	5.71	0.46	0.84	2.58	2.15	0.73	7.18
$2*10^{-3}$	4.2	4.68	0.5	0.8	2.49	2.13	1.46	8.7
$5*10^{-3}$	3.18	3.33	0.33	0.43	1.76	1.57	1.42	11.75
Mean	5.38	6.87	0.54	1.11	2.4	2.18	1.12	6.66

C: Growth response of Loblolly Pine (LP) and Red Maple (RM) versus Fluoride Concentration.

mole/l	Height (cm)		Stem dry weight (g)		Root dry weight (g)		Root/shoot ratio	
	LP	RM	LP	RM	LP	RM	LP	RM
0	6.74	9.31	0.29	0.76	0.76	1.97	3.02	2.79
$2*10^{-4}$	6.56	9.07	0.3	0.68	0.69	1.88	2.28	2.83
10^{-3}	6.68	7.99	0.3	0.68	0.7	2.45	2.89	3.71
$2*10^{-3}$	6.15	8.35	0.27	0.58	0.59	1.76	2.26	2.78
$2*10^{-2}$	3.74	3.91	0.19	0.31	0.32	0.55	1.77	1.84
Mean	5.97	7.73	0.27	0.6	0.61	1.72	2.44	2.79

Lethal doses 50% Mortality (LD$_{50}$)

Component	Concentration (mg per kg body)	Species	
Ag	100	Mouse	(oral)
Ag as oxide	2820	Rat	-
Al	770	Mouse	-
Al	3700	Rat	-
As	9	Mouse	-
As as As$_2$O$_5$	8	Rat	-
As as As$_2$O$_3$	45	Rat	-
B as borax	4500	Rat	-
B as boric acid	2660	Rat	-
Ba as chloride	500	Mouse	-
Ba as chloride	150	Rat	-
Ba as carbonate	800	Rat	-
Be as chloride	86	Rat	-
Bi	13	Rat	(intraven.)
Ca as acetate	4280	Rat	(oral)
Cu as chloride	4000	Rat	-
CCl$_4$	4620	Mouse	-
Cd as oxide	72	Rat	-
Ce as nitrate	4200	Rat	-
Co as chloride	80	Mouse	-
Cr as chloride	1870	Rat	-
Cu as chloride	140	Rat	-
CN$^-$ as Na-salt	3	Mouse	-
Fe(III) as nitrate	3250	Rat	-
Fe(II) as sulphate	1480	Rat	-
Ge as oxide	750	Rat	-
Hf as chloride	112	Mouse	(intraven.)
Hg(II) as chloride	37	Rat	(oral)
La	35	Rat	(intraven.)
Li as carbonate	710	Rat	(oral)
Mg as chloride	2800	Rat	-
Ni as fluoride	130	Mouse	(intraven.)
Pb as acetate	120	Rat	(oral)
Se as sulphide	38	Rat	-
Sn(II) as chloride	41	Mouse	(intraven.)
Strychnine	0.98	Mouse	(oral)
Te as Na-salt	20	Mouse	(oral)
Th as chloride	114	Mouse	(intraven.)
Tl as oxide	44	Rat	(oral)
U as oxide	6	Mouse	(intraven.)
V(II) as chloride	540	Rat	(oral)
V(IV) as chloride	160	Rat	-
Zn as acetate	2460	Rat	-

Lethal Concentration 50% Mortality (LD_{50})

Component	Concentration $\mu g\ l^{-1}$	Duration	Species
Ag as nitrate	30	4 d	Daphnia magna
Al as chloride	3900	2 d	- -
Alkyl benzene sulphonate	25,000	38 h	Tilapia (fish)
Ammonia	280	24 h	Salmo salar
Ba as chloride	14,500	2 d	Daphnia magna
Cd as chloride	65	2 d	- -
Cd (hard water)	17	5 d	Salmo gairdneri
Chloramine (NH_2Cl)	100	24 h	Phytoplankton
Chlorine	100	24 h	- -
Co as chloride	1100	2 d	Daphnia magna
Cr(IV)	50	2 d	- -
Cr(IV)	32-6000	2 d	Phytoplankton
Cu as chloride	9.8	2 d	Daphnia magna
Dibutyl-phthalate	50,000	2 d	Goldfish
Hg as chloride	5	2 d	Daphnia magna
Mn(II) as chloride	9800	2 d	- -
Polyelectrolytes	345,000	48 h	- -
Polyelectrolytes	>8000	48 h	Salmonoid fish
Polyoxyethylene	14,500	48 h	- -
Sn(II) as chloride	55,000	2 d	Daphnia magna
Sr as chloride	125,000	2 d	- -
Zn as chloride	100	2 d	- -
Zn as chloride	6000	15 h	Salmonoid fish

APPENDIX 6

Water Quality Standards for Domestic Water

Water use	Sub-stances	Units	Int.WHO accept-able	Int.WHO allow-able	Euro-pean WHO	U.S.A.	Sweden	France	Bulgaria	Tanza-nia
Toxic effects	Lead	µg/l	50	50	100	50	20/50		100	100
Toxic effects	Arsenic	µg/l	50	50	50	50	10/50		50	50
Toxic effects	Selenium	µg/l	10	10	10	10	10/50		50	50
Toxic effects	Chromium	µg/l	50	50	50	50	20		50	50
Toxic effects	Cyanide	µg/l	200	200	50	10	10/20		10	200
Toxic effects	Cadmium	µg/l	10	10	10	10	10		50	50
Toxic effects	Barium	µg/l	1000	1000	1000	1000			1000	1000
Toxic effects	Mercury	µg/l					1/5			
Toxic effects	Silver	µg/l				50				
Human health	Fluoride	mg/l	1.5	1.5	0.7	0.8-1.7	1.5		0.7-1.0	8.0
Human health	Nitrate	mg/l	30.0	30.0	50/100	45	30	44	30	(100)
General use	Colour	mg pt/l	5	50		15	10		15	50
General use	Turbidity	mg SiO_2/l	5	25		3	weak		30 cm	30
General use	pH		7.0-8.5	6.5-9.2			6.0-8.0		6.5-8.5	6.5-9.2
General use	Tot. dis. matter	mg/l	500	1500			200	2000		2000
General use	Tot. hard-ness	mg $CaCO_3$/l			500			300	450	600
General use	Calcium	mg/l	75	200					150	
General use	Magnesium	mg/l	50	150	125			125	50	
General use	Sulphate	mg/l	200	400	250	250	25/250	250	250	600
General use	Chloride	mg/l	200	600	600	250	25/250	250	250	800
General use	Iron	mg/l	0.3	1.0	1.0	0.3	0.2	0.2	0.2	1.0
General use	Manganese	mg/l	0.1	0.5	0.05	0.05	0.05	0.1	0.1	0.5
General use	Copper	mg/l	1.0	1.5	0.05	1.0	0.05/1.0	0.2	0.2	3.0
General use	Zinc	mg/l	5.0	15.0	5.0	5.0	0.3/5.0	3	3	15.0
General use	Phenol	µg/l	1	2	1	1			1	2

APPENDIX 7

Elements: Abundance and Biological Activity

Symbols used:

a = elements formed by radioactive decay of uranium and thorium. Have short physical half-lives and their crustal abundance are too low to be measured accurately.
b = very low, unmeasureable
ra = radioactive
cs = carcinogenic, suspected only.

s = stimulatory
cp = carcinogenic, proven
en = essential nutrient, established
ep = essential nutrient, probably or required under special conditions
t1 = toxic
t2 = very toxic

Element	Symbol	Atomic number	Crustal abundance weight (%)	Abundance in hydrosphere (mg/l)	Abundance in atmosphere (vol ppm)	Biological activity	Threshold limit (mg/m^3 in air in 8 hours)
Actinium	Ac	89	manmade	manmade		ra	
Aluminium	Al	13	8	0.01		cs	
Americum	Am	93	manmade	manmade		ra	
Antimony	Sb	51	0.00002	0.0005		s t2	
Argon	Ar	18	0	0.6	9300		
Arsenic	As	33	0.00020	0.003		cs s t2	
Astatine	At	85	manmade	manmade		ra	
Barium	Ba	56	0.0380	0.03		s t1	0.5
Berkelium	Bk	97	manmade	manmade		ra	
Beryllium	Be	4	0.0002	b		cp (s) t2	
Bismuth	Bi	83	4E-7	2E-5		t1	
Boron	B	5	0.0007	4.6			
Bromine	Br	35	0.00040	65		t1=Br2	
Cadmium	Cd	48	0.000018	0.001			0.2
Calcium	Ca	20	5.06	400		en	
Californium	Cf	98	manmade	manmade		ra	
Carbon	C	6	0.02	28	CO_2=330	en	
Cerium	Ce	58	0.0083	0.0004		s	
Cesium	Cs	55	0.00016	0.0005			
Chlorine	Cl	17	0.019	18,980		Cl(-)=en Cl_2=t1	
Chromium	Cr	24	0.0096	5E-5		en cp s t1	0.1(CrO_3)
Cobalt	Co	27	0.0028	0.0005		en cp	
Copper	Cu	29	0.0056	0.003		en s t1	
Curium	Cm	96	manmade	manmade		ra	
Dysprosium	Dy	66	0.00085	b		s	
Einsteinium	Es	99	manmade	manmade		ra	
Erbium	Er	68	0.00036	b		s	
Europium	Eu	63	0.00022	b			
Fermium	Fm	100	manmade	manmade		ra	
Fluorine	F	9	0.0460	1.3		ep s	

Elements: Abundance and Biological Activity (continued)

Element	Symbol	Atomic number	Crustal abundance weight (%)	Abundance in hydrosphere (mg/l)	Abundance in atmosphere (vol ppm)	Biological activity	Threshold limit (mg/ m^3 in air in 8 hours)
Francium	Fr	87	manmade	manmade		r a	
Gadolinium	Gd	64	0.00063	b			
Gallium	Ga	31	0.00063	b			
Germanium	Ge	32	0.00013	7E-5		s	
Gold	Au	79	2E-7	4E-6		s t1	
Hafnium	Hf	72	0.0004	b			
Helium	He	2	0	5E-6	5.2		
Holmium	Ho	67	0.00016	b		s	
Hydrogen	H	1	0.14	H2O	CH_4=1.5 H_2=0.5	en	
Indium	In	49	0.00002	0.02		s t2	
Iodine	I	53	0.00005	0.06		I(-)=en I_2=t1	
Iridium	Ir	77	2E-8	b		t1	
Iron	Fe	26	5.80	0.01		en	
Krypton	Kr	36	0	0.0003	1		
Lanthanium	La	57	0.0050	0.0003			
Lead	Pb	82	0.0010	3E-5		t2 cp s	0.2
Lithium	Li	3	0.0020	0.17		s	
Lutetium	Lu	71	8E-5	b			
Magnesium	Mg	12	2.77	1350		en	
Manganese	Mn	25	0.100	0.002		en cs	5
Mendelevium	Md	101	manmade	manmade		r a	
Mercury	Hg	50	2E-6	3E-5		s t2	
Molybdenium	Mo	42	0.00012	0.01		en	5-15
Neodymium	Nd	60	0.0044	b			
Neon	Ne	10	0	0.0001	18		
Neptunium	Np	93	manmade	manmade		r a	
Nickel	Ni	28	0.0072	0.002		ep cp s	
Niobium	Nb	28	0.0072	0.002		ep cp s	
Nitrogen	N	7	0.0020	0.5	780,900	en	
Nobelium	No	102	manmade	manmade		r a	
Osmium	Os	76	2E-8	b		r a	
Oxygen	O	8	45.2	H_2O	209,500	en	
Palladium	Pd	46	3E-7	b		cs	
Phosphorus	P	15	0.1010	0.07		en	
Platinum	Pt	78	5E-7	b		t1	0.002
Plutonium	Pu	94	manmade	manmade		r a	
Polonium	Po	84	a	b		r a	
Potassium	K	19	1.68	380		en	
Praseody- mium	Pr	59	0.0015				
Promethium	Pm	61	manmade				Protach- tinium
Radium	Ra	88	a	1E-10		r a	
Radon	Rn	86	a	0.6E-15		r a	

Elements: **Abundance and Biological Activity** (continued)

Element	Symbol	Atomic number	Crustal abundance weight (%)	Abundance in hydros- phere (mg/l)	Abundance in atmos- phere (vol ppm)	Biological activity	Threshold limit (mg/ m^3 in air in 8 hours)
Rhenium	Re	75	4E-8	b			
Rhodium	Rh	45	1E-8	b		cs	0.001
Rubidium	Rb	37	0.0070	0.12		s	
Rutherium	Ru	44	1E-8	b			
Samarium	Sm	62	0.00077	b			
Scandium	Sc	21	0.0022	4E-5		cs	
Selenium	Se	34	5E-6	0.004		cp s en t2	
Silicon	Si	14	27.2	3		ep cs	
Silver	Ag	47	8E-7	0.0003		cs t1	0.01
Sodium	Na	11	2.32	10,556		en	
Strontium	Sr	38	0.045	8			
Sulphur	S	16	0.030	885		en	
Tantalum	Ta	73	0.00024	b			5
Technetium	Tc	43	manmade	manmade		r a	

APPENDIX 8

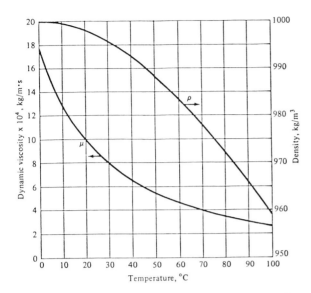

Density and dynamic viscosity of liquid water as a function of temperature

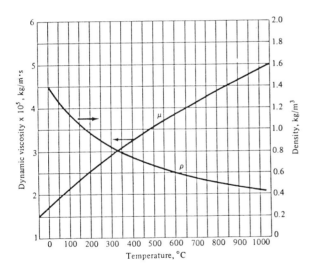

Density and dynamic viscosity of pure air at 1.0 atm pressure as a function of temperature

APPENDIX 9

Ranger (1981) compared fertilized and non-fertilized stands at the ages of 15 and 18 years. The stem density was 4,500 stems per hectare. Half of the stand was fertilized at planting with additions of 9.3 g m^{-2} of nitrogen (27.5 g m^{-2} of urea), 16.3 g m^{-2} of phosphorus (38.0 g m^{-2} of P$_2$O$_5$) 57.1 g m^{-2} of calcium (80.0 g m^{-2} CaO), and 13.3 g m^{-2} of potassium (16.0 g m^{-2} of K$_2$O).

Fertilization largely increased the productivity of the stand (+ 122%) (Table 1). This increment was correlated with a higher uptake of nutrients. This effect changed with time. Between 15 and 18 years the biomass increased 39% in the unfertilized plot and only 30% in the fertilized one. At 18 years the biomass of the fertilized plot was only 107% greater than the control stand, whereas at 15 years it was 122% greater.

The same pattern was evident for nutrients. The fertilized stand, which produced 122% more biomass, used 153% more of the five nutrients studied than did the control plot. There was no direct relationship between biomass production and nutrient uptake. On the contrary, the increment of productivity was accompanied by an "over-consumption" of some, but not all, nutrients. Potassium was not "over-consumed", and more surprising, neither was nitrogen.

Table 2 indicates the amounts of elements included in a thousand kilograms of woody biomass and in a thousand kilograms of 1-year-old needles, at 15 and 18 years, and also the amounts of these same nutrients found in the same quantity of litter.

Table 1
Biomass and the included nutrients in a fertilized (F) and control (C) plot of *Pinus laricio* after 15 and 18 years

Age of the stand	Site	Biomass kg m^{-2}	Mineral elements in biomass g m^{-2}				
			N	P	K	Ca	Mg
15	F	11.14	17.02	2.21	14.57	19.07	4.06
	C	5.01	7.05	0.46	7.15	6.40	1.47
	F-C	6.13	9.97	1.75	7.42	12.67	2.59
18	F	14.46	21.47	2.70	20.00	23.94	5.15
	C	6.98	9.61	0.62	9.90	8.64	2.00
	F-C	7.48	11.86	2.08	10.10	15.30	3.15

Table 2
Nutrients required to build up 1,000 kg of biomass in fertilized (F) and control (C) plots and in the yearly litter fall of a *Pinus laricio* stand (kg of nutrients)

Age	Type of material	Site	N	P	K	Ca	Mg	Total
15	Woody biomass	F	1.53	0.20	1.31	1.71	0.36	5.11
		C	1.41	0.09	1.43	1.28	0.29	4.50
	1-year-old needles	F	13.16	1.20	6.78	4.01	1.06	26.21
		C	13.74	0.77	6.81	3.10	1.16	25.58
18	Woody biomass	F	1.48	0.19	1.38	1.66	0.36	5.07
		C	1.38	0.09	1.42	1.24	0.29	4.43
	1-year-old needles	F	13.22	1.20	6.85	4.06	1.08	26.41
		C	13.89	0.76	6.87	3.20	1.16	25.88
	Litter	F	7.21	0.49	1.90	6.59	0.95	17.14
		C	8.00	0.23	1.31	5.35	1.00	15.89

Table 3
Losses of nutrients by various harvesting methods in a control and a fertilized stand of *Pinus laricio* (g m^{-2}). F - Fertilized plot; C - Control plot; I - Input

		N	P	K	Ca	Mg
Input with fertilizers		9.2	16.3	13.3	57.1	
Atmospheric inputs		0.5-1.0	0	0.2	0.2-0.5	0.1
Total harving	F	49.5	5.1	34.6	35.2	7.4
(boles, branches,	F-I	40.3	(+11.2)	21.3	(+21.9)	7.4
roots and leaves)	C	20.9	1.2	15.1	11.6	2.8
Harvesting total	F	46.1	4.8	32.5	32.8	6.7
aerial biomass	F-I	-36.9	(+11.5)	19.2	(+24.3)	6.7
	C	19.3	0.9	13.8	10.6	2.8
Boles harvested	F	5.1	0.9	6.7	5.2	1.5
(without bark)	F-I	(+4.1)	(+15.2)	(+6.6)	(+51.9)	1.5
	C	2.6	0.1	3.6	1.9	0.5

In contrast to forest fertilization, which involves an input of nutrients into the system, forestry practices have involved the increased utilization of forest products by harvesting all of the biomass. This management practice increases the output from the system.

Foresters have long respected the biological processes and mineral requirements of the forest ecosystems, and have used only a fraction of the biomass, such as boles or big branches, most often excluding leaves, small branches, bark and stumps.

Some utilization of forest litter as fertilizers of agricultural soils has been made in the past in Bavaria. After a few years, the productivity of the forests decreased so much that this method was prohibited.

Some recent projects for exploiting forest products have attempted to harvest not only the aerial part of forest, but also the root system, which can make up an appreciable part of the forest production. To study the effects of such practices, Ranger (1981) used the data of the stand discussed above to analyze the effect of different harvesting processes on the mineral budget. He compared total harvesting of aerial and root biomass, harvesting of the total aerial part and harvesting, by the old methods, of the boles, without bark and branches.

Table 3 indicates that on the control plot, the effect of the three harvesting procedures is always on the output of nutrients from the ecosystem, and that output could possible be matched for some nutrients, on a several year basis, by inputs from the atmosphere, by rain and by dry and wet deposition. However, when total input and output are considered, the final results is always a loss for succeeding plantations.

APPENDIX 10

1. **TEST GUIDELINES TO MEASURE PHYSICAL-CHEMICAL PROPERTIES**

1.1 Test used to identity the chemical

  Relative Molecular Mass
  UV-VIS Absorption Spectra
  Melting Point/Melting Range
  Boiling Point/Boiling Range

1.2 Test used

  (a) to obtain pre-requisite information for tests on degradation, accumulation, ecotoxicity, toxicity
  (b) to evaluate environmental mobility and transport

  Vapour Pressure Curve
  * Water Solubility
  * Adsorption/Desorption
  * Partition Coefficient
  Volatility from Aqueous Solutions
  Complex Formation Ability
  * Density of Liquids and Solids
  * Particle Size Distribution
  * Dissociation Constants
  Viscosity of Liquids
  Surface Tension of Aqueous Solution
  Fat Solubility of Solids and Liquids
  Permeability
  Corrosiveness

1.3 Tests used to evaluate abiotic degradation

  * Hydrolysis as a Function of pH
  Storage Stability   (thermal stability)

1.4 Tests used to obtain safety data

  Determination of explosive properties
  Determination of oxidizing effects of gases yielding oxygen
  Determination of pyrophoric behaviour of solids & liquids
  Determination of pyrophoric behaviour of gases & liquids
  Determination of high inflammability of powdered granular and pasty
    substances
  Determination of high inflammability
    Inflammability of liquids
  Determination of high inflammability of gases
  Determination of substances, which give off highly inflammable gases
    in dangerous amounts on contact with water

2. GUIDELINES TO EVALUATE DEGRADATION AND ACCUMULATION IN THE ENVIRONMENT

2.1 Bio-Degradation in Water

Ready biodegradability (level I)
The following five tests are considered to be equally sensitive:
- French AFNOR test T 90/302
- Modified OECD screening test with DOC analysis
- Modified Sturm test, based on the measurement of CO_2 evolution
- MITI test
- Closed bottle test according to Fischer

Inherent biodegradability (level II)
- Modified semicontinuous activated sludge test for determination of inherent biodegradability (SCAS)
- Modified static method for testing the inherent biodegradability (Zahn-Wellens test)
- Test guideline for inherent biodegradability (MITI)

Simulation tests (level III)
OECD condirmatory test modified for the application of unspecific analysis (coupled units test)

2.2 Photodegradation

Laboratory gas phase photodegradation test
Laboratory solid surface photomineralisation test
Test on photochemical transformation in water

2.3 Bioaccumulation in organisms

Static tests
Bioaccumulation test with mussels
Static fish test on bioaccumulation

Semi-static test
Semi-static procedure for measuring bioconcentration of chemicals in fish
Sequential static procedure for measuring the bioaccumulation potential of chemicals with fishes

Dynamic flow-through test
Test guideline for testing the degree of accumulation of chemical substances in fish body

2.4 Degradation and accumulation in soil and sediments

Degradability of chemicals in soil
Leaching behaviour of chemicals in soil
Residue behaviour of chemicals in soil

3. **GUIDELINES TO EVALUATE ECOTOXICOLOGICAL EFFECTS STUDIES TO EVALUATE ENVIRONMENTAL EFFECTS**

* Acute toxicity LC_{50} of fish (96 hours, static)
* Reproduction study and LC_{50} with daphnia magna (14 days)
* Growth inhibition study with unicellular algae (4 days)

4. **GUIDELINES TO EVALUATE TOXICOLOGICAL EFFECTS**

4.1 Acute toxicity studies

* Acute oral toxicity
* Acute dermal toxicity
* Acute inhalation toxicity
* Acute dermal irritation/corrosivity
* Skin sensitisation

4.2 Subchronic toxicity studies

* Subchronic oral toxicity - rodent: 28 day or 14 day study
  Subchronic oral toxicity - rodent: 90 day study
  Subchronic oral toxicity - non-rodent: 90 day study
  Subchronic dermal toxicity: 21/28 day study
  Subchronic dermal toxicity: 90 day study
  Subchronic inhalation toxicity: 28 day or 14 day study
  Subchronic inhalation toxicity: 90 day study
  Subchronic neurotoxicity: 90 day study
  Teratogenicity

4.3 * Studies for the Evaluation of the Mutagenic and/or Carcino- genic Potential of Chemicals

4.4 Toxicolinetics

4.5 Chronic Toxicity Studies

Chronic toxicity
Combined Chronic Toxicity/Carcinogenicity
Carcinogenicity

2.3 Bioaccumulation in Organisms

Bioaccumulation is presented by the bioaccumulation factor (the ratio of concentration of a chemical in a test organism compared to that in the ambient medium at steady state conditions). Among the species tested, fish show the highest bioaccumulation factors for a given chemical. In view of the importance of water as a major carrier of envirnmental chemicals, it is recommended that fish should be **the** representative animal species in bioaccumulation testing.

Test procedures may be static (initial concentration of test chemical may decrease due to uptake by test organism) or **dynamic** (constant concentration of test chemical maintained in flow through system).

Eight **species**, for which there is considerable experience, are recommended (pelagic: zebrafish, carp, guppy, rainbow trout, fathead minnow, bluegill sunfish; bottom/filter feeders: mussel, catfish). Bottom and filter feeders ara the recommended test species for chemicals which have an octanol/water partition coefficient less than 1000, but are highly adsorbed on suspended matter and sediments/soils.

The dicision not to test a chemical for bioaccumulation will be based upon the assessment of its relevant physical-chemical properties and its degradability.

There will generally be no need for the determination of bioaccumulation of unionised organic chemicals, if
- the water solubility is greater than 2 g/litre;
- the n-octanol/water partition coefficient is less than 1000;
- the water/air partition coefficient (=volatility) is less than 10;
- the chemical is ready biodegradable

The following test stages are recommended:

Level I (Screening phase): Identifies chemicals with a significant bioaccumulation potential by performing phsysical-chemical tests (preferably the partition coefficient) and by taking into account their environmental stability. There may be cases where the physical-chemical testing yields an unequivocal result, in which case, a static test could be used as a screening test.

Level II and III: Confirmation of bioaccumulation as disclosed in the screening phase by means of studies with representative living organisms. The decision whether to use a dynamic or static test procedure depends on the reliability with which the bioaccumulation factor and the kinetics (uptake/depletion rate) can be measured.

Static tests

Bioaccumulation test with mussels.

Static fish test on bioaccumulation.

Semi-static tests

Semi-static procedure for measuring the bioaccumulation potential of chemicals in fishes.

Sequential static procedure for measuring the bioaccumulation potential of chemicals with fishes.

Dynamic flow-through test

Test Guideline for testing the degree of accumulation of chemical substances in fish body.

2.4 Degradation and Accumulation in Soils and Sediments

Soil tests should only be carried out on chemicals likely to reach the soil. Tests on soil/sediment should be run under aerobic/anaerobic conditions, respectively.

Chemicals readily biodegradable in water do not need to be tested for biodegradation in soils or sediments.

The likelihood of leaching of chemicals to deeper soil layers where anaerobic conditions prevail must be considered.

Due to the difficulties inherent in obtaining standardised soils, the use of the U.S. soil classification for tests on adsorption, biodegradation and/or leaching is recommended. Three types are described, which are common in temperature zones (but not representative of arid/tropical zones):

Nature	pH	Clay	Organic $CaCO_3$	Matter	Example
1. very strongly to strongly acid: sandy	4.5-5.5	5%		1-6%	Spodosol
2. moderately or slightly acid: loamy	5.6-6.5	15-25%		1-4%	Alfisol
3. neutral to slightly alkaline: loamy	6.6-8.0	11-25%	1-10%	1-4%	Entisol

In order to simulate sediments with anaerobic conditions, use of water-logged soils flushed with an inert gas may be used.

1. Biodegradability in soil

Level I: Ready Biodegradability
Tests on soils would be restricted to specialised substances as, for example, to chemicals directly applied to soil. Normally procedures for tests on soil/sediments will start at a level similar to level II for aquatic conditions.

Level II: Inherent Biodegradability
Such tests are recommended for chemicals which are not readily biodegradable in an aquatic screening test, and which may be expected to contaminate soils as a result of their anticipated use/disposal pattern.
A preliminary test guideline (C121/79) for degradability in soil requires use of ^{14}C labelled organic chemicals and is based on a technique developed for pesticides, monitoring $^{14}CO_2$ evolution for up to 64 days. No pass level is quoted. Techniques monitoring unlabelled CO_2 are usually not sensitive enough.

Level III: Simulation Tests
Currently available test methods relate to sewage treatment conditions. Tests at this level may be indicated for chemicals which are found to be nonbiodegradable, or have a low rate of biodegradation, or which are leaching and do not degrade easily anaerobically. Chemicals which degrade to recalcitrant metabolites are a special group for consideration.
In most cases, use of radiolabelled chemicals will be advantageous.

2. Abiotic Degradation in Soils

Where chemicals are disposed of in soils with low biological activity, tests for degradability by non-metabolic processes may be indicated. In such studies, use of sterilised soils could be involved. Sterilisation of soil by irradiation or autoclaving is recommended. It should be noted that the remainder of the test must be carried out under aseptic conditions.

3. Accumulation in Soils   (Adsorption/Desorption)

The n-octanol-water partition coefficient for unionised organic chemicals correlates empirically to leaching characteristics as a measure of the accumulation tendency. Direct measurements of adsorption coefficient between water and soil may be used as a confirmatory test and also as an indicator of accumulation in soil of ionic materials.

4. Mobility of Chemicals in Soils/Sediments

Such tests are indicated for chemicals which are non-biodegradable and/or require simulation tests (see above); the tests allow an evaluation of the combined effects of specific adsorption properties and inherent degradability/persistence.

(a) Leaching
A test guideline presents a method (used for pesticides registration in Germany for assessment of leaching). Essentially the method consists of pouring a solution of the test chemical on to a column of water-saturated soil, followed by leaching water and analysing the eluate to detect the amount of test chemical washed through the soil column.

(b) Adsorption
Adsorption coefficient may also be used to predict leaching behaviour of a chemical in soil.

(c) Residue Behaviour
The test measures the degree to which the test substances is irreversibly adsorbed on soil.

3. Guidelines to Evaluate Ecotoxicological Effects

Three levels of testing are envisaged:

Level I (basic): Simple tests to indicate possible effects on a few functionally important types of organism. Tests with several species are judged to be more important than a single very accurate test with one organism.

[**Level II** (confirmatory): Tests which give more information than level I tests and confirm their results.

Any chemical which meets the following criteria should be submitted for level II tests:

1. Is not readily biodegradable
2. Bioaccumulates significantly
3. Is rated highly toxic in short/longterm toxicology tests and/or has an acute LC_{50} to an organism of less than 1 mg/litre
4. Is a positive mutagen
5. Does not show asymptotes in LC_{50} determinations
6. Undergoes change(s) in production volume/use/disposal pattern such that predicted environmental concentration (PEC) increases by an order of magnitude
7. Has a no-observed-effect concentration (NOEC) for fish, daphnia, algae less than 10 times the PEC
8. Demonstrates increased toxicity following chemical, physical or biological change in the environment.

For the time being, no test guidelines are presented for level II tests.]

[**Level III** (definitive): Restricted to special cases, e.g. where appreciable concentration of the chemical in the environment may exist, or possible environmental hazard has been identified.]

The objectives of level I testing are to indicate general types of ecotoxicologically significant effects (e.g., toxic effects, inhibition of growth, reproduction, photosynthesis) in a range of organisms with widely different physiological and biochemical properties. The tests recommended under level I are:

Fish LC_{50} (4 days)   LC = lethal concentration
Daphnia LC_{50} (14 days)
Algae IC_{50} (4 days)   IC = inhibitory concentration
                              EC = effect concentration

Acute Toxicity LC_{50} of fish   (96h, static)

Duration   4 days (but may be extended to 14 days);

Applicability   Can be used for any chemical, volatile or non-volatile, which enters the fresh water environment;

Test species   Guppy, but one of several other suitable test species may be used;

Concentrations Five concentrations   (selection guided by solubility and following range-finding test) plus blank, each one tested in duplicate;

No. of Fish   10 per concentration;

Food   Only feed if duration exceeds 4 days;

Non-volatiles   Aerate solutions gently; transfer fish to freshly prepared solutions every 48 hours;

Volatiles   Do not aerate; transfer to fresh solutions every 24 hours;

Monitor   pH, and concentrations of oxygen and test substances throughout test (not necessarily on all solutions);

Record   Number dead after 3, 4, 24, 48, 72, 96 hours (dead fish are removed). If the last two observations suggest that mortality will continue, the test may be extended up to 14 days. Calculate LC_{50} for as many observation times as possible. Observe for effects other than mortality and calculate effect concentrations (EC_{50}) is possible.

Reproduction study and LC_{50} with Daphnia Magna (14 days)

Duration   14 days (but may be extended)

Applicability   Can be used for any chemical, volatile or non-volatile, which enters the fresh water environment. Test substances which give highly coloured solutions, or consume oxygen at a rate necessitating intensive aeration, will be more difficult to evaluate by this technique.

Concentrations   Three to five concentrations (selected following 48 hour range-finding test), plus blank, each one tested in duplicate.

No. of Daphnids   25 per concentration, less than 24 hours old ("P generation").

Renew   Medium, standard water, test compound once every 48 hours (or Monday, Wednesday, Friday)

Food   Add specified algal suspension daily

Non-volatiles	Test in beakers or flasks
Volatiles	Tests in stoppered flasks
Monitor	pH/O_2 concentration before each renewal; concentration once per test;
Reproduction	When the mother daphnids are about 7 days old first new "brood" (F1 generation) appears, with further batches appearing every 2-3 days. Extend test duration past 14 days, if necessary for 3 broods of the F1 generation to appear in the blank concentrations
Record	Mortality: LC_{50} (at least) at 2, 4, 7, 14 days
	Effects: EC_{50} (where possible) for any other effects
	Reproduction: average young/female, and total young/female/test solution and calculate reproduction indicator "r".

For both the above tests, "conditions for the validity of the test" are presented, and recommendations are made to (1) include reference substances in the tests and (2) repeat the whole test (in the case of daphnia starting with the third batch of F1 animals).

Growth inhibition study with unicellular algae   (4 days)

Duration   4 days;

Applicability   Can be used for all compounds that do not interfere with the counting of algae;

Test Organism   One of three specified green algal species   (other may be used, with justification given);

Concentrations   Usually five concentrations   (selected following range-finding test), plus blank, each one tested in duplicate.   Highest concentration should at least give distinct inhibition of growth, lowest concentration should be no different from blank.
Chemicals of low solubility can be predissolved in organic solvent;

Algal Suspension $2*10^4$ cells/ml   for readily water-soluble substances; 10^4 cells/ml   for sparingly soluble substances;

Measure   algal concentration   (by Coulter counter, counting chamber fluorimeter, spectro-photometer) at 1, 2, 3, 4 days in cells/ml;

Calculate   Growth rate, then per cent inhibition, derive IC_{50} and no-observed-effect concentration from graph of per cent inhibition against concentration.

4.   Guidelines to Evaluate Toxicological Effects Relative to Human Health

Acute Tests assess responses of a test organism to a single dose/exposure. They identify chemicals of high toxicity and provide information on potential hazards to humans which could result from exposure to single doses.   The data enable the hazard associated with exposure to a given chemical to be positioned relative to that of other chemicals.

1. Acute Toxicity: Guidelines are presented for tests involving administration by oral, dermal or inhalation routes.   Such tests enable the hazard of poisoning by a single exposure to a chemical to be assessed.
Test guidelines allow an LD_{50} or LC_{50} value to be calculated.   For less

toxic chemicals test guidelines allow the assessment of lethality be a "limit value" test in which proposed dose levels are 5000 mg/kg (oral), 2000 mg/kg (dermal), 5 mg/litre (inhalation), and record all toxic responses observed during the 14 days following administration (note that properly conducted tests do not merely determine lethality.

Preferred species are rat (oral), rat, rabbit or guinea pig (dermal), rat (inhalation); test groups include equal numbers of both sexes.

2. <u>Acute Dermal or Eye Irritation/Corrosivity:</u>   Irritation/corrosivity are defined as reversible/irreversible effects.

Effects of single applications to skin or eye should be assessed for chemicals which are likely to come into contact with skin or eye. After exposure, residual material is removed from skin and the application site is graded for up to 72 hours, although observation can continue for 14-21 days to evaluate e.g. reversibility of effects. Not all chemicals need to be assessed for eye irritation; substances found to be irritant to skin can be prejudged to be irritant to the eye.

Rabbits are the preferred species for both tests; sex is not considered to be important.

3. <u>Sensitisation</u>:   Is not an acute test because it involves more than one exposure, but the period over which exposures take place is short.

Sensitisation tests are, however, usually carried out in the initial assessment phase. Any chemical which is likely to come into repeated contact with skin should be assessed for sensitisation potential.

Six test methods are considered acceptable, all of which use guinea pigs (sex not important).

Periodic use of a positive control substance is recommended to assess the reliability of the test system.

4.2   Subchronic Toxicity Studies

<u>Subchronic Tests:</u>   Assess the toxic effects from repeated doses/exposures of a chemical for part of a lifespan. Tests which in duration do not exceed 10% of an average lifetime are termed "subchronic" in these Guidelines. The tests involve treatment of separate groups with different dose levels and provide information on the occurrence of abnormalities, target organs, dose response curves. and an estimate of a no-effect-level. Such studies should be helpful in selecting dose levels for longer term (i.e., chronic) studies and for establishing safety criteria for human exposure.

1. <u>Guidelines are presented for oral, dermal and inhalation routes.</u> Physical-chemical properties and likely human exposure route(s) need to be

considered in selecting the exposure route(s) to be tested. Recommended durations (14, 28, 90 days for oral and inhalation, 21/28 and 90 days for dermal) are those which are supported by experience or existing regulatory requirements. The longer the study, the greater is the amount of information likely to be obtained. For some chemicals the subchronic study will be the only repeated dose study to be carried out, therefore, it is important to obtain the maximum amount of information. The Guidelines recommend more detailed clinical chemistry, and histopathology in the 90 day compared with the 28 day studies, and the longer duration studies involve more animals per group. In the first instance, histopathology is restricted to control and high-dose groups. Use of satellite groups in which additional animals are treated with the highest dose level for the test duration and then maintained for a period without dosing to observe for reversibility, persistence, or delayed occurrence of toxic effects is recommended, as is the use of control groups which are handled in an identical manner to the test groups except for treatment with the test substance.

Preferred species are rat (oral), rat, rabbit or guinea pig (dermal), rat (inhalation). A separate guideline is included for a 90 day oral study in dogs sicne occasionally studies in non-rodent (in addition to rodent) species are advisable. In all cases, groups consist of equal numbers of each sex in order to establish whether there is a difference in toxic response between sexes.

2. Neurotoxicity: Has special importance because effects may be irreversible, and a guideline is included for a 90 day study involving oral administration to hens.

3. Teratogenicity: Is the property of a chemical which causes permanent structural or functional abnormalities during the period of embryonic development. It is assessed by administering the chemical daily to a pregnant animal through the period of organogenesis. Route of administration is usually oral but other routes more representative of likely human exposure can be used. Different dose levels are administered to separate groups of animals, with a control (and if necessary a vehicle control) group included in the test design. The study should detect any teratogenic potential; it this exists, the data should enable a no-effect level to be established so that appropriate measures may be taken.

4.3 Studies for the Evaluation of the Mutagenic and/or the Carcinogenic Potential of Chemicals

Mutagenicity: Is the property of a chemical which causes either mutations in genes or changes in chromosomes. The groups have presented a report outlining principles of testing for mutagenic potential, but expect to elaborate on the subject, and the use of short-term tests to detect carcinogenic potential, following a meeting in October 1980.]

4.4 Toxicokinetics

Studies of toxicokinetics give information on absorption, distribution, excretion and metabolism of a test chemical. Such data help to evaluate results of other toxicity studies. The time at which a toxicokinetic study may be carried out will vary according to the need for additional data to help assess the safety of a chemical. It may be done soon after tests for acute toxicity, but should definitely be carried out before chronic studies are initiated because it will provide data useful in selecting dose levels for long-term studies. If the study is to be used in such a way (or to evaluate other toxicity data), obviously the species tested must be the same. Alternatively, toxicokinetics studies in a range of species may help identify a species which metabolises the chemical in a similar pattern to man. This could help select a species for chronic studies which may be most predictive of effects in man.

4.5 Chronic Studies

Involve prolonged and repeated exposure of the test substances to animals for the major part of their lifespan and should enable effects which require a long latent period, or are cumulative, to be detected. For long-term studies the route(s) of administration will depend on the physical-chemical properties of teh test substance and teh route(s) typifying human exposure. The oral route is preferred, providing that it can be shown that the chemical is absorbed from the gastrointestinal tract. In all cases, selection of dose levels should be made following well-designed subchronic studies. For all the longterm studies, histopathology is restricted in the first instance to control and high-dose group animals only.

1. Chronic Toxicity: Guidelines are presented for chronic toxicity by oral, dermal or inhalation routes of exposure for at least 12 months. Such studies should identify the majority of chronic effects to establish dose-response relationships.
General toxicity (including neurological, physiological, biochemical, haemotological and pathological) effects should be detected. Guidelines for

both rodent and non-rodent species are presented, though it is noted that testing with a single species may provide sufficient data for assessing the hazard of a chemical. 20 rodents (4 non-rodents) of each sex are recommended group sizes.

2. Carcinogenicity: Guidelines are presented for carcinogenicity by oral, dermal or inhalation routes of exposure of animals to various doses of a test substance by an appropriate route of administration for the major part of their lifespan while observing for the development of neoplastic lesions during or after exposure. A test substance of unknown activity should be tested in both sexes in each of the animal species; of the three rodent species of choice the mouse and the rat have been more widely used than the hamster.

The number of animals tested have to be sufficient so that, at the end of the study, enough animals are available in every group for thorough biological and statistical evaluation; this is essential to support a negative conclusion. 50 animals of each sex are recommended for each test and control group; if interim sacrifice(s) are planned, then the initial number should be increased by the number of animals scheduled for interim sacrifice. Study duration is recommended to by 18 months for mice and hamsters and 24 months for rats, in animal strains of greater longevity and/or low spontaneous tunour rate, termination should be at 24/30 months for mice and hamsters/rats. Termination is acceptable when survival rate in control and lower-dose groups drops to 25%, but for a test which generates negative results to be acceptable, survival rate in all groups must exceed 50% after 18/24 months for mice and hamsters/rats.

3. Combined Chronic Toxicity/Carcinogenicity: Guidelines are presented for combined chronic toxicity/carcinogenicity study by oral, dermal or inhalation routes of exposure for 18-30 months. These guidelines for one species (typically the rat but other species may be used) are essentially a combination of those in (1) and (2) above, group size is 50 animals per sex in treatment and control groups for carcinogenicity assessment, with satellites of treated (20 per sex) and control (10 per sex) groups for toxicity assessment. Study duration constraints for carcinogenicity are as in paragraph 2 above; the satellite groups should be retained in the study for at least 12 months.

REFERENCES

Abbott, W.E., 1948. Water Sewage Works, 95: 424.

Ahl, T. and Weiderholm, T., 1977. Svenska vattenkvalitetskriterier. Eurofierande ammen. SNV PM (Swed)., 918.

Alabaster, J.S. and Lloyd, R., 1980. Water quality criteria for freshwater fish. pp. 21-45. FAO, Butterworths, London.

Albertson, O.E. and Sherwood, R.J., 1968. Improved sludge digestion is possible. Water Wastes Eng., 5: 8, 43.

Almer, B.U., 1972. Inf. from Freshwater Lab., Drottningholm, Sweden, No. 12.

Aly, O.M. and Faust, S.D., 1963. J. Am. Water Works Assoc., 55: 639.

Ambühl, H., 1969. Die neueste Entwicklung der Vierwaldstättersees. Inst. Verein. theor. angew. Limnologie, 17: 210-230.

Ames, L.L., 1969. Evaluation of operating parameters of alumina columns for the selective removal of phosphorus from waste water and the ultimate disposal of phosphorus as calcium phosphate. Final report FWPCA Contract No. 14-12: 413.

Ames, Lloyd L. jr. and Dean, Robert B., 1970. Phosphorus removal from effluents in alumina columns. J. Wat. Pol. Contr. Fed., 42: 161-172.

Anderson, B.G., 1946. Sewage Works J., 18: 82.

Anderson, B.G., 1950. Trans. Am. Fisheries Soc., 78: 96.

Anderson, J.W., 1977. Responses to Sub-lethal Levels of Petroleum Hydrocarbons: Are They Sensitive Indicators and Do They Correlate with Tissue Contamination? In: D.A. Wolfe (ed.), Fate and Effects of Petroleum Hydrocarbons in Marine Organisms and Ecosystems. Pergamon Press, New York. p. 95.

Anderson, N.L. and Smith, K.E., 1977. Dynamics of mercury at coal-fired power plant and adjacent Cooling Lake. Environ. Sci. Technol., 11(1): 75-80.

Anderson, P.D. and D' Apollonia, S., 1978. Aquatic Animals. In: G.C. Butler (ed.), Principles of Ecotoxicology, SCOPE 12. John Wiley % Sons, New York. p. 187.

Anon, 1958. Ozone counter waste cyanide's lethal punch. Chem. Eng., 65: 63.

Arrhenius, G., 1963. The Sea, vol. 3, p. 655. (N.M. Hill, ed.), Interscience, New York.

Aston, R.N., 1947. Chlorine dioxide use in plants on the Niagara boarder. J. Amer. Water Works Assoc., 39: 687.

Aten, A.H.W., 1961. Health Phys., 6: 114.

Augustsson, T. and Ramanathan, V., 1977. A radioactive-convective model study of the CO_2 climate problem. J. Atmos. Sci., 34: 448-451.

Axelrod, D.I. and Bailey, H.P., 1968. Cretaceous dinosaur extinction. Evolution, 22: 595-611.

Babu, S.P. (Ed.), 1975. Trace Elements in Fuels. Adv. Chem. Ser. No. 141, A.C.S.

Baes, C.F., Goeller, H.E., Olson, J.S. and Rotty, R.M., 1976. The global CO_2 problem. ORNL-5194, Oak Ridge National Laboratory, Oak Ridge, Tennessee.

Baes, C.F., Goeller, H.E., Olson, J.S. and Rotty, R.M., 1977. Carbon dioxide and climate: The uncontrolled experiment. American Scientists, 65f: 310.

Balmer, Peter, Blomqvist, Magnus and Lindholm, Margareta, 1968. Simultanfällning i en högbelastet aktivslamproces. Vatten, 24: No. 2, 112-116. (Simultaneous precipitation at highloaded activated sludge treatment).

Barry, R.G., 1978. Cryospheric responses to a global temperature increase. In: Carbon Dioxide, Climate and Society - Proceedings of an IIASA Workshop, Febr. 21-24, 1978 (Ed. Jim Williams). Pergamon Press, Oxford.

Bauer, Robert C. and Vernon, L. Snoeyink., 1973. Reaction of chloramines with active carbon. J.W.P.Cr. Fed., 45: 2290.

Baumester, W., 1958. Encyclopedia of Plant Physiology, vol. 4, pp. 5, 482. (W. Ruhland, Ed.), Springer, Berlin.

Bazzaz, F.A., 1968. Succession on abandoned fields in the Shawnee Hills, Southern Illinois. Ecology, 49: 924-936.

Beharrell, J., 1942. Nature, London, 149: 306.

Benci. J.F., et al., 1975. Effects of hypothetical climate changes on production and yield of corn, in impacts of climatic change on the biosphere. CIAP Monograph 5, pt. 2, Climatic Effects, 4-3 to 4-36.

Bencko, V. and Symon, K., 1977. Health aspects of burning coal with a high arsenic contents. Environ. Res., 13: 378-95.

Benijts-Claus, C. and Benijts, F., 1975. In: Koeman, J.H. et al., (Eds); Sublethal effect of toxic chemicals on aquatic animals. Elsevier, Amsterdam, pp 43-52.

Bennett, J., 1980. A comparison of selective methods and a test of the pre-adaption hypothesis. Heredity, 15: 65-77.

Berck, B., 1953. Anal. Chem., 25: 1253.

Berg, A.P., Kirman, R.C., Thomas, W.C., Freund, G. and Bird, E.D., 1965. Use of iodine for disinfection. J. Amer. Water Works Assoc., 57: 1401.

Bernhart, E.L., 1975. Nitrification in industrial treatment works. 2nd International Congress on Industrial Waste Water and Wastes, Stockholm, Febr. 4-7, 1975.

Beroza, M. and Bowman, M.C., 1966. J. Assoc. Offic. Agr. Chem., 49: 1007.

Berry, James W. et al., 1974. Chemical Villains: A Biology of Pollution. St. Louis: Mosby.

Bertrand, D., 1942. Annls. Inst. Pasteur, Paris, 68: 58.

Bertrand, D., 1950. Bull. Am. Mus. Nat. Hist., 94: 409.

Bertrand, G. and Levy, G., 1931. C. r. hebd. Sénanc. Acad. Sci., Paris, 192: 525.

Billings, C.E. and Matson, W.R., 1972. Mercury emissions from coal combustion. Science, 176: 1232-33.

Black, A.P., 1960. Basic mechanism of coagulation. J. Am. Water Works Ass., 52: No. 4, 492-501.

Black, W.A.P. and Mitchell, R.L., 1952. J. mar. biol. Ass. U.K., 30: 575.

Block, C. and Dams, R., 1976. Study of fly ash emission during combustion of coal. Environ. Sci. Technol., 10: (10), 1011-17.

Boeghlin, 1972. Les polyelectrolytes wt leur utilitation dans l'amelioration des traitements de clarification des eaux. Residuaire et dans les procedes de deshydratation mecanique des bones. Chemie et Industrie. Génie Chimique, 105: 527-534.

Boesch, D.F., 1974. Diversity, stability and responses to human disturbances in estuarine ecosystems. Proc. First Int. Congr. Ecol., pp. 109-114.

Boirie, C., Boss, D., Hugot, G. and Platzer, R., 1962. Acta chim., hung., 33: 281.

Boliden Information FKK 120S (30/10-1967).

Boliden AVR, September 1969.

Bolin, B., Degens, E.T., Kempe S. and Ketner, P., 1979. The global carbon cycle. SCOPE Report 13. John Wiley and Sons, New York.

Bonde, G.J., 1966. Health Lab. Sci., 3: 124.

Bowen, H.J.M., 1960. Biochem. J., 77: 79.

Bowen, H.J.M., 1963. U.K. Atomic Energy Authority Report AERE-R 4196.

Bowen, H.J.M., 1966. Trace Elements in Biochemistry. A.P. New York, 241 pp.

Bowen, H.J.M., and Cawse, P.A., 1963. Analyst, London, 88: 721.

Bowen, H.J.M. and Dymond, J.A., 1955. Proc. R. Soc., B. 144, 355.

Bowman, K.O., Hutcheson, K., Odum, E.P. and Shenton, L.R., 1970. Comments on the distribution.

Bouilding, R. 1976. What is pure coal? Environment 13(1): 12-36.

Boyt, F.L., Bayley, S.E. and Zoltek, J. Jr., 1977. Removal of nutrients from treated municipal waste water by wetland vegetations. J. Wat. Pol. Cont.

Fed. 49: 789-799.

Bretsky, P.W. and Lorenz, D.M., 1970. Adaptive response to environmental stability: a unifying concept in paleoecology. Proc. N. America.

Brett, J.R., 1960. Thermal Requirements of Fish - Three Decades of Study, 1940-70. In: C.M. Tarzwell, ed. Biological Problems in Water Pollution, Transactions of the 1959 Seminar. R.A. Taft, Sanit. Eng. Center Tech. Report W60-3, Cincinatti, p. 110.

Bridges, B.A., 1980. Introduction of Enzymes Involved in DNA Repair and Mutagenesis. In: Ciba Foundation Symposium 76, Environmental Chemicals, Enzyme Function and Human Disease. Excerpta Medica, Amsterdam, pp. 67-81.

Brooksbank, W.A. and Leddicotte, G.W., 1953. J. Phys. Chem., 57: 819.

Brosset, C., 1973. Air-borne acid. Ambio, 2: 1-9.

Brown, C.C., 1978. The Statistical Analysis of Dose-Effect Relationships. In: G.C. Butler (ed.), Principles of Ecotoxicology, SCOPE 12. John Wiley & Sons, New York.

Bucksteeg, Müller, Mörtl, Adelt, Sceer, Labitzky, Sendl, Hörmann and Kästner, 1985. Erste erfahrungen mit 12 sumpfpflanzenkläranlagen. Korrespondenz Abwasser, no. 5, 1985.

Burd, R.S., 1968. Study of sludge handling and disposal. Water Poll. Contr. Res. Ser., Fed. Water Poll. Contr. Adm. Publ. WP-20-4, Washington, D.C.

Burton, I.D, Milbourne, G.M. and Russel, R.S., 1960. Relationship between the rate of fall-out and the concentration of strontium-90 in human diet in the U.K. Nature, 185: 498.

Butler, G.C., 1972. Retention and excretion equations for different patterns of uptake. In: Assessment of Radioactive Contamination in Man, IAEA, Vienna, STI/PUB/290, p. 495.

Buzzell, James C. and Sawyer, Clair N., 1967. Removal of algal nutrients from raw waste with lime. Wat. Poll. Contr. Fed., 39: R 16.

Camp, T.R., 1946. Sedimentation and design of settling tanks. Trans. Amer. Soc. Chem. Eng., 111: 895.

Canham, R.A., 1955. Some problems encountered in spray irrigation of canning plant waste. Proc. k10th Ind. Wast. Conf.

Cannon, H.L., 1960. Science, N.Y., 132: 591.

Cannon, H.L., 1963. Soil Sci., 96: 196.

Caruso, S.C., Bramer, H.C. and Hoak, R.D., 1968. In: Developments in Applied Spectroscopy (W.K. Bair and E.L. Grove, eds.). Vol. 6, Plenum, New York, pp. 323-338.

Caughley, Graeme, 1970. Eruption of ungulate populations with emphasis on Himalayan thor in New Zealand. Ecology, 51: 53-72.

Chadwick, H.W. and Dalke, P.D., 1965. Plant succession on dune sands in Fremont County, Idaho. Ecology, 46: 765-780.

Chamberlin. T.A. et al., 1975. Color removal from bleached kraft. TAPPI Env. Conf., May.

Chau, Y.K. and Riley, J.P., 1965. Analytica chem. Acta, 33: 36.

Chemical Week, 1976. Cyanide tamer looks promsin. 118: 44.

Chemistry, 1968. The lead we breathe. Chemistry, 41: 7.

Chilean Iodine Educational Bureau, 1952. Iodine Content of Foods, London.

Chiou, C.T., Freed, V.H, Schmedding, D.W. and Kohnert, R.L., 1977. Partition coefficient and bioaccumulation of selected organic chemicals. Environmental Science and Technology 11: 475-478.

Clark, H.F., Geldreich, E.E., Jeter, H.L. and Kabler, P.W., 1951. Public Health Rept. (U.S.), 66: 951.

Clark, W.E., 1962. Prediction of ultrafiltration membrane performance. Science, 138: 148.

Cloud, Preston E., Jr., 1971. Resources, Population, and Quality of Life. In: S.

Fred Singer, ed., Is There an Optimum Level of Population? New York, McGraw-Hill.

Coackley, P., 1960. Principles of Vacuum Filtration and Their Application to Sludge-Drying Problems. Waste Treat., Isaac, P (Ed.). Pergamon Press, London.

Cody, M.L., 1966. A general theory of clutch size. Evolution, 20: 174.184.

Cole, C.A., Paul, P.E. and Brewer, H.P., 1976. Odour control by hydrogen peroxide. J. Wat. Poll. Contr. Fed., 48: 297.

Coles, D.G. et al., 1979. Chemical studies of stack fly ash from a coal-fired power plant. Environ. Sci. Technol., 13: (4) 455-59.

Commorev, Burry, 1971. The Closing Circle. Bantam Books, Inc. New York.

Cook, L.M., Askey, R.R. and Bishop, J.A., 1970. Increasing frequency of the typical form of the peppered moth in Manchester. Nature, 227: 1155.

Cooke, G. Dennis, 1967. The pattern of autotrophic succession in laboratory microecosystems. Bioscience, 17: 717-721.

Counsil on Environmental Quality: Twelfth Annual Report of the Council on Environmental Quality, Washington D.C., 1982.

Cowgill, U.M. and Hutchinson, G.E., 1964. Cultural eutrophication in Lago Monterosi during Roman antiquity. Proc. Int. Assoc. Theor. Appl. Limnol., 15(2): 644-645.

Cox, J.L., 1970. Accumulation of DDT residues in Triphoturus mexicanus from the Gulf of California. Nature, 227: 192-193.

Crosby, N.T. and Laws, E.Q., 1964. Analyst, 89: 318.

Crouch, S.R. and Malmstadt, H.V., 1967. Anal. Chem., 39: 1084.

Cueto, C. and Biros, F.J., 1967. Toxicol. Pharmacol., 10: 261.

Cueto, C. and Hayes, W.J., Jr., 1962. J. Agr. Food Chem., 10: 366.

Culp, Gordon, 1967. Chemical treatment of raw sewage 1. Water Wastes Eng., July 61-63, vol. 4.

Culp, Gordon, 1967. Chemical treatment of raw sewage 2. Water Wastes Eng., Oct. 54-57, vol. 4.

Culp, G., Hansen, S. and Richardson, G., 1968. High-rate sedimentation in water treatment works. J. Am. Water Works Ass., 60: 681.

Dague, R.R. and Baumann, E.R., 1961. Hydraulics of circular settling tanks determined by models. Paper presented at the 1961 annual meeting, Iowa Wat. Poll. Contr. Ass.

Dalton, F.E., Stein and Lynam, H.T., 1968. Land reclamation - a complete solution to the sludge and solids disposal problem. J. Water Poll. Contr. Fed., 40: 789.

Davidson, Göran and Ullman, Peter, 1971. Use of prickling acid for precipitation of municipal waste water. Vatten, 27: No. 1, pp. 95-106.

Davidson, R.L. et al., 1974. Trace elements in fly ash. Environ. Sci. Technol., 8(13): 1107-13.

Davis, G.E., Foster, J. and Warren, C.E., 1963. The influence of oxygen concentration on the swimming performance of juvenile Pacific salmon at various temperatures. Trans. Am. Fish Soc. 92, 111.

Davis, Wayne, H., 1970. Overpopulated America. New Republic, January 10.

Debach, P.V., 1974. Biological control by Natural Enemies. Cambridge University Press.

de Freitas, A.S.W., Gidney, M.A.J., McKinnon, A.E. and Norstrom, R.J., 1975. Factors affecting whole body retention of methylmercury in fish. Proc. 15th Hanford Life Science Symposium on the Biological Implications of Metals in the Environment, Richland, Wash., Publ. in the Energy Research and Development Administration Symposium Series.

de Freitas, A.S.W., and Hart, J.S., 1975. Effect of body weight on uptake of methylmercury in fish. Water Quality Parameters, ASTM STP 573, Amer. Soc. Testing Materials, p. 356.

Dibbs, H.P. and Marier, P., 1975. Some environmental effects, sources, and control methods for fine particulates. A.I.Ch.E. Symp. Ser. 71(147): 60.69.

Dillon, P.J. and Rigler, F.H., 1974. A test of a simple nutrient budget model predicting the phosphorus concentration in lake water. J. Fisk. Res. Board Can., 31: 1771-1778.

Dillon. P.J. and Kirchner, W.B., 1975. The effects of geology and land use on the export of phosphorus from watersheds. Water Res., 9:, 135-148.

Dingle, H., 1974. The experimental analysis of migration and life history strategies in insects. In: L. Barton-Browne (ed.), Experimental Analysis of Insect Behaviour, Springer-Verlag, Berlin, pp. 329-42.

Dismukes, E.B., 1975. Conditioning of fly ash with ammonia. J. Air Pollut. Control Assoc., 25: 152-155.

Douglas, G.S. (ed.), 1967. Radioassay Procedures for Environmental Samples. U.S. Public Health Service. Publ. No. 999-RH-27, U.S. Dept. Health Education and Welfare, Washington, D.C.

Downing, A., 1966. Advances in water quality improvement, 1. University of Texas Press, Austin, Texas, April, 1966.

Dryden, F.D. and Stern, G., 1968. Renovated waste water creates recreational environ. Science Tech., 2: 1079.

Dye, W.B., Bretthauer, E., Seim, H.J. and Blincoe, C., 1963. Analyt. Chem., 35: 1687.

Eberle, S.H., Donnert, D. and Stöber, H., 1976. Möglichkeiten des Einsatzes von Aluminium Oxide zur Reinigung organisch belasteter Abwässer. Chem. - Ing. - Tech., 48: 731.

Eckenfelder, W.W., Lawler, J.P. and Walsh, J.T., 1958. Study of fruit and vegetable processing waste disposal methods in the eastern region. U.S. Department of Agriculture, Final Report, September.

Eckenfelder, W.W., Jr. and O'Connor, D.J., 1961. Biological waste treatment. Pergamon Press, New York.

Ehrlich, Paul P. and Raven, Peter H., 1964. Butterflies and plants: a study in coevolution. Evolution, 18: 586-606.

Ehrlich, Paul R. and Raven, Peter H., 1969. Differentiation of populations. Science, 165: 1228-1232.

Eisenhauer, H.R., 1968. The ozonation of phenolic wastes. J. W. P. Contr. Fed., 40: 1887, 1896.

El-Refai, H.R. and Giuffrida, L., 1965. J. Assoc. Offic. Agr. Chem., 48: 374.

Epps, E.A., Bonner, F.L., Newsom, L.D., Carlton, R. and Smitherman, R.O., 1967. Bull. Environ. Contamination Toxicol., 2: 333.

Erichsen-Jones, J.R., 1964. Fish and River Pollution. Butterworths, London.

Ericsson, Bernt and Westberg, Nils, 1968. Use of flocculants by reduction of phosphorus in waste water. Vatten, 24: No. 2, 125-31.

Eriksson, C. and Mortimer, D.C., 1975. Mercury uptake in rooted higher plants - laboratory studies. Verh. Internat. Verein. Limnol., 19: 2087.

Eriksson, and Jernelöv, 1978. Lufttransport och reemission an Kvicksilver. Energikommissionen, rapport No. 6, Stockholm.

Fair, G.M., Geyer J.C. and Okun, D.A., 1968. Water and Waste Engineering. J. Wiley and Sons, Inc., New York.

Faust, S.D. and Hunter, J.W. (eds.), 1967. Principles and Applications of Water Chemistry. J. Wiley and Sons, New York.

Federal Water Pollution Control Administration, 1968. Report of the Committee on Water Quality Criteria. Section 3, Fish, other aquatic life and wildlife. pp. 27-110. Govt. Ptg. Off., Washington, D.C.

Fenger, J., 1979. Luftforurening - en introduktion. MST LUFT B 46, 16 pp.

Ferguson, W.S. and Armitage, E.R., 1944. J. agric. Sci., 34: 165.

Finalyson, C.M. and Chick, A.J., 1983. Testing the Potential of Aquatic Plants

to Trat Abattoir Effluent. Water Research Vol. 17. no. 4 p. 415-422.
Fishman, M.J. and Skougstad, M.W., 1963. Anal. Chem., 35: 146.
Forbes, R.M., Cooper, A.R. and Mitchell, H.H., 1954. J. Biol. Chem., 209: 857.
Ford, D. and Reynolds, T., 1956. Proc. 3rd Ind. Waste Conf., Dallas, Texas.
Force, H. and Morton, R.A., 1952. Biochem. J., 51: 598, 600.
Francis, C.W. and Callahan, M.W., 1975. Biological denitrification and its application in treatment of high-nitrate waste water. J. Environ. Qual., 4: 153-163.
Friberg L. (Ed.), 1977. Toxicology of metals. U.S. EPA, 600/1-77-022.
Fukai, R. and Meinki, W.W., 1959. Nature, London, 184: 815.
Fukai, R. and Meinki, W.W., 1962. Limnol, Oceanorg., 7: 186.

Gaffney, P.E. and Heukelekian, H., 1958. Sewage Ind. Wastes, 30: 503.
Gale, R.S., 1968. Some aspects of the mechanical dewatering of sewage sludges. Filtr. Separ., 5: 2, 133.
Garber, K., 1967. Luftverunreinigung und ihre Wirkungen. Gebrüder Bornträger, Berlin. 2799 pp.
Garp, 1975. The physical basis of climate and climate modeling. Garp Publication series, No. 14, WMO-ISCU, Geneva.
Garrels, R.M. and Christ, C.L., 1965. Solutions, Minerals, and Equilibria. Harper and Row, New York.
Gauvin, W.H., 1955. Tappi, 40: 11, 866-877 (November 1947), Chem. Can.
Gellman, I. and Blosser, R.O., 1959. Proc. 14th Ind. Waste Conf., Purdue University.
Gillespie, D.M., 1965. Proc. Mont. Acad. Sci., 24: 11.
Gladney, E.S. et al., 1978. Coal combustion: Sources of toxic elements in urban air? J. Environ. Sci. Health, A 13'7) 481-91.
Gleason, G.H. and Loonam, A.C., 1933. The development of a chemical process for treatment of sewage. Sewage Works Journal p. 61.
Godwin, H., 1923. Dispersal of pond floras. J. Ecol., 11: 160-64.
Golley, Frank B., 1960. Energy dynamics of a foodchain of an old-field community. Ecol. Monogr., 30: 187-206.
Golterman, H.L. and Clymo, R.S. (eds.), 1969. Methods for chemical analysis of fresh waters. International Biological Programme Handbook, No. 8. Blackwell, Oxford. p. 17.
Gooch, N.P. and Francis, N.L., 1975. A theoretically based mathematical model for calculation of electrostatic precipitator performance. J. Air Pollut. Contr. Assoc., 25: 108-113.
Goodenkauf, A. and Erdei, J., 1964. J. Am. Water Works Assoc., 56: 600.
Goodman, 1974. The validity of the diversity - stability hypothesis. Proc. First Int. Congr. Ecol., pp. 75-79.
Gooring, G., 1962. Soil Sci., 93: 211.
Gould, J.P. and Weber, W.J., 1976. Oxidation of phenols by ozone. J. Wat. Poll. Contr. Fed., 48.
Goulden, C.E., 1969. Temporal changes in diversity. Brookhaven Symp. Biol., 22: 96-100.
Gouw, T.H., 1968. In: Progress in Separation and Purification (E.S. Perry, ed.), vol. 1. Wiley (Interscience), New York, pp. 57-82.
Grahn, O., Hultberg, H. and Landner, L., 1974. Oligotrophication - a self-Accelerating Process in Lakes Subjected to Excessive Supply of Acid Substances. Ambio vol. 3, No. 2, pp. 93-94.
Granstrom, M.L. and Lee, G.F., 1958. Generation and use of chlorine dioxide. J. Amer. Water Works Assoc., 50: 1453.
Gromiec, M.J., 1983. Biochemical Oxygen Demand - Dissolved Oxygen. River Models. p. 131-226. In: Application of Ecological Modelling in Environmental Management. Part A. Elsevier, Amsterdam.
Gustafson, Bengt and Westberg, Nils, 1968. Information från Statens

Naturvårdsverk, vol. 3, p. 13-24.
Gutenmann, N.H. et al., 1976. Selenium in fly ash. Science, 191: 966-67.

Halfon, E., 1976. Relative stability of ecosystem linear models. Ecol. Modelling, 2: 279-296.
Hamaguchi, H., Kuroda, R. and Hosohara, K., 1960. J. atom. Energy Soc., Japan, 2: 317.
Hamence, J.H., Hall, P.S. and Caverly, J.T., 1965. Analyst, 90: 649.
Hamilton, R.D. and Holm-Hansen, O., 1967. Limnol. Oceanogr., 12: 319.
Hanf, E.B., 1970. A guide to scrubber selection. Environ. Sci. Technol., 4: 110-115.
Hansen, J.A. and Tjell, J.C., 1978. Guidelines and sludge utilization - Practice in Scandinavia. Paper presented on "Utilization of Sewage Sludge on Land". Oxford, April 10-13.
Harley, J.H. (ed.). Manual of Standard Procedures. Health and Safety Laboratory.
Harrison, A.D., 1962. Hydrobiological studies of all saline and acid still waters in Western Cape Province. Trans. Roy. Soc., South Africa, No. 36, p. 213.
Harrison, G.E. and Sutton, A., 1963. Nature, London, 197: 809.
Hasebe, S. and Yamamoto, K., 1970. Studies of the removal of cadmium ions from mine water by utilizing xanthate as selective precipitate. Int. congr. on Ind. Waste Water, Stockholm.
Hassler, J.W., 1951. Active Carbon. Chemical Publishing, Brooklyn, N.Y., p. 252.
Hauck, A.R. and Sourirajan, S., 1972. Reverse osmosis treatment of diluted nickel plating solution. J. Wat. Poll. Contr. Fed., 44: 372.
Hawes, C.G. and Davidson, R.R., 1971. Kinetic of ozone decomposition and reaction with organics in water. Chem. Kin. Eng. J., 17: 141.
Hawkes, H.A., 1962. In: River Pollution II. Causes and Effects. (L. Klein, ed.). Butterworths, London. pp. 311-432.
Henderson, C., 1957. Application factors to be applied to bioassays for the safe disposal of toxic waters. In: Biological Problems in Water Pollution (C.M. Trazwell, ed.). Robert A Taft, Sanitary Engineering Center, Cincinnati. pp. 31-37.
Henderson, E.H., Parker, A. and Webb, M.S.W., 1962. U.K. Atomic Energy Authority Report AERE-R 4035.
Hill, J. et al., 1976. Dynamic Behavior of Vinyl Chloride in Aquatic Ecosystems. USEPA, ORD, ERL, Athens, Georgia, EPA-600/3-76-001 p. 64.
Hill, J.B. and Kessler, G., 1961. J. Lab. Clin. Med., 57: 970.
Hindin, E., May, D.S. and Dunstan, G.H., 1964. Residue Rev., 7: 130.
Holden, A.V. and Marsden, K., 1966. Inst. Sewage Purif., J. Proc., Part 3: 3.
Holluta, t., 1963. Das Ozon in der Wasserchemie. GWF, 104: 1261.
Howard, Reiquam, 1971. Paper presented to the American Assoc. for Advancement of Science, 1971 meeting.
Huffaker, C.B., Kennett, C.E. and Finney, G.L., 1962. Biological control of the olive scale Parlatoria oleae (Colvee) in California by imported Aphytis maculicornis (Masi) (Hymenoptera: Aphelinidae). Hilgardia, 32: 541-636.
Hummel, J.R. and Reck, R.A., 1978. Development of a global surface Albedo Model, presented at the Meteorology Section, American Geophysical Union, San Francisco, December 1977. GMR-2607, General Motors Research Laboratories, Warren, Michigan.
Hunter, J.G., 1942. Nature, London, 150: 578.
Hunter, J.G., 1953. J. Sci. Fd. Agric., 4: 10.
Hunter, J.W., 1962. Ph. D. Thesis Rutgers University, New Brunswick, N.J.
Hurwitz, E., Barnett, G.R., Beaudoin, R.E. and Kramer, H.P., 1947. Sewage Works J., 19: 995.

Hutcheson, Kermit, 1970. A tes for comparing diversities based on the Shannon formula. J. Theor. Biol., 29: 151-154.
Hynes, H.B.N., 1960. The Biology of Polluted Waters. Liverpool University Press, Liverpool, England.
Hynes, H.B.N., 1971. Ecology of Running Water. Liverpool University Press, Liverpool, England.
Häfele, E., 1978. A Perspective of Energy Systems and Carbon Dioxide. In: Carbon Dioxide, Climate and Society. Proceedings of a IIASA Workshop, Febr. 21-24, 1978. (Jim Williams, ed.). Pergamon Press, Oxford. p. 21-34.
Hägg, G., 1979. Kemisk Reaktionslære. Almquist and Wiksell, Stockholm.

IAWPR Conference, 1970. Efficiency of polyelectrolytes as flocculating agent. Session 2, Part 1, Section "Control of Water Pollution".
ICRP, 1977. Principles and Methods for Use in Radiation Protection Assessments.
Ingols, R.S. and Fetner, R.H., 195f7. Proc. Soc. Water Treatment Exam., 6: 8.
International Commission on Radiological Protection, Publication 6. Recommendations of the committee on permissible doses for internal radiation, 1964. Pergamon Press, Oxford.
International Standards for Drinking Water, 2nd ed. World Health Organization, Geneva, Switzerland, 1963.
Isgard, Erik, 1971. Chemical Methods in Present Swedish Purification. Effluent and Water Treatment Convention (24-25 June, 1971). Pillar Hall, Olympia, London.

Jacobs, J., 1974. Diversity, stability and maturity in ecosystems influenced by human activities. Proc. First Int. Cingr. Ecol., p. 94.
Jenkins, David, Ferguson, John F. and Menar, Arnolds B., 1971. Chemical processes for phosphate removal. Water Research, vol. 5, No. 7, p. 369-390.
Jensen, K.W. and Snevik, E., 1972. Low pH levels wipe out salmon and trout populations in southernmost Norway. AMBIO, vol. 1, No. 6, pp. 223-225.
Jensen, Soren, 1966. Report of a new chemical hazard. New Scientist, 32: 612.
Jervis, R.E., Perkons, A.K., Mackintosh, W.D. and Kerr, K.F., 1961. Modern trens in activation analysis. Proc. Int. Conf., Texas, p. 107.
Joensuu, O., 1971. Fossil fuels as a source of mercury pollution. Science, 172: 1027-1028.
Johnsen, I. and Søchting, U., 1973. Influence of air pollution on the epiphytic lichen vegetation and bark properties of deciduous trees in the Copenhagen area. Oikos, 24: 344-351.
Hohnson, J.M. and Butler, G.W., 1957. Physiologia Pl., 10: 100.
Johnston, D.W. and Odum, E.P., 1956. Breeding bird populations in relation to plant succession on the Piedmont of Georgia. Ecology, 37: 50-62.
Johnston, Harold, 1972. The effect of supersonic transport planes on the stratospheric ozone shield. Environmental Affairs, vol. 1, No. 4, 735-781.
Joyner, M.L., Healy, D., Chakravarti and Koyanagi, T., 1967. Environ. Sci. Tech., 1: 417.
Juhola, A.J. and Tupper, F., 1969. Laboratory investigation of the regeneration spent granular activated carbon. FWPCA Report, No. TWRC-7.
Jørgensen, E.G., 1969. The adaption of plankton algae. IV. Light adaption in different algal species. Physiol. Plant., 22: 1307-15.
Jørgensen, S.E., 1968. The purification of waste water containing protein and carbohydrate. Vatten 24(4): 332-338.
Jørgensen, S.E., 1969. Purification of highly polluted waste water by protein precipitation and ion exchange. Vatten, 25(3): 278-288.
Jørgensen, S.E., 1970. Ion exchange of waste water from the food industry.

Vatten, 26(4): 350-357.
Jørgensen, S.E., 1971. Precipitation of proteins in waste water. Vatten, 27(1): 58-72.
Jørgensen, S.E., 1973. A new advanced waste water treatment plant in Ätran, Falkenberg, Sweden. Vatten, 29(1): 36-40.
Jørgensen, S.E., 1973. The combination precipitation - ion exchange for waste water from the food industry. Vatten, 29(1): 40-42.
Jørgensen, S.E., 1973. Industrial Waste Water Treatment by Precipitation and Ion Exchange. In: Environmental Engineering, Eds. G. Lindner and K. Nyberg. D. Reidel Publ. Co., Holland, U.S.A., 239 pp.
Jørgensen, S.E., 1974. Recirculation of waste water from the textile industry. Vatten, 29: 364.
Jørgensen, S.E., 1975. Do heavy metals prevent the agricultural use of municipal sludge. Water Res., 9: 163-170.
Jørgensen, S.E., 1975. Recovery of ammonia from industrial waste water. Water Res., 9: 1187.
Jørgensen, S.E., 1976. A eutrophication model for a lake. J. Ecol. Model., 2: 147-165.
Jørgensen, S.E., 1976. An ecological model for heavy metal contamination of crops and ground water. Ecol. Modelling, 2: 59-67.
Jørgensen, S.E., 1976. Reinigung häuslicher Abwässer durch Kombination eines chemischen Fällungs- und Ionenaustausch Verfahrens. (Doctor thesis, Karlsruhe).
Jørgensen, S.E., 1976. A model of fish growth. J. Ecol. Model., 2: 303-313.
Jørgensen, S.E., 1976. Recovery of phenols from industrial waste water. Prog. Wat. Tech., 8: 2/3, 65-79.
Jørgensen, S.E., 1978. The application of cellulose ion exchanger to industrial waste water management. Water Supply Management, 12.
Jørgensen, S.E., 1978. Exchange of heavy metals sediment-water, a review. Interactions Between Sediment and Water. 6th Nordic Symposium on Sediments, 9-12 March, 1978, Hurdal, Norway. Published by Nordforsk, December 1978.
Jørgensen, S.E., 1979. A holistic approach to ecological modelling. Ecol. Model., 7: 169-189.
Jørgensen, S.E., 1979. Modelling the distriburion and effect of heavy metals in an aquatic ecosystem. Ecol. Model., 6: 199-222.
Jørgensen, S.E., 1979. Removal of phosphorus by ion exchange. Vatten, 34: 179-182.
Jørgensen, S.E., 1980. Lake Management. Pergamon Press, London. 180 pp.
Jørgensen, S.E., 1980. The application of intensive measurement for the calibration of eutrophication models. Ecol. Model., 8.
Jørgensen, S.E., 1980. Ecological submodels for Upper Nile Lake Sustem. WMO Report.
Jørgensen, S.E., 1981. Application of Ecological Modelling in Environmental Management. Elsevier.
Jørgensen, S.E., Bengtsson, L., Mejer, J. and Friis, M., 1979. A case study of lake modelling, Södra Bergundasjön. ISEM Journal: in press.
Jørgensen, S.E., Friis, M.B., Henriksen, J., Jørgensen, L.A. and Mejer, H.F., 1979. Handbook of Environmental Data and Ecological Parameters (S.E. Jørgensen, ed.). ISEM, Copenhagen.
Jørgensen, S.E. and Harleman, D.R.F., 1978. Hydrophysical and ecological modelling of deep lakes and reservoirs. Summary Report of an IIASA Workshop, December 12-5, 1977. Published by IIASA.
Jørgensen, S.E., Jacobsen, O.S. and Hoi, I., 1973. A prognosis for a lake. Vatten, 29: 382-404.
Jørgensen, S.E., Kamp-Nielsen, L. and Jacobsen, O.S., 1975. A submodel for anaerobic mud-water exchange of phosphate. Ecol. Model., 1: 133-146.

Jørgensen, S.E. and Mejer, J.F., 1977. Ecological buffer capacity. Ecol. Model., 3: 39-61.

Jørgensen, S.E., Mejer, H.F. and Friis, M., 1978. Examination of a lake model. Ecol. Model., 4: 253-279.

Jørgensen, S.E. and Mitsch, W., 1988. Ecological Engineering and Introduction to Ecotechnology. John Wiley, New York.

Kaakinen, J.W. et al., 1975. Trace element behavior in coal-fired power plant. Environ. Sci. Technol., 9(9): 862-869.

Kalb, G.W., 1975. Total mercury mass balance at a coal-fired power plant. Adv. Chem. Ser. No. 141: 154-74.

Kamp-Nielsen, L., 1974. Mud-water exchange of phosphate and other ions in undisturbed sediment cores and factors effecting the exchange rate. Arch. Hydrobiol., 73(2): 218-237.

Kautz, K. et al., 1975. Über Spurenelementgehalte in Steinkohlen und den daraus entstehenden Reingasstäuben. VGB Kraftwerkstechnic, 55(10): 672-76.

Kawahara, F.K., Eichelberger, J.W., Reid, B.H. and Stierly, H., 1967. J. Wat. Poll. Contr. Fed., 39: 572.

Kehoe, R.A., Cholak, J. and Story, R.V., 1940. J. Nutr., 19: 579.

Kennedy, J.S., 1975. Insect dispersal. In: D. Pimentel (ed.), Insects, Science and Society. pp. 103-119. New York Academic Press.

Kettlewell, H.B.D., 1956. Further selection experiments on industrial melanism in the Lepidoptera. Heredity, 10: 287-301.

Keyfitz, N., 1977. Population of the World and its Regions. Internal paper, International Institute for Applied Systems Analysis, Laxenburg, Austria.

Khandelwal, K.K., Barduhn, A.J. and Grove, C.S., Jr., 1959. Kinetics of ozonation of cyanides. Ozone Chemistry and Technology, Adv. Chem. Ser., 21, 78. Amer. Chem. Soc., Washington, D.C.

Kilgore, W.W. and Doutt, R.L. (eds.), 1967. Pest Control: Biological, Physical and Selected Chemical Methods. Academic Press, New York.

King, E.J. and Belt, T.H., 1938. Physiol. Rev., 18: 329.

King, R.C., 1957. Am. Nat., 41: 319.

Kira, T. and Shidei, T., 1967. Primary production and turnover of organic matter in different forest ecosystems of the western Pacific. Japanese J. Ecol., 17: 70-87.

Kirk-Othmer, 1967. Encyclopedia of Chemical Technology. John Wiley, New York. Sec. Ed., 14: 410-43.

Klein, D.H. et al., 1975. Trace elements discharges from coal combustion for power production. Water, Air and Soil Poll., 5: 71-77.

Klein, D.H. and Russell, P., 1973. Heavy metals: Fall-out around a power plant. Environ. Sci. Technol., 7(4): 357-58.

Klein, L., 1966. River Pollution 3. Control Butterworth and Co. Ltd., Washington, D.C.

Koch, R.C. and Keisch, B., 1963. U.S. Atomic Energy Commission Report, NSEC-100.

Koch, R.C. and Roesmer, J., 1962. J. Fd. Sci., 27: 309.

Kopp, J.F. and Kroner, R.C., 1965. Appl. Spectry., 19(5): 155.

Kopp, J.F., Kroner, R.C. and Barnett, D.L., 1967. Paper presented at 18th Pittsburg Conference on Analytical Chemistry and Applied Spectroscopy, Pittsburgh, Pa., March, 1967.

Koczy, F.F. and Titze, H., 1958. J. mar. Res., 17: 302.

Kratochvil, B., Boyer, S.L. and Hicks, G.P., 1967. Anal. Chem., 39: 45.

Kraus, K.A., Shore, A.J. and Johnson, J.S., 1967. Hyperfiltration studies. Desalination, 2: 243.

Kringsley, D., 1959. Nature, London, 183: 770.

Krogh Andersen, Karsten, 1973. Recirculation of paper wastes. (Recirkulation

af papiraffald). The Technical University of Denmark.
Kukla, G.J. and Kukla, H.J., 1974. Increased surface albedo in the northern hemisphere. Science, 183: 709-714.

Laitinen, H.A., 1960. Chemical Analysis. McGraw-Hill, New York. Chapt. 10.
Lamar, W.L., Goerlitz, D.F. and Low, L.M., 1965. U.S. Geol. Surv. Water Supply Paper 1817B. Washington, D.C.
Langloid, G., 1955. Adsorption of fatty acids, which are partially dissociated in solution. Mem. Serb. Chim. Etat., Paris; 40: 83.
Lawrence, A. and McCarty, P.L., 1967. Kinetics of methane fermentation in anaerobic waste treatment. Tech. Rep. 75, Department of Civil Engineering, Stanford University, Stanford, California, Feb. 1967.
Leddicotte, G.W., 1959. International Commission on Radiological Protection. Report of committee on permissible doses for internal radiation. Pergamon Press, New York and Oxford.
Lee, G.F., 1965. Water Chemistry Seminars. University of Wisconsin, Madison.
Lee, G.F. and Kluesener, J.W., 1971. Nutrient sources for Lake Wingra, Madison, Wisconsin. Rep. Univ. Wisconsin Water Chemistry Program, December 1971.
Leith, D. And Licht, W., 1972. The collection of cyclone type particle collectors - a new theoretical approach. AICHE Symp. Ser. 68(126): 196-206.
Leopold, Aldo, 1943. Deer irruptions. Wisconsin Cons. Bull., Aug., 1943. Reprinted in Wisconsin Cons. Dept. Publ., 321: 3-11.
Leridy, L.W. and Koksoy, M., 1962. Econ. Geol., 57: 107.
Lieth, H. and Whittaker, R.H., 1975. Primary Productivity of the Biosplhere. Springer-Verlag. pp. 203-215.
Lindberg, S.E. et al., 1975. Mass balance of trace elements in Walker Branch Watershed: Relation to coal-fired steam plants. Environ. Health Persp., 12: 9-18.
Linton, R.N. et al., 1976. Surface predominance of trace elements in airborne particles. Science, 191: 852-54.
Lloyd, M. and Ghelardi, R.J., 1964. A table for calculating the equitability component of species diversity. J. Anim. Ecol., 33: 421-425.
Loehr, R.C., 1974. Characteristics and comparative magnitude of nonpoint sources. J. Wat. Poll. Contr. Fed., 46: 1849-1872.
Logsdon, G.S. and Symons, J.M., 1973. Mercury removal by concentional water treatment techniques. J. AWWA: 554-562.
Long, C., 1961. Biochemists Handbook. Spon. London.
Lotka, A.J., 1925. Elements of Physical Biology. Williams and Wilkins, Baltimore. 460 pp. Reprinted by Dover Publ., New York, 1956.
Lounamaa, J., 1956. Suomal. eläin-ja kasvit. Seur. van Julk, 29: No. 4.
Low, E.M., 1949. J. Mar. Res., 8: 97.
Lu, J.C.S. and Chen, K.Y., 1977. Migration of Trace Metals in Interface of Seawater and Polluted Superficial Sediments. Env. Science and Technology. Vol. 1. pp. 174-182.
Ludzack, F.J. and Ettinger, M.B., 1960. Chemical structures resistance to aerobic biological stabilization. J. Wat. Poll. Contr. Fed., 32: 1173.
Luley, H.G., 1963. Spray irrigation of vegetable and fruit processing wastes. J. Wat. Contr. Fed., 35: 1252.
Lumb, C., 1951. Heat treatment as an aid to sludge dewatering - ten years full-scale operation. J. Proc. Inst. Sew. Purif. part 1, 5.
Lundelius, E.F., 1920. Adsorption and solubility. Kolloid Z., 26: 145.
Lux, H., 1938. Z. anorg. allg. Chem., 240: 21.
Lønholdt, Jens, 1973. The BOD_5, P and N content in raw waste water. Stads- og Havneingeniøren, 7: 1-6.
Lønholdt, Jens, 1976. Nutrient engineering WMO Training Course on Coastal

Pollution (DANIDA): 244-261.

MacArthur, R.H. and Wilson, E.O., 1967. The theory of island biogeography. Princeton Univ. Press, Princeton, N.J. 203 pp.

Mackereth, F.J.H., 1965. Chemical investigations of lake sediments and their interpretation. Proc. Roy. Soc., London, Series B, 161: 295-309.

Mackle, W., Scott, E.W. and Treon, J., 1939. Am. J. Hyg., 29A: 139.

Magee, E.M. et al., 1973. Potential pollutants in fossil fuels. EPA-R2-73-249.

Manabe, S. and Wetherald, R.T., 1975. The effect of doubling the CO_2 concentration on the climate of a general circulation model. J. Atmos. Sci., 32: 3-15.

Margalef, Ramon, 1958. Information theory in ecology. Gen. Syst., 3: 36-71.

Margalef, Ramon, 1963. Successions of populations. Adv. Frontiers of Plant Sci. (Instit. Adv. Sci. and Culture, New Delhi, India), 2: 137-188.

Margalef, Ramon, 1968. Perspectives in Ecological Theory. University of Chicago Press, Chicago. 122 pp.

Margalef, Ramon, 1969. Diversity and stability: A practical proposal and a model for interdependence. In: Diversity and Stability in Ecological Systems. (Woodwell & Smith, eds.).

Margalef, Ramon and Ryther, J.H., 1960. Pigment composition and productivity as related to succession in experimental populations of phytoplankton. Biol. Bull. No. 119: 326-327.

Mather, K., 1953. The genetical structure of populations. Symp. Soc. Exp. Biol., 7: 66-95.

Matsumura, S., Kokubu, N., Watanabe, S. and Sameshima, Y., 1955. Mem. Fac. Sci., Kyushu University Ser. C2: 81.

May, R.M., 1974. Stability in ecosystems: Some comments. Proc. First Int. Congr. Ecol., p. 67.

May, R.M., 1975. Patterns of species abundance and diversity. Chapter 4 (pp. 81-120). In: M.L. Cody and J.M. Diamond, eds. Ecology and evolution of communities. Harvard Univ. Press, Cambridge, Mass.

May, R.M., 1975. Stability and Complexity in Model Ecosystems. (Second edition). Princeton, Princeton University Press.

Mayer, A.M. and Gorham, E., 1957. Ann. Bot., 15: 247.

McCance, R.A. and Widdowson, E.P., 1960. The Composition of Foods. H.M. Stationary Office.

McCarty, P.L., 1964. Anaerobic waste treatment fundamentals. Publ. Works, 95: 9, 107; 95: 10, 123; 95: 11, 19; 12, 95.

McConnell, K.P., 1961. Modern Trends in Acitvation Analysis. Proc. Int. Conf., Texas, p. 137.

McErlean, A.J., Mihursky, J.A. and Brinkley, H.J., 1969. Determination of upper temperature tolerance triangles for aquatic organisms. Chesapeake Sci. 10, 293.

Meadows, Dennis L. and Meadows, Donella H., 1973. Toward Global Equilibrium: Collected Papers. Cambridge, Mass. Wright-Allen Press.

Meadows, Donella H. et al., 1972. The Limits to Growth. New York: Universe Books for Potomac Assosiates.

Menhinick, Edward F., 1964. A comparison of some species diversity indices applied to samples of field insects. Ecology, 45: 859-61.

Mitchell, J. Jr., 1966. In: Treatise on Analytical Chemistry (I.M. Koltoff and P.J. Elving, eds.). Part II, vol. 13. Wiley (Interscience), New York. p. 1.

Mitchell, R.L., 1944. Proc. Nutr. Soc., 1: 183.

Miyaji, I. and Cato, K., 1975. Biological treatment of industrial waste water by using nitrate as an oxygen source. Wat. Research, 9: 95.

Mohr, Eugen, 1969. Konventionelle oder Ionenaustauschanlage. Galvanotechnik, 8: 60.

Moiseenko, U.I., 1959. Geochemistry, 117.

Monier-Williams, G.W., 1950. Trace Elements in Food. Wiley and Sons, New York.

Moody, G.J. and Thomas, J.D.R., 1968. Analyst, 93: 557.

Moon, F.E. and Pall, A.K., 1944. J. Agric. Sci., Camb., 34: 165.

Moore, E.W., 1951. Fundamentals of chlorination of sewage and wastes. Water and Sewage Works, 98(3): 130.

Morgan, J.J. and Stumm, W., 1965. J. Am. Water Works Assoc., 57: 107.

Moriarty, Frank, 1972. Pollutants and food chains. New Scientist, March 16, pp. 594-596.

Morowitz, H.J., 1968. Energy flow in biology. Biological Organization as a Problem in Thermal Physics. Academic Press, N.Y. 179 pp. (See review by H.T. Odum, Science, 164: 683-84 (1969)).

Mortimer, D.C. and Kundo, A., 1975. Interaction between aquatic plants and bed sediments in mercury uptake from flowing water. J. Environ. Qual., 4: 491.

Mount, D.I. and Stephan, C.E., 1967. Trans. Am. Fisheries Soc., 96: 21.

Mueller, H.F., Larson, T.E. and Ferreti, M., 1960. Anal. Chem., 32: 687.

Mulla, M.S., 1966. J. Econ. Entomol., 59: 1085.

Mulligan, Thomas J. and Fox, Robert D., 1976. Chem. Eng.: 49-66.

Mullin, J.B. and Riley, J.P., 1956. J. mar. Res., 15: 103.

Muth, H., Rajewsky, B., Hantke, H.J. and Aurand, K., 1960. Health Phys., 2: 239.

Myhrstad, J.A. and Sandal, J.E., 1969. Behavior and determination of chlorine dioxide. J. Amer. Water Works Assoc., 61: 205.

National Academi of Sciences, 1969. Insect-pest management and control. Chapters 6,9,10 and 15. In: Principles of Plant and Animal Pest Control, vol. 3. Natl. Acad. Sci. Washington, D.C. Publn. 1695, 508 pp.

National Academy of Sciences, 1972. Biological Impacts of Increased Intensities of solar Ultraviolet Radiation. Washington, D.C.

National Air Pollution Control Administration. Air Quality Criteria for Sulfur Oxides. Publ. no. AP-50. Superintendent of Documents. U.S. Government Printing Office, Washington D.C., 1970.

National Technical Advisory Committee, 1968. Water Quality Criteria. Federal Water Pollution Control Administration, Washington, D.C.

Natusch, D.F.S., 1978. Potential carcinogenic species emitted to the atmosphere by fossil-fueled power plants. Environ. Health Persp., 22: 79-90.

Natusch, D.F.S. et al., 1974. Toxic trace elements: Preferential concentration in respirable particles. Science, 183: 202-204.

Neubauer, W.K., 1966. Waste alum sludge characteristics and treatment. N.Y. State Dept. Health Res. Rep. 15.

Neufeld, A.H., 1936. Can. J. Res., 14B: 160.

Neufeld, R.D. and Hermann, F.G., 1975. Heavy metal removal by acclimated activated sludge. J.W.P. Contr. Fed., 47: 310.

Neufeld, R.D. and Thodos, g., 1969. Removal of orthophosphates from aqueous solutions with aluminia. Environ. Sci. Tech., vol. 3, p. 661.

Newman, W.F., 1949. Pharmacology and Toxicology of Uranium Compounds. McGraw-Hill, New York.

Nilsson, Rolf, 1969. Precipitation of phosphates at municipal waste water plants. Stads- og Havneingeniøren, 4: 1-8.

Nordforsk, 1972. Sludge Problems. Naturvårdsverket, Solna, Sweden.

Nordforsk, 1975. Intercalibrering av slamanalyser. (Intercalibration of sludge analyses).

Nordstrom, R.J., McKinnon, A.E. and de Freitas, A.S.W., 1975. A bioenergetics-band model for pollutant accumulation by fish. Simulation of PCB and methylmercury levels in Ottawa River perch (Perca flevescens). J. Fish. Res. Bd. Canada, 33: 248.

Northington, C.W., Chang, S.L. and McCabe, L.J., 1970. In: Water Quality Improvements by Physical-Chemical Processes (N.F. Gloyna and W.W. Eikenfelder, Jr., eds.). Univ. Texas Press, Austin. pp. 49-56.

Odum, E.P., 1950. Bird populations of the Highlands (North Carolina) Plateau in relation to plant succession and avian invasion. Ecology, 31: 587-605.
Odum, E.P., 1969. The strategy of ecosystem development. Science, 164: 262-270.
Odum. E.P., 1971. Fundamentals of Ecology. W.B. Saunders Co., Philadelphia.
Odum, H.T., 1956. Primary production in flowing waters. Limnol. Oceanogr., 1: 102-117.
Odum, H.T., Cantlon, J.E. and Kornicker, L.S., 1960. An organizational hierarchy postulate for the interpretation of species-individuals distribution, species entropy and ecosystem evolution and the meaning of a species-variety index. Ecology, 41: 395-399.
Odum, H.T. and Pinkerton, R.C., 1955. Times speed regulator, the optimum efficiency for maximum output in physical and biological systems. Amer. Sci., 43: 331-343.
Onnen, J.H., 1972. Wet scrubbers tackle pollution. Environ. Sci. Technol., 6: 994-998.
Orians, G.H., 1974. Diversity, stability and maturity in natural ecosystems. Proc. First Int. Congr. Ecol., pp. 64-65.
Orlob, G., 1981. State of the Art of Water Quality Modelling. IIASA, Laxenburg, Austria. (In press).

Paine, R.T., 1966. Food web diversity and species diversity. Amer. Nat., 100: 65-75.
Painter, H.A. and Jones, K., 1963. J. Appl. Bacteriol., 26: 471.
Park, Richard A. et al., 1978. The aquatic Ecosystem Model MS. CLEANER. Proc. Int. conf. on Ecol. Modelling, 28 Aug.-2 Sep. 1978, Copenhagen (ISEM), p. 579.
Parker, David D., Cliffor, W. Randall and King, Paul H., 1972. Biological conditioning for improved sludge. Filer ability. J. Wat. Poll. contr. Fed., 44: 266-277.
Parr, R.M. and Taylor, D.M., 1963. Physics Med. Biol., 8: 43.
Parr, R.M. and Taylor, D.M., 1964. Biochem. J., 91: 424.
Patterson, J.W., 1970. Ph. D. Thesis. University of Florida. Gainsville.
Patterson, J.W., Brezonik, P.L. and Putnam, H.D., 1970. Environ. Sci. Technol., 4: 569.
Pauling, L., 1960. The Nature of Chemical Bond. Cornell University Press, Itacha, N.Y., U.S.A.
Pavlova, A.K., 1956. Vestsi Akad. Navuk. BSSR, No. 3: 83.
Penny, C. and Adams, C., 1964. 4th Rept. Roy. Comm. Pollution of Rivers in Scotland, vol. 2, Evidence, London, 1863, pp. 377-391. In: J.R.E. Jones: Fish and River Pollution. Butterworth, Washington, D.C.
Perkins, H.C., 1974. Air Pollution. I.S.E., Tokyo. 407 pp.
Perry, H. and Landsberg, H.H., 1977. Projected World Energy Consumption. In: Energy and Climate. National Academy of Sciences, Washington, D.C.
Pielou, E.C., 1966. Species-diversity and pattern diversity in the study of ecological succession. J. Theoret. Biol., 10: 370-83.
Pionke, H.B., Konrad, J.G., Chesters, G. and Armstrong, D.E., 1968. Analyst, 93: 363.
Piperno, E., 1975. Trace element emissions: Aspects of environmental toxicology. Adv. Chem. Ser. No. 141: 192-209.
Porter, J.R., 1946. Bacterial Chemistry and Physiology. Wiley and Sons, New York.
Posselt, H.S., 1966. Sodium and potassium ferrate (VI). A Bibliography and

Literature Review. Unpublished Research Report, Carus Chemical Company, Inc., LaSalle, I11.

Prakash, C.B. and Murray, F.E., 1975. Particle conditioning by steam condensation. A.I.Ch.E. Symp. Ser., 71(147): 81-88.

Preston, F.W., 1969. Diversity and stability in the biological world. In: Diversity and Stability in Ecological Systems. Woodwell and Smith, eds.

Pring, R.T., 1972. Specification considerations for fabric sollectors. Pollution Eng., 4(12): 22-24.

Push, C.H. and Walcha, W., 1975. Fractionation of metal salts by hyperfiltration. 2nd Int. Congr. on Industrial Waste Water, Stockholm.

Pytkowicz, R.M., 1967. Geochim cosmochim Acta, 31: 63.

Quastel, J.H. and Scholefield, P.G., 1951. Bacteriol. Rev., 15: 1.

Quirk, T.P., 1964. Economic aspects of incineration vs. incineration-drying. J. Wat. Poll. contr. Fed., 36: 11, 1355.

Ragaini, R.C. and Ondov, J.M., 1975. Trace contaminants from coal-fired fower plants. Paper, Int. Conf. Env. Sens. Assessm., Las Vegas.

Ramanathan, V., 1975. Greenhouse effect due to chlorofluorocarbons: Climatic Implications. Science, 190: 50-52.

Rasmussen, D.I., 1941. Biotic communities of Kaibah Plateau, Arizona. Ecol. Mongr., 11: 229-275.

Reck, R.A., 1978. Global Temperature Changes: Relative Importance of different Parameters as calculated with a Radioactive-Convective Model. In: Carbon Dioxide, Climate and Society. Proceedings of an IIASA Workshop, Febr. 21-24, 1978. (Jim Williams, ed.). Pergamon Press, Oxford. p. 193-200.

Reck, R.A. and Fry, D.L. The Direct Effect of Chlorofluoromethanes on the Atmospheric Temperature. GMR-2564. General Motors Research Laboratories, Warren, Michigan.

Reddy, K.R., Campbell, K.L., Graetz, D.A. and Portier, K.M., 1982. use of biological filters for treating agricultural drainage effluents. J. Environ. Qual., Vol 11, no. 4, 1982, p. 591.

Rhodes, A.J. and van Rooyen, C.E., 1968. Textbook of Virology. Williams and Wilkins, Baltimore, Md. p. 966.

Rice, J.R. and Dishburger, H.J., 1968. J. Agr. Food Chem., 16: 867.

Riley, J.P., 1965. In: Chemical Oceanography (J.P. Riley and G. Skirrow, eds.). Vol. 2. Academic Press, New York, Chapter 21.

Rinaldi, S., Soncini-Sessa, R., Stehfest, H. and Tamura, H., 1979. Modelling and Control of River Quality. McGraw-Hill, Great Britain.

Rizzo, J.C. and Shepherd, A.R., 1977. ChemIlll. Eng.: 95-100.

Roberts, E.J., 1949. Thickening - art or science? Mining Eng., 1: 61.

Roberts, Kelvin, Wenneberg, Ann-Marie and Friberg, Stig, 1970. Section "Control of Water Pollution". IAWPR Conference, 1970.

Rohde, H., 1978. Kompendium i luftforureningsmeteorologi. Meteorologiska Instittuonen, Stockholm University. 75 pp.

Rose, H.E. and Wood, A.J., 1966. An introduction to electrostatic precipitation in theory and practice. 2nd ed., p. 81. Constable and Co., Ltd. London.

Rosendahl, Arne, 1970. Undersökelser i halvteknisk målestok vedrörende fjerning av fosfor från mekanisk och biologisk renset kommunalt avlöpsvann ved kjemisk rensning. 6. Nordiske Symposiet om Vattenforskning, Scandicon 21-23 April 1970.

Rovel, J.M., 1972. Chemical regeneration of activated carbon. Proc. Wat. Techn., 1: 187-190.

Rullman, D.H., 1976. Baghouse technology: a perspective. J. Air Pollut. Control Assoc., 26: 16-18.

Rüb, Friedmond, 1969. Aufbereitung von spielwässern der Metalindustrie nach dem Ionenaustauscherverfahren. Wasser, Luft und Betreib, 8: 292-296.

Samsahl, K. and Soremark, R., 1961. Modern Trends in Activation Analysis. Proc. Int. Conf., Texas.

Sanborn, N.P., 1953. Disposal of food processing water by spray irrigation. Sewage and Int. Wastes, 25: 1034.

Sanders, L.H., 1968. Marine benthic diversity: A comparative study. Amer. Nat., 102: 243-282.

Sawyer, C.N., 1960. Chemistry for Sanitary Engineers (R. Eliassen, ed.). McGraw-Hill, New York. pp. 274.

Sawyer, C.N. and Bradney, L., 1946. Sewage Works J., 18: 1113.

Scaramelli, A.B. and Dibiano, F.A., 1973. Upgrading the activated sludge system by addition of powdered carbon. Water and Sewage Works: 90-94.

Schaufler, Gerhard, 1969. Ion exchangers in industrial waste water clarification. Chem. Tech. Ind.: 787-790.

Schindler, D.W. and Nighswander, J.E., 1970. Nutrient supply and primary production in Clear Lake, Eastern Ontario. J. Fish. Res. North Canada, 27: 2009-2036.

Schjødtz-Hansen, P., 1968. Dansk Teknisk Tidsskrift, 5.

Schjødtz-Hansen, P. and Krogh, O., 1968. Nordisk Mejeritidsskrift, 34: 194.

Schofield, A. and Haskin, L., 1964. Geochim. cosmochim. Acta, 28: 437.

Schwarz, K. and Foltz, C.M., 1958. Fed. Proc. Fedn. Am. Socs. exp. Biol., 17: 492.

Schwitzgebel, K. et al., 1975. Trace element discharge from coal-fired power plants. Symposium Proceedings, vol. II, part 2, pp. 533-51. International Conference on Heavy Metals in the Environment, Toronto.

Scorer, R., 1968. Air Pollution. Pergamon Press, Oxford.

Seip, Knut Lehre, 1979. Mathematical model for uptake of heavy metals in benthic algae. J. Ecol. Model., 6: 183-198.

Shacklette, H.T., 1965. Bull. U.S. geol. Surv. 1198D.

Shannon, C.E. and Weaver, W., 1963. The mathematical theory of communication. University of Illinois Press, Urbana. 117 pp.

Shibuya, M. and Nakai, T., 1963. Proc. 5th Conf. Radioisotopes, Japan.

Shimp, N.F., Connor, J., Prince, A.L. and Bear, F.E., 195f7. Soil Sci., 83: 51.

Sillen, I.G. and Martell, A.J., 1964. Stability Constants of Metal - Ion Complexes. The Chemical Society, London.

Sillen, L.G., 1961. Oceanography Publ. No. 67 AAAS, Washington D.C.: 549-581.

Sillen, L.G., 1967. Adv. Chem. Serv., 67: 45-57.

Simpson, E.H., 1949. Measurement of diversity. Nature, 163: 688.

Singer, Philip C. and Zilly, William B., 1975. Ozonation of ammonium in waste water. Water Research, 9: 127.

Skinder, Brian J., 1969. Earth Resources. Englewood Cliffs, N.J.: Prentice-Hall.

Slobodkin, L.B. and Sanders, H.L., 1969. On the contribution of environmental predictability to species diversity. In: Diversity and Stability in Ecological Systems (Woodwell and Smith, eds.).

Smales, A.A. and Pate, B.D., 1952. Analyst, London, 77: 196.

Smales, A.A. and Salmon, L., 1955. Analyst, London, 80: 37.

Smidth, F.L./MT, 1973. Report on the eutrophication of Lake Lyngby.

Smith, F.E., 1963. Population dynamics in Daphnia magna and a new model for population growth. Ecology, 44: 651-663.

Smith, M. (ed.), 1968. Recommended Guide for the Prediction of the Dispersion of Airborne Effluents. Am. Soc. Mech. Eng.

Smith, R.D. et al., 1979. Concentration dependence upon particle size of volatilized elements in fly ash. Environ. Sci. Technol., 13(5): 553-58.

Smith, R.D. et al., 1979. Characterizational formation of submicron particles in coal-fired plants. Atmospheric Environment, 13: 607-17.

Snider, Erich H. and Porter, John J., 1974. Ozone treatment of dye waste, J.W.P. Contr. Fed., 46: 887.

Snyder, L.R., 1961. J. Chromatogr. 6: 22.

Soremark, R. and Bergman, B., 1962. Acta Isotopica, 2: 1.

Sorensen, T., 1948. A method of estabilishing groups of equal amplitude in plant society based on similarity of species content. Kgl. Danske Vidensk. Selskab, 5: 1-34.

Southern, H.N., 1970. The natural control of a population of Tawny Owls (Strix aluco). J. Zool., London, 162: 197-285.

Southwood, T.T.E., May, R.M., Hassell, M.P. and Conway, G.R., 1974. Ecological strategies and population parameters. Am. Nat., 108: 791-804.

Sowden, E.M. and Stitch, S.R., 1957. Biochem. J., 67: 104.

Spanier, G., 1969. Ionenaustauscher für die Wasser un Abwasseraufbereitung in der Metallveredlung. Oberfläche, 10: 365-373.

Spector, W.S., 1956. Handbook of Biological Data. Saunders, Philadelphia.

Sprague, J.B., 1970. Water Res., 4(1): 3-32.

Stahl, E., 1965. Thin-Layer Chromatography. A Laboratory Handbook. Academic, New York. 31 pp.

Stamm, M.D. and Fernandez, F., 1958. Revta. esp. Fisiol., 14: 177,185.

Standard Methods for the Examination of Water and Waste Water, 1965. 12th ed. Americal Public Health Assoc., New York. 287 pp.

Stansel, J. and Huke, R.E., 1975. Rice. In: Impacts of Climatic Change on the Biosphere. CIAP Monograph 5, pt. 2, Climatic Effects, 4-90 to 4-132.

Statens Naturvårdsverk Publikationer, 1969. Experiences by Chemical Purification, 10: 12-13.

Stern, A.C. (ed.), 1977. Air Pollution II. A.P. New York. 684 pp.

Stitch, S.R., 1957. Biochem. Z., 304: 73.

Storherr, R.W. and Watts, R.R., 1965. J. Assoc. Offic. Agr. Chem., 48: 1154.

Stumm, W., 1958. Ozone as a disinfectant for water sewage. J. Boston Soc. Civ. Eng., 45(1): 68.

Stumm, W. (ed.), 1967. Equilibrium Concepts in Natural Water Systems. Advan. Chem. Ser., 67. Americal Chemical Society, Washington D.C.

Stumm, W. and Leckie, J.O., 1970. Phosphate exchange with sediments; Its role in the productivity of surface waters. Advances in water pollution research. Proc. of 5th Int. Conf. in San Francisco and Hawaii, III 26/1 to 16.

Stumm, W. and Morgan, 1970. Aquatic Chemistry. Wiley Interscience, New York.

Suzuki, T. and Hamada, I., 1956. J. Chem. Soc., Japan, Pure Chem. Sect., 77: 125.

Swaine, D.J., 1977. Trace elements in coal. In: Trace substances in Environmental Health XI. (H. Hemphill, ed.), Missouri. p. 107-115.

Svensson, B.H. and Söderlund, R. (eds.), 1976. Ecological Bulletin, No. 22. Nitrogen, phosphorus and sulphur - Global Cycles. SCOPE Report 7, Örsundsbro, Sweden.

Sylvester, R.O., 1960. Nutrient Content of Drainage Water from Forested, Urban and Agricultural Areas. Trans. Sem. Algae Metropolitan Wastes, U.S. Public Health Service. R.A. Taft Center, Cincinnati, Ohio.

Särkkä, Mirja, 1970. Praktiska erfarenheter av närsaltreduktion. 6. Nordiske Symposiet om Vattenforskning, Scandicon 21-23 April 1970: 177-187.

Søchting, U. and Johnsen, I., 1978. Lichen transplants as biological indicators of SO_2 air pollution in Copenhagen. Bull. Environ. Contam. & Toxicol., 19: 1-7.

Tarzwell, C.M. and Gaufin, A.R., 1953. Proc. Ind. Waste Conf. 8th. Purdue University, Lafayette Indiana, 1953. p. 295.

Teasley, J.I. and Cox, W.S., 1963. J. am. Water Works Assoc., 55: 1093.
Thomann, R.V., 1978. Size Dependent Model of Harzardous Substances in Aquatic Food Chain. EPA-600/3-78-036. April 1978.
Thomas, E.A., 1965. Phosphat-Elimination in der Bebetschlammanlage von Mannedorf und Phosphat Fixation in See und Klarschlamm. Vierteljahrish Naturf. Ges. Zürick, Bd. 110: 419.
Thomas, H.A., Jr. and McKee, J.E., Congitudinal mixing in aeration tanks. Sew. Works J., 16: 42-56.
Thomas, M.D., Hendricks, R.H. and Hill, G.R., 1950. Soil Sci., 70: 9.
Thomas, M.J. and Theis, T.L., 1976. Effect of selected ions on the removal of chrome(III) hydroxide. J. Wat. Poll. Contr. Fed., 43: 2032.
Thompson, L.M., 1975. Weather variablity, climatic change, and grain production. Science, 188: 553-541.
Thompson, T.G. and Chow, T.J., 1956. Univ. Washington Publs. Oceanog. No. 184: 20.
Tipton, I.H. and Cook, M.J., 1963. Health Phys., 9: 103.
Trubnick, E.H. and Mueller, P.K., 1958. Sludge dewatering practice. Sew. Ind. Wastes, 30: 11, 1364.
Turner, D.B., 1970. Workbook of Atmospheric Dispersion Estimates. U.S. Public Health Service Publication 999-AP-26, revised 1970 ed.
Turner, R.C., Radley, J.M. and Mayneord, W.V., 1958. Health Phys. 1: 268.
Turner, R.C., Radley, J.M. and Mayneord, W.V., 1958. Br. J. Radiol., 31: 397.
Tyutina, N.A., Aleskovsky, V.B. and Vasilev, P.I., 1959. Geochemistry: 668.
Tönseth, E.I. and Berridge, H.B., 1968. Removal of proteins from industrial waste water. Effluent and Water Treatment J., 8: 124-128.

Untersteiner, N., 1975. Sea ice and ice sheets and their role in climatic variations. In: The Physical Basis of Climate and Climatic Modelling. GARP Publ. Ser. 16. World Meteorological Organization, Geneva: 206-226.
U.S. Department of Health, Education and Welfare, 1969. Tall Stacks Various Atmospheric Phenomena, and Related Aspects. Report APTD 69-12, pp. 1-12, Washington D.C.

Van Hook, R.I., 1978. Potential health and environmental effects of trace elements and reaionuclides from increased coal utilization. Oak Ridge National Laboratory (ORNL-5367).
Veger, J., 1962. Vodn. Hospodarstvi, 12: 172. Through Chem Abstr., 57: 10947d.
Veith, G.D., Defoe, D.L. and Bergstedt, B.V., 1979. Measuring and estimating the bioconcentration factor of chemicals in fish. J. Fish. Res. Board Can., 36: 1040-1048.
Vinogradov, A.P., 1953. The elementary chemical composition of marine organisms. Sears Foundation, New Haven, Conn.
Vinogradova, Z.A. and Kobalsky, V.V., 1962. Dokl. Acad. Nauk SSSR, 147: 1458.
Vollenweider, R.A., 1968. The scientific basis of Lake and Stream Eutrophication with particular reference to phosphorus and nitrogen as eutrophication factors. Tech. Rep. OECD, Paris. DAS/DSI/68, 27: 1-182.
Vollenweider, R.A., 1969. Möglichkeiten und Grenzen elementarer Modelle der Stoffbilanz von Seen. Arch. Hydrobiol., 66: 1-136.
Vollenweider, R.A., 1975. Input-output models with special reference to the phosphorus loading concept in limnology. Schweiz. Z. Hydrol., 37: 53-83.
Von Lehmden, D.J. et al., 1974. Determination of trace elements in coal, fly ash, fuel oil, and gasoline - A preliminary comparison of selected analytical techniques. Anal. Chem., 46: 239-245.
Voss, K.D., Burries, F.O. Jr. and Riley, R.L., 1966. Kinetic study of hydrolysis of cellulose acetate in the pH range 2-10. J. Appl. Poly. Sci., 10: 825.

Waid, D.E., 1972. Controlling pollutants via thermal incineration. Chem. Eng. Prog., 68(8): 57-58.

Waid, D.E., 1974. Thermal oxidation or incineration. Proc. Special - Pollut. Control Assoc., Pittsburgh, PA: 62-79.

Wakita, H. and Kigoshi, K., 1964. J. Chem. Sic., Japan, Pure Chem. Sect., 85: 476.

Walker, J.D. and Drier, D.E., 1966. Aerobic digestion of sewage sludge solids. Paper presented at 35th Annual Meeting of Georgia Water Pollution Association, Walker Process Equipment, Inc., Publ. 26.130.

Wang, W.C., Yung, Y.L., Lacis, A.A, Mo, T. and Hansen, J.E., 1976. Greenhouse effects due to manmade perturbations of trace gases. Science, 194: 685.

Wangersky, P.J. and Cunningham, 1956. On time lags in equations of growth. Proc nat. Acad. Sci., 42: 699-702.

Wangersky, P.J. and Cunningham, 1957. Time lag in population models. Cold Spring Harbor Symp. Quant. Biol., 42: 329-338.

Warnick, S.L. and Gaufin, A.R., 1965. J. Am. Water Works Assoc., 57: 1023.

Warren, C.E. and Doudoroff, P., 1958. The development of methods for using bioassays in the control of pulp mill waste disposal. TAPPI, 41: 8, 211A.

Watt, Kenneth E.F., 1972. Tambora and Krakatau: Volcanoes and the cooling of the world. Saturday Review, Dec. 23: 43-44.

Weatherholtz, W.M., Cornwell, G.W., Young, R.W. and Webb, R.E., 1967. J. Agr. Food Chem., 15: 667.

Weber, W.J. Jr. and Morris, J.C., 1963. Kinetics of adsorption on carbon from solution. J. Sanit. Eng. Div. Amer. Sic. Civ. Eng., 89: SA 2, 31-59.

Weber, W.J. Jr. and Morris, J.C., 1964. Equilibria and capacities for adsorption of carbon. J. Sanit. Eng. Div. Amer. Soc. Civ. Eng., 90: SA 3, 79-107.

Weber, W.J. Jr. and Morris, J.C., 1965. Adsorption of biochemically resistant materials from solution. J. Wat. Poll. Contr. Fed., 37: 425.

Wester, P.O., 1965. Biochem. Biophys. Acta, 109: 268.

Weibel, S.R. et al., 1964. Urban land runoff as a factor in stream pollution. J. Wat. Poll. Contr. Fed., 36(7): 914.

Weichselbaum, T.E., Hagerty, J.C. and Mark, H.B. Jr., 1969. Annal. Chem., 41: 848.

Weiderholm, T., 1980. Use of benthos in lake monitoring. J. Water Pollut. Control Fed. 52, 537.

Wheatly, G.A. and Hardman, J.A., 1965. Nature, 207: 486.

Whealer, G.L., 1976. Chlorine dioxide a selective oxidant for industrial waste water treatment. 4th Annual Ind. Poll. Conf. April 1976.

White, H.J., 1974. Resistivity problems in electrostatic procipiattion. J. Air Poll. Control. Assoc., 24: 313-338.

Whittaker, R.H., 1970. Communities and ecosystems. Macmillan, New York. 162 pp.

Wicke, M., 1971. Collection efficiency and operation behaviour of wet scrubbers. Paper EN 16H, pp. 713-718, in Proceedings of the 2nd Int. Clean Air Congress, H.M. Englund and W.T. Beery, eds. Academic Press, Inc., New York.

Williams, Jim, (ed.), 1978. Carbon dioxide, climate and society. Proceedings of a IIASA Workshop cosponllsored by WMO, UNEP, and SCOPE, Febr. 21-24, 1978. Pergamon Press, Oxford.

Williamson, G., 1959. Proc. 6th Ind. Waste Conf., Ontario Canada, p. 17.

Wing, A. Bruce and Steinfeld, William M., 1970. Comparison of stone-packed and plastic-paced trickling filters. J.W. Poll. Contr. Fed., 42: 255.

Wisneiwski, T.F., Wiley, A.J. and Lueck, B.F., 1956. TAPPI, 39: 2,65.

Wolf, Phillip C., 1971. Carbon Monoxide Measurement and Monitoring in Urban Air. Env. Sci. Tech. 5(3): 231.

Wong, J.B., Ranz, W.E. and Johnstone, H.F., 1956. Collection of aerosols by fiber mats. Technical Report No. 11. Engineering Experiment Station,

University of Illinois.

Woodwell, G.M. et al., 1967. DDT residues in an East Coast esturay: A case of biological concentration of a persistent insecticide. Science, 156: 821-824.

Wren, J.J., 1960. Chromatogr. Rev., 3: 111.

Wright, F.C., Gilbert, B.N. and Riner, J.C., 1967. J. Agr. Food Chem., 15: 1038.

Wright, R.F. and Gjessing, E.T., 1976. Changes in the chemical composition of lakes. Ambiom 5f(5-6): 219-223.

WRL, 1977. Heavy metal ion exchanger WRL/500CX. Copenhagen.

Wuhrmann, K., 1956. In: Biological Treatment of Sewage and Industrial Waste, vol. I. (J. McCabe and W.W. Eckenfelder, eds.). Reinhold, New York.

Wynn, C.S., Kirk, B.S. and McNabney, R., 1972. Pilot plant for tertiary treatment of waste water with ozone. Water, 69: 42.

Yamagata, N., 1950. J. Chem. Soc., Japan, 71: 228.

Yamagata, N., 1962. J. Radiat. Res., 3: 4.

Yamagata, N., Murata, S. and Torii, T., 1962. J. Radiat. Res., 3: 4.

Yee, W.C., 1966. Selective removal of mixed phosphate by activated alumina. J. Am. Water Works Assoc., p. 239.

Zahn, R., 1961. Staub. 21, 56-60.

Zimen, K.E., 1978. Source function for CO_2 in the atmosphere. In: Carbon Dioxide, Climate and Society. Proceedings of a IIASA Workshop, Febr. 21-24, 1978. (Jim Williams, ed.). Pergamon Press Oxford. p. 89-96.

Zimen, K.E., Offermann, P. and Hartmann, G., 1977. Source functions of CO_2 and furture CO_2 burden in the atmosphere. Z. Naturforschung, 32a: 1544.

Zimmerman, F.J., 1958. Chem. Eng., 65: 17, 117.

Zuckermann, M.M. and Molof, A.H., 1970. High quality reuse water by chemical physical waste water treatment. J. Wat. Poll. Contr. Fed., p. 437. (Included the discussion by Weber).

INDEX

Atomic absorption spectrophoto-
metry (AAS), 505
Atomic fluorescence spectrophoto-
metry, 505
Atomized suspension technique,
433, 434
ATP, 47, 500
Automatic sampling devices, 494
Autotrophic organisms, 299
AVR, 330, 331
AWT system, 318

Background concentration in
atmosphere, 37-39
Bacteria, 80
Bacterial analysis, 506
Bathypelagic, 247
Bentonite, 365-366
Baffle scrubbers, 470
Benzene, 367, 488, 502
Beryllium, 268
Beta-mesosaprobic water, 75, 79
Bioaccumulation, 132
Bioadsorption, 309
Bioassay techniques, 508
Bioconcentration, 132
Biodegradability, 192, 261, 271,
363, 364, 439
Biodegradable
- compounds, 40
Biological
- concentration factor, 127
- half-life time, 40
- index, 535
- magnification 95, 96, 158, 261
- modelling, 133
- monitoring, 533-538
Biomagnification, 132, 139
Biomanipulation, 255
Biotower, 309, 310
Birth rate, 17
Bismuth, 267-269
Bleaching powder, 412
BOD, 499
BOD$_5$, 32-33, 65, 82, 142, 246,
287, 288, 289, 300, 303, 307,
309, 310, 315, 316, 318, 326,
327, 338, 359, 414, 415, 439,
497, 498, 499
Bogs, 248
Boric acid, 240
Break point, 371, 410
Breeze inversion, 520
Bromine, 413
Brown algae, 140
Buffer capacity (see also ecologi-
cal buffer capacity), 188, 239,
256

Buffering capacity, 240, 243-246
Butadiene, 416

Cadmium, 34, 97, 122, 140, 164,
165, 262, 267-269, 337, 382,
524, 529
Calcium, 382
- carbonate, 348, 400
- hydroxide, 242, 255, 328, 333,
334, 335, 337, 378, 384, 385,
389, 401, 415, 427, 487
- oxide, 337, 400
- sulphate, 140
Carbamates, 262
Carbohydrate, 289, 312
Carbon, 35, 90
- activated, 316, 317, 318, 349,
351, 370, 372, 373, 492
- cycle, 60
- dioxide, 58, 60-62, 169, 172,
173, 174, 175, 177, 180, 225,
237, 312, 337, 345, 401, 402,
430, 439, 475, 477, 486, 491,
492
- hydrides, 260, 475, 476, 477
- monoxide, 117, 259, 449, 475,
477, 479, 517-518, 530
- pools, 173
Carbonanceous oxygen demand, 304
Carbonate ions, 240
Carcinogenic metals, 265, 269
Carnivores, 186
Carnivorous, 58
Carrying capacity, 166, 196, 199
Catalytic
- afterburners, 478
- oxidation, 478, 482, 492
Catastrophies, 271
Cation exchange capacity, 235
Caustic soda factory, 264
Cellular growth, 300
Cellulose, 485
- acetate membrane, 396
- ion exchangers, 316, 359
- polyethyleneimine, 389
Cement, 485
Centrifugal force, 459
Centrifugation, 336, 385, 422,
429, 497
Chamber scrubbers, 470
Charged-droplet scrubbers, 472
Chemical precipitation, 317, 325,
327, 335, 338, 358, 359, 365,
381, 382, 424, 448
Chemical regeneration, 374
Chemotrophic organisms, 299
Chick's las, 404, 405
Chloramines, 349, 409, 413

Chlorinated, 95
Chlorinated hydrocarbons, 96, 262
Chlorination, 349, 351, 410
Chlorine, 350, 375, 408, 409, 412,
  413, 414, 416, 487
  - dioxide, 375, 378, 380, 404
  - free available, 408
  - species, 411
Chlorobenzene, 367
Chlorobiphenyl, 97
Chlorphenol, 186
Chromate, 385, 387, 389, 390
Chromatography, 495
Chromium, 263, 267-269, 337, 382,
  387, 389
  - (III)hydroxide, 387, 388
Chromosomes, 190
Circulation, 253
Citric acid, 416
Classification of ecosystems, 249
Claus process, 481
Clay, 353
Clean air, 518
Cleaning up, 501
Climate, 169, 173, 174, 177
  - diversity, 209
Climatic factors, 177
  - variations, 177
Clinoptilolite, 353, 357
Coagulant, 429
Cobalt, 97, 267-269, 363
COD, 344, 345, 431, 499
Codistillation method, 501
Coevolution, 222, 223
COHb, 117
Coli, E., 398, 405, 406, 414, 416
Collecting electrodes, 466
Colloidal
  - particles, 496
  - substances, 370
Colour, 376, 398, 413
Combustion, 422, 427, 434, 446,
  476, 491
Commensalism, 202
Compartment models, 133
Competition, 202, 206
Competitive adsorption, 368
Complexity, 34, 167
Composting, 422, 448
  - plant, 441
  - time, 440
Compounds, 446
Compressibility, 424
Concentration factor, 127, 130
Condensation, 474, 475
Conditioner, 282, 163, 164, 165,
  166
Conditioning, 422

Conduction, 181
Coniferous, 247
Coning, 455
Contact stabilization process, 309
Continous mixed flow, 40-42
Convection, 182
Cooling system, 285
Cope's rule, 199
Copper, 30-32, 97, 124, 267-269,
  337, 363, 382, 389, 416
Corn yield, 177
Corrosion problem, 480
Countercurrent extraction, 501
Coverage of sediment, 252
Creatinine, 289
Crop yields, 52, 155
Cross-linking, 345
Cryosphere, 173, 178, 179
Cumulative, 135
Cyanate, 378
Cyanides, 363, 376, 378, 380
Cyclone, 435, 459, 461
Cyclonic scrubbers, 470

D_{50}, 473, 492
DDD, 262
DDE, 262
DDT, 36, 40, 95, 96, 97, 125, 140,
  165, 191, 192, 203, 258, 262,
  502
Death rate, 17
Deciduous forest, 247
Decomposition, 65, 70, 87
Decomposition, 300
  - chain, 66
  - rate, 236
Degradation of energy, 149
Degradation rate, 44
Degradation zone, 75
Denitrification, 88, 341, 344,
  345
Density, 222
Deposition, 113, 135, 282, 348
Desert, 247
Desorption, 489
Detergents, 495
Detritus, 88
Dewatering characteristics, 425
Dialysis, 393, 497
Dichlorophenol, 416
Dieldrin, 502
Differential themal analysis, 497
Diffusion, 69, 366, 462, 486
  - coefficient, 43, 113, 486
Digestion
  - methods, 515
  - of sludge, 430-434
Dilution, 68

Fertilization, 233
Fertilizers, 63, 156, 271, 447
 - control, 253-254
Fertilizing value, 447
Filter, 462, 466
 - analysis, 529
 - media, 295
Filtration, 295, 296, 318, 336,
 351, 365, 370, 385, 398, 422,
 425, 429
 - time, 425
Fire, 202
First law of thermodynamics,
 143, 187
First order kinetic, 66
Fish ponds, 448
Fixed adsorption, 371
Flame photometry, 505
Flocculation, 253, 304, 339, 364,
 384, 385, 398
Flooding, 184
Flotation, 296, 297, 364
Fluctuation, 200
Flue gas cleaning, 481
Fluor, 535
Fluorescense spectra, 500
Fluoride, 535, 537, 538
Flyash, 445
Food
 - additives, 258, 275
 - chain, 95, 148, 157, 186
 - processing industry, 288,
 316, 319
 - waste, 437
 - web, 157, 187
Forest, 247
Formaldehyde, 479, 496
Fossil energy, 171
 - fuel, 18, 61, 150, 171, 172,
 228, 229, 442
 - record, 199
Free available chlorine, 408
Freeze concentration, 495
Freeze-drying, 406
Freezing, 348, 406, 427
Freundlich's adsorption isotherm,
 331, 333, 488
Freundlich isotherm, 367
Front inversion, 520
Fuel cells, 479
Fumes, 450
Fumigation, 112, 455
Fungicides, 265
Fusion energy, 278

Garden waste, 437
Gasification, 482
Gaussian function, 108-110

Gel filtration, 497
Generation time, 163, 186, 195,
 200
Genetic variation, 189
Geothermal energy, 278
Glass and ceramic products, 437
Glucose, 500
 - trisulphate, 365-366
Gradual agents, 34
Grazing, 87, 88, 123
Green fields, 447, 448
Gross production, 218
Ground water, 105, 107, 399, 438,
 320, 503
Growth
 - coefficients, 196, 198, 255
 - efficiency, 159
 - rate, 194, 301
 - rate, Atlantic Salmon, 52
Guggenheim, 326
 - process, 317

Habitat modification, 205
Half life, 166, 377
Half saturation constants, 54
Halogens, 446
Hardness, 398, 400
Heat trapping, 184, 185
Heavy metals, 36, 46, 95, 98, 99,
 256, 262-271, 272, 380-397, 445,
 446, 495, 518, 529, 533
Henry's Constant, 59, 61
Henry's Law, 59, 68
Herbicides, 156
Herbivores, 223
Herbivorous, 58
Herring filetting, 319
Heterotrophic organisms, 299
Hexachloriphenyl, 97
Hexane, 502
Homoeostatis, 188
Hospital waste, 445
Hot capping, 184
Humidity, 186, 463
 - relative, 183
Hydraulic
 - loading, 309
 - stability, 293
Hydrocarbons, 95, 117, 495, 530
Hydrochloric acid, 485
Hydrogen, 90, 312, 374, 481
 - bond, 356
 - carbonate, 225, 237, 402
 - cyanide, 492
 - cycle, 352
 - peroxide, 375
 - sulphide, 142, 312, 345, 365,

Mercury, 34, 97, 131, 140, 141, 226, 262, 263, 264, 267-269, 363, 416, 446, 529
- levels, 266
Meridional profile, 175
Mesopelagic, 247
Mesotrophic, 91, 95
Metabolic processes, 128
Metabolic rate, 129, 162, 163
Metabolism, 135
Meta-hydrogen sulphite, 385
Metals, 437
- oxides, 482
Meteorological conditions, 519
Methane, 312, 517-518, 530
- formers, 430
Methane-producing organisms, 312
Methoxychlor, 140
Methyl mercaptane, 488
Methyle methacrylate, 478
Michaelis-Menten equation, 301, 302, 361
Migration, 200
Minerals, 35
- waste, 274
Mineralization, 87, 88
Mining wastes, 274, 419, 445
Mixed forests, 215
Model, 139
Modified logistic equation, 198
Moisture content, 320
Monobromamine, 413
Monochloramine, 350, 411
Mortality, 87, 138, 194, 197, 205
- rate, 199
Motor vehicle, 476-479
Mulching, 184
Multidimensional niche, 215
Municipial sludge, 267
Mutalism, 202
Mutation, 190, 223

NADPH, 500
Natality, 193, 197
Net primary productivity, 151, 152, 154, 155
Net production, 218, 248
Neutralism, 202
Neutralization, 46, 365
Niche specialization, 217
Nickel, 140, 263, 267-269, 363, 382
- plating rinsing water, 397
Nitrate, 66, 80, 82, 88, 300, 313, 343, 345, 439, 483, 498
Nitric
- acid, 485
- oxide, 410

Nitrification, 56, 67, 72, 88, 300, 313, 341, 342, 415, 498
- of ammonia, 65, 314
- process, 66
Nitrifying microorganisms, 66
Nitrilo Three Acetate (NTA), 337
Nitrite, 82, 88, 378
Nitrobacter, 300, 342
Nitrobenzene, 367
Nitrochloro-benzene, 368
Nitrogen, 35, 83, 86, 90, 323, 324, 343, 344, 350, 447, 448, 517-518, 526
- cycle, 63, 64, 88
- dioxide, 491
- fixation, 88
- oxides (see also NO_x), 117, 476
Nitrogenous gases (see also NO_x), 482-484
Nitrosomonas, 300, 342
Noise, 258, 272, 273
Non-biodegradable material, 327
Non-carbonate calcium, 402
Non-competitive irreversible process, 136
NO_x, 259, 260, 517-518, 524, 526, 533
Noxious animals, 439
NTA (Nitrilo Three Acetate), 337, 339
Nuclear energy, 277
Nucleation centres, 474
Nutrient, 284, 288, 322, 323, 352, 500
- discharge, 54
- traps, 254
- uptake, 123
Nylon, 463

Odorous water, 511
Odour, 376, 380, 413
OECD, 127
Oil, 390
- pollution, 247
Oligo-saprobic water, 75, 79
Oligotrophic, 91, 92, 95
Optical microscopy, 497
Organic
- acids, 476
- phosphates, 262, 327
- polymer (see also polyfloc-culant), 427
- residues, 284
Organism size, 161, 165
Organotrophic organisms, 299
Orthophosphate, 327
Osmosis, reverse, 317, 381, 393
Osmotic pressure, 393, 394

Oxidation, 68, 365
- ditch, 307, 308
- number, 375
Oxidation-reduction reaction
(redox), 374
Oxygen, 70, 74, 82, 85, 90, 256,
284, 300, 304, 318, 364, 374,
375, 439, 491, 513, 517-518, 526
- bridge, 363
- concentration, 52, 65, 69, 70,
72, 76, 84
- concentration, critical, 73
- consumption, 70
- profile, 72, 84
- uptake, 56
Ozone, 375, 377, 404, 413, 415,
517-518, 536
- decomposition, 377

Packed-bed scrubbers, 472, 473
Paper, 437
- finres, 297, 298, 437
Parasitism, 201-203
Parathion, 502
Particle
- capture mechanism, 463
- distribution, 451
- nucleation, 474
- size, 100
Particle matter, 117
Particulate pollution, 453
Particulate polycyclic organic
matter, (PPOM), 450
Partition
- chromatography, 501
- coefficient, 127-128
Pathogenic organisms in sludge,
427
PCB, 165, 261, 272, 501
Pelagic, 247
Pentanone, 416
Permanente hardness, 400
Permanganate, 379, 487
- concentration, 378
Permeability, 361, 464
- coefficient, 320, 321
Peroxyacetyl nitrate, 259
Peroxyacyl nitrates, 476
Peroxybutyl nitrate, 259
Peroxypropanyl nitrate, 259
Persistence, 34, 261, 262
Persistent pollutants, 36
Pesticides, 40, 46, 96, 97, 98,
192, 203, 204, 258, 260-267,
272, 286, 495, 501
Petrochemical, 300
pH, 58
- adjustment, 336

- buffering capacity, 224-228
- effects of, 226
Phenol, 367, 368, 372, 376, 380,
486, 492, 495
Phenotype, 189, 191, 193
Phenoxy herbicides, 262
Phosphate, 328, 329, 330, 358
- organic, 262, 328
Phosphoric acid, 485
Phosphorus, 35, 83, 86, 90, 243,
322, 323, 324, 325, 330, 331,
332, 334, 338, 359
- cycle, 88
Photochemical
- oxidants, 117
- processes, 45, 46
- smog, 259, 476
Photosynthesis, 55, 58, 68, 87,
93, 123
Phototrophic organisms, 299
Pinene, 488
Plant composition, 85
- growth, 55
- resistance, 205
- temperature, 183
Plastics, 437
Plug flow, 42-45, 292
Plume, 107
- dispersion model, 107
Plutonium, 277
Pollution indicators, 75, 77
Polyacrylamide, 340, 341, 365-366
Polyacrylic acid, 340
Polyamide membrane, 397
Polycyclic organic matter, 450
Polyelectrolytes, 339, 340, 388
Polyethylene oxide, 340
Poly-saprobic water, 75, 79
Polystyrene sulphonate, 340
Polyvinyl-chloride, 416
Population density, 166, 193
- growth, 16-18, 201, 282
Potassium permanganate, 375, 487
- number, 316, 398
Power Plant, 113
Prairie, 247
Precipitation, 105, 177, 178, 317,
322, 323, 333, 338, 351, 360,
387, 390, 422, 495
Precoating, 195
Predation, 201-203
Predator, 201
Preservation
- methods of, 496
- of samples, 493, 496, 514
Primary
- amin, 354
- consumer, 160

- production, 153
- productivity, 57, 151, 248
152, 186
- sludge, 424
Productivity, 92, 95, 151, 322
Propylene, 478
Protein, 66, 312, 316, 359
Protozoan, 77, 80
Purge gas stripping, 490
Pyridine, 488, 492
Pyrolysis, 422, 445
Pyrophosphate, 327, 387

Quasi threshold, 135
Quaternary ammonium group, 354

Radiation, 182
- balance, 175
Radioactive fuel, 277
Rainfall, 209
Rain water, 323
Raman spectra, 500
Reaeration, 56, 68, 69, 71
Recalcination, 403
Recarbonization, 351
Recombination, 190, 223
Recoverable reserves, 171
Recovery, 420-422, 487
- of ammonia, 349
- of fat, 316
- of metals, 269-271
- of paper, 421
- of proteins, 318
- zone, 75
Recruitment, 199
Redox (oxidation-reduction
reaction)
Reduction, 365
Refractive index, 500
Refractory
- material, 327, 363
- organic compound, 316
Regeneration, 489
Regulation of hydrology, 253
Reject, 397
Relative humidity, 183
- mortality, 194
- natality, 193
Removal of macrophytes, 251
Removal of superficial sediment,
251
Reproduction, 216
- rate, 195
Resin matrix, 356
Resistance, 191, 192, 197, 205,
207, 261
Resistivity, 468
Respiration, 56, 68, 87, 147, 151,

300, 302
- aerobic, 299
- anaerobic, 299
Respiratory system, 115
- uptake, 129
Response/dose relationship, 121
Responses, types, 135
Resuspension, 68
Retention, 131
- time, 292, 298, 305, 344
Returnable bottles, 420
Reverse osmosis, 317, 381, 393
Reynolds numbers, 458
River Rhine, 267
RNA, 406
Root-zone plant, 360
R-strategy, 198-200

Sampling procedure, 494
Sand traps, 290
Sanitary landfills, 275
Saprobic-system, 75, 80, 82
Saturation, 298
Savannah, 247
Screening, 289
Scrubbers, 487
Searching ability, 206
Seasonal variations, 494
Secondary amine, 354
Secondary productivity, 151
Second Law of Thermodynamics,
147, 187
Sedimentation, 68, 320, 364, 497
Sediments, 129, 139, 251, 515
Selection, 191, 193, 217
Selectivity, 393, 488
- coefficient, 355-357
Selenium, 268
Self-maintenance, 188
Self-regulation, 188
Semi-permeable membrane, 393, 394
Separation, 422, 437-438
- plant, 438
Separators, 364
Settling, 68, 87, 318, 385, 398
- chambers, 457, 458
- tank, 294
Shading, 184
Shannon index, 208, 210, 213
Shelter belts, 184
Shock loadings, 316
Side-effects, 33
Silica, 35
- cycle, 89
Silicates, 242
Silver, 263, 267-269, 363, 382,
397
- nitrate, 397

Thermal tolerance, 189
Thickener, 424, 427
Thickening, 422, 423, 427
Thin layer chromatography, 501
Thiosulphate, 350
Threshold, 135
 - limits, 511
TL_{50}, 509
TOC, 499
Tolerance, 189, 190
Toluene, 367
Toxicant receptor, 135
Toxicity, 262
Transparency, 58, 92, 233
Transpiration, 183
Treatability constant, 309
Trickling filter, 309, 310, 314,
 315, 424
Tripolyphosphate, 327
Trophic niche, 215
Tropical regions, 360
Tube settler, 292
Tundra, 247
Turbidity, 83, 398
Types of responses, 135
Typical emission factors, 266, 270

Ultrafiltration, 393
Ultraviolet radiation, 407
Uptake, 130, 164
 - efficiency, 98, 131
 - rate, 164
Uranium, 277
Urbanization, 33
Uric acid, 370
UV, visible and infrared
 spectroscopy, 497

Vacuum
 - desorption, 490
 - filtration, 428
Van Hoff factor, 393
Venturi scrubber, 471
Vinyl chloride, 367
Viral analysis, 506
Vitamins, 35, 288
Volatile acids, 312
Vollenweider, 91, 92

Warming system, 519
Water, 169
 - balance, 105
 - cycle, 107
 - management, 283
 - resources, 105
Watershed, 322
Weak lapse (coning), 455
Wet grinding, 437
 - scrubbers, 469
Wetlands, 254
WHO, 102, 123, 398
Wind
 - effects, 291
 - power, 279
 - speed, 107-109, 183
Windhoek, 352
Wood, 437

Xanthates, 365-366
X-ray
 - diffusion, 497
 - spectroscopy, 505

Zeolite, 317
Zimmerman process, 434, 435
Zinc, 131, 267-269, 337, 382,
 389, 416, 516, 524
Zonation, 214